Climate Change and Community Resilience

A. K. Enamul Haque · Pranab Mukhopadhyay ·
Mani Nepal · Md Rumi Shammin
Editors

Climate Change and Community Resilience

Insights from South Asia

Editors
A. K. Enamul Haque 🅾
Department of Economics,
East West University,
Dhaka, Bangladesh

Mani Nepal 🅾
South Asian Network for Development
and Environmental Economics (SANDEE),
International Centre for Integrated
Mountain Development (ICIMOD),
Lalitpur, Nepal

Pranab Mukhopadhyay 🅾
Goa Business School,
Goa University,
Taleigao, Goa, India

Md Rumi Shammin 🅾
Environmental Studies Program,
Oberlin College,
Oberlin, OH, USA

ISBN 978-981-16-0682-3   ISBN 978-981-16-0680-9 (eBook)
https://doi.org/10.1007/978-981-16-0680-9

In Memory of Karl-Göran Mäler (1939–2020)

Karl-Göran Mäler the Visionary

This book is dedicated to Karl-Göran Mäler for his contribution in building research capacity and to the growth of environmental economics in South Asia.

Karl-Göran Mäler became professor in 1975 at the Stockholm School of Economics where he started his activities in building research capacity in economics.

By the end of the 1980s, the Royal Swedish Academy of Sciences wanted to give a new start to one of its research institutes. Members of the academy were invited to write proposals on alternative scientific fields for the new institute. Karl-Göran (Academy Member since 1981) teamed up with Bengt-Owe Jansson (Swedish Ecologist and Member of the Academy), and their proposal was chosen. Thus was born 'the Beijer International Institute of Ecological Economics'.

Karl-Göran became Beijer's first director in 1991, and this provided him with a prestigious institutional set-up where he could ground his ideas and foster environmental economics. During his years as Beijer's Director, he promoted bridge building between disciplines and between scientists. Communication and collaboration between top scientists within economics and ecology have been prized as a major achievement of the institute.

Even more relevant here is his work in *Teaching the Teachers*, the capacity-building programme that played a substantial role in the creation of SANDEE in 1999. Karl-Göran and Professor Sir Partha Dasgupta had met in 1979 and were in close contact ever since. But it was after Partha was appointed Beijer's first Chairman that their professional collaboration developed, and it became so iconic that a colleague, former student, and friend wrote about them 'Watson and Crick, Lennon and McCartney, Dasgupta and Mäler' (S. Barrett).

Teaching the Teachers was designed and implemented by Karl-Göran and Partha. They had already launched the programme in 1991, at the Conference of the World Institute for Development Economics Research (WIDER), in Finland. However, soon after the programme moved with them to the Beijer Institute which hosted and developed it.

Environmental economics had been relatively absent from university courses. *Teaching the Teachers* addressed economics faculties in third world universities and had a comprehensive scope. It included teaching, research, support for publication of research results, and enhancement of libraries in universities where the participants worked. The programme had a limited span of time and sought to ignite teacher's curiosity for environmental economics. The programme deepened participants' information about environmental issues and shared ideas about how to use the economist toolbox to analyse them.

The ultimate aim of *Teaching the Teachers* was to catalyse the creation of regional networks, a space for teachers to continue meeting, investigating, and setting their own priorities and broaden capacity building within each region. *Goal reached! Networks were created in Asia, Africa and Latin America.*

This book reinforces and sheds light on essential issues within environmental economics, from a South Asian perspective. Karl-Göran's own research has been considered foundational, making him a pioneer of this scientific field. He persistently spread his own research and fostered others. He was influential in the creation of SANDEE and came back regularly to give lectures. He supported ongoing research and participated in the advisory board, hardly considering this as 'work', so much as paying a visit to family.

SANDEE, with 21 years of uninterrupted activities today, proves the seed was planted in fertile land.

Sara Aniyar

A Life in Dedication to Environmental Economics

Professor Emeritus Karl-Göran Mäler was a pioneering, leading environmental economist. He was one of the founders of the Beijer Institute of Ecological Economics in 1991 and the institute's first director until his retirement in 2002. He was an intellectual giant, mentor, and dear friend to colleagues around the world.

Karl-Göran's curiosity and continuous desire to understand matters at a deeper level were key to the success of the Beijer Institute's early endeavour to build bridges between disciplines that did not usually collaborate. The Beijer Institute had an inspiring journey with Karl-Göran, searching for new understandings, always focused on the problems, exploring and investigating with an open mind and deep commitment.

Under Karl-Göran's leadership, the Beijer Institute established several regional networks of environmental economists and founded the journal *Environment and Development Economics*, which encourages submissions from researchers in the field in both developed and developing countries. Both have had a great impact on research and policy in developing countries and have been very important to many researchers from the network regions. The South Asian Network for Development and Environmental Economics started in 1999 and is one of the most influential networks.

The Beijer Institute also organised a series of teaching and training workshops in the network regions aiming to teach economics teachers in universities so that they could themselves start teaching environmental economics. Karl-Göran Mäler and Sir Partha Dasgupta led the work, while other leading ecologists and economists were part of the teaching teams. Karl-Göran regarded this work as one of the most important accomplishments in his career and probably the one he cherished the most. It was in one of the teaching workshops, in Jamaica, that he met his beloved wife Sara.

Many of Karl-Göran Mäler's former students and workshop attendees testify to his genuine interest in sharing his knowledge and in learning from his pupils. He would dive into the technicalities of a student's draft paper to discuss underlying assumptions, alternative approaches, potential improvements, or previous publications that this work related to. One could tell that he really enjoyed those conversations.

Christina Leijonhufvud, Office Manager at the Beijer Institute, who organised the teaching workshops and courses and often travelled with Karl-Göran, says:

'Karl-Göran was very much appreciated among the participants with his straightforwardness, patience and support. He showed such interest in people and ideas and loved to discuss things from different angles. He shared his wisdom and encouraged everyone to think, dig deeper and never give up. His broad and deep knowledge about the most diverse matters was impressive and his love for Nature was remarkable'.

Above all, Karl-Göran was kind, a helping soul who had amazing stories, jokes, and laughs to share. The stories were often about the people he met in his travels, mixed with the experiences of Nature and scientific discussions they had enjoyed together. He was an inspiring role model and father figure. He will be deeply missed, not least by those of us who had the privilege of working under his leadership for many years.

Carl Folke, Professor
Director of the Beijer Institute of Ecological Economics

Anne-Sophie Crépin, Associate Professor
Deputy Director of the Beijer Institute of Ecological Economics

Remembering Karl-Göran Mäler

Karl-Göran Mäler helped create and build the South Asian Network for Development and Environmental Economics (SANDEE). In the 1990s, as Director of the Beijer Institute of Ecological Economics, Dr. Karl-Göran Mäler, along with Sir Partha Dasgupta, who was the institute's Board Chair, envisioned the growth of a new type of scholarship in environment and development economics. The Beijer Institute is a 'boundary' organisation that works at the frontier of ecology, economics, and other disciplines. It examines global changes, understanding that humanity both is embedded in and shapes the biosphere. While this systems approach is integral to the institute's work, with Karl-Göran Mäler as Director, the institute took a big step towards strengthening enquiry into the role of economic development in shaping humanity's continued interactions with earth systems.

The partnership between Karl-Göran Mäler and Sir Partha Dasgupta was enormously productive and opened the world of environment and ecological economics to many scholars in developing countries. Together, these two friends envisaged a need for environment and development economics, a discipline that would be grounded in economic theory but build on a lived understanding of the synergies and trade-offs between economic development and environmental change. They rightly saw the importance of research and a body of evidence driven by scholars from low- and middle-income countries, so that problems were correctly identified, and solutions were home grown. This belief, his confidence in the usefulness of economic thinking for solving global problems, and his passion for the environment led Prof. Mäler to help create a number of environmental economics networks around the world— SANDEE in South Asia, LACEEP in Latin America, and CEPA and RANESA in Africa. Each of these networks has played an important part in many low- and middle-income countries in building the capacity of researchers to think about, collect data, and analyse environmental problems. Professors Mäler and Dasgupta also helped launch the journal *Environment and Development Economics*, which continues to build evidence on the nexus between environment and development.

Karl-Göran Mäler loved being an economist. His Ph.D. thesis became, *Environmental Economics—A Theoretical Inquiry*, a much-cited book that examines and builds the conceptual underpinnings for studying a range of environmental problems. By developing the methods for undertaking a monetary evaluation of changes in environmental quality, the book helped grow the now large body of the literature on environmental valuation. Karl-Göran continued to focus on solving complex problems and is best known for his work on acid rain, inclusive wealth accounts, and pollution in small lakes. His research on comprehensive wealth (inclusive of natural capital), as an indicator of sustainable development, is another productive area of work with Sir Partha. Many of us have fond memories of KGM discussing the importance of wealth and natural capital or drawing graphs to show us how lakes could switch from being just polluted to plain dead. Karl-Göran for many years was also part of the committee that selected recipients of the Nobel Prize in Economics. His love for knowledge, however, far exceeded the bounds of economics. He could possibly surpass Google with his ability to present minute historical facts on all sorts of topics. His was an utterly curious mind, wherein perhaps lies the explanation for his wealth of knowledge and his deep scholarship on linked human and environmental systems.

To Karl-Göran Mäler, it was important to understand and honour the magical complexity of the natural world. As Director of the Beijer Institute, he brought many ecologists and economists together to synthesise and draw lessons from different strands of science. He was also an avid birdwatcher and carried his binoculars everywhere he travelled. In fact, the very first time I met him, he convinced all the conference participants at a meeting in Malta to climb up a steep cliff and lie down on top to look over the edge for birds. This was how persuasive he could be when it came to birds. I also vividly remember many stories of his safaris and his reverence for the wildlife he encountered in Africa. This awe for biodiversity extended to all of life. The many regional dances that we watched on our trips to Nepal simply reinforced for Karl-Göran the importance of diversity.

Professor Mäler spent many years teaching and nurturing SANDEE. For the first several years of SANDEE's evolution, Karl-Göran served as a research advisor. Scholars from South Asia will never forget his many questions, which sometimes seemed simple, but were always foundational. As SANDEE matured, so did Karl-Göran's relationship with the network. He took on more teaching and was always willing to join SANDEE's summer workshops in Bangkok, which gave him an opportunity to teach, ask questions, and share his intellectual interests with his many SANDEE friends. It meant a great deal to Karl-Göran to see SANDEE grows and its members publish and thrive. He was deeply interested in the many local economic problems presented. Perhaps, his single most important contribution was to push everyone to think carefully about the complexities underlying these problems. '*Think harder*' was his mantra, the key that would open all doors.

Karl-Göran lived a good life. His wonderful sense of humour helped, and he was ever willing to crack a joke, hum a song, or raise a toast. He showed us, by example, how to enjoy both the pursuit of intellectual activity and the friendships that come along the way. Building trust and creating institutions was important to Karl-Göran Mäler. Social capital, the relationships, and networks we accumulate as we move through life were his secret sauce to living the good life. He built the Beijer Institute with this recipe, making it easier for all of us from SANDEE to follow in his footsteps.

Thank you, Prof. Mäler—for being our mentor, our guru, and our friend.

Priya Shyamsundar
Lead Economist, The Nature Conservancy

Foreword by Sir Partha Dasgupta

Economics, like I imagine other scientific disciplines, normally moves in incremental steps and always without a central guide. Much like practitioners of other disciplines, we economists work with models of those features of the world we want to study in detail. That involves keeping all else in the far background. Models are thus parables, and some say they are caricatures, which is of course their point.[1]

Economics is also a quantitative subject. Finance ministers need estimates of tax revenues if they are to meet intended government expenditure; environment ministers today cannot but ask how much farmers should be paid to set aside land for 'greening' the landscape, and whether fossil fuel subsidies should be eliminated; health ministers look to convince cabinet colleagues that investment in health is good for economic growth; and so on. Which is why economic models are almost invariably cast in mathematical terms.

Which is also why the models that appear in economics journals can appear esoteric, unreal, and even self-indulgent. Many would argue as well that to model human behaviour formally, let alone mathematically, is to tarnish the human experience, with all its richness. And yet, economists in governments, international organisations, and private corporations find those models and their adaptations essential for collecting and analysing data, forecasting economic trajectories, evaluating options, and designing policy. Perhaps, then, it should be no surprise that those same models go on to shape the conception we build of our economic possibilities. In turn, our acceptance that the economic possibilities those models say are open to us encourages academic economists to refine and develop them further along their tested contours. And that in turn further contributes to our beliefs about what is achievable in our economic future. The mutual influence is synergistic.[2]

---

[1] The reflections here have been adapted from *The Dasgupta Review on the Economics of Biodiversity*, prepared at the invitation of the UK Treasury and published in February 2021.

[2] It will be asked who is represented in the collective 'we' and 'our' I am using here. It is not everyone in the world and certainly not restricted to those who agree with the claims I am making about the mutual influence of academic economic models and a general reading of economic possibilities. The group I have in mind is not fixed by designation but through invitation—for example, people

That has had at least one unintended and costly consequence. Not so long ago, when the world was very different from what it is now, the economic questions that needed urgent response could be studied most productively by excluding Nature from economic models. At the end of the Second World War, absolute poverty was endemic in much of Africa, Asia, and Latin America, and Europe needed reconstruction. It was natural to focus on the accumulation of produced capital (roads, machines, buildings, factories, and ports) and what we today call human capital (health and education). To introduce Nature, or *natural* capital, into economic models would have been to add unnecessary luggage to the exercise.[3]

Nature entered macroeconomic models of growth and development in the 1970s, but in an inessential form.[4] The thought was that human ingenuity could overcome Nature's scarcity over time and ultimately (formally, in the limit) allow humanity to be free of Nature's constraints. But the practice of building economic models on the backs of those that had most recently been designed meant that the macroeconomics of growth and development continued to be built without Nature's appearance as an essential entity in our economic lives. Historians of science and technology call that feature of the process of selection 'path dependence'.[5] That may be why economic and finance ministries and international organisations today *graft* particular features of Nature, such as the global climate, onto their models as and when the need arises, but otherwise continue to assume the biosphere to be *external* to the human economy. In turn, the practice continues to influence our conception of economic possibilities for the future. We may have increasingly queried the absence of Nature from official conceptions of economic possibilities, but among economists at large, the worry has been left for Sundays. On weekdays, our thinking has remained as usual.

Nature has features that differ subtly from produced capital goods. The financier may be moving assets around in his portfolio, but that is only a figure of speech. His portfolio represents factories and ports, plantations and agricultural land, and mines and oil fields. Reasonably, he takes them to be immobile. In contrast, Nature is in large measure mobile. Insects and birds fly, fish swim, the wind blows, rivers flow, and the oceans circulate. Economists have long realised that Nature's *mobility* is one reason the citizen investor will not take the market prices of natural capital to represent their social worth even when markets for them exist. The wedge between the prices we pay for Nature's goods and services and their social values (their social values are called 'accounting prices') is usually studied in terms of what economists call 'externalities'. Over the years, a rich and extensive literature has identified the institutional measures that can be deployed for closing that wedge.

---

who read this volume of essays—to consider why and how we need to break the cycle and revise the conception we hold of humanity's place in the biosphere.

[3] I am referring to the evolution of economic thinking in the West. However, to the best of my knowledge the economic models that shaped state policy in the Soviet Union, and the ones developed by prominent academics in Latin America, also did not include Nature.

[4] See, for example, the special issue in the *Review of Economic Studies* (1974) on the economics of exhaustible resources.

[5] A clear statement is in P. A. David, 'Clio and the Economics of QWERTY', *American Economic Review*, 1985; 75(2), 332–337.

But in addition to mobility, Nature has two properties that make the economics of biodiversity markedly different from the economics that informs our intuitions about the character of produced capital. Many of the processes that shape our natural world are *silent* and *invisible*. The soils are a seat of a bewildering number of processes with all three attributes. Taken together, the attributes are the reason it is not possible to trace many of the harms inflicted on Nature (and by extension, on humanity too) to those who are responsible. Just who is responsible for a particular harm is often neither observable nor verifiable. No social mechanism can meet this problem in its entirety, meaning that no institution can be devised to enforce socially responsible conduct. Ultimately, we each will have to serve as judge and jury over our own conduct. That can happen only if we felt impelled to account for the personal impact we have on Nature. The economics of Nature, like the economics of so much else, ranges from the (global) macro- to the (very local) micro-reasoning.

Humanity is embedded in Nature, and we are not external to it. Nature's goods and services are essential for our existence (the air we breathe and the water we drink are immediate examples), of direct use as consumption goods (fisheries), of indirect use as inputs in production (timber and fibres), and for our emotional well-being (green landscape and sacred sites). Many have multiple uses (forests, rivers, the oceans). And there are parts of Nature communities regard as sacred.

But like so much academic economics, the economics of Nature until relatively recently addressed life in the affluent West. Unknowingly, that acted as a constraint on the development of environmental and resource economics. For example, until relatively recently influential writers on economic development saw Nature's goods and services mostly as luxuries. An unnecessary debate took place between those who expressed environmental concerns in low-income countries and those who saw the need for economic growth there above all else. Well-meaning writers tried to reconcile the two viewpoints. An editorial in the UK's *Independent* (4 December 1999), for example, observed that '... (economic) growth is good for the environment because countries need to put poverty behind them in order to care', and a column in *The Economist* (4 December 1999: 17) insisted '... trade improves the environment, because it raises incomes, and the richer people are, the more willing they are to devote resources to cleaning up their living space'.

Given this background, it is hard to overemphasise the significance of SANDEE. Since its inception, the network has seen natural capital as both ends and a means to human ends, and it has studied Nature from the vantage point of both rural communities and urban citizens of South Asia. In a remarkable collection of essays, SANDEE scholars provided quantitative estimates of externalities that are embedded in the production and consumption of a wide range of natural assets, ranging from wetlands to urban airsheds.[6] The estimates were understood as being the wedge between their market prices and their accounting prices. I know of no comparable set of studies on the local natural resource base in the developing world. The studies have proved

---

[6] A. K. E. Haque, M. N. Murty, and P. Shyamsundar, eds. (2011), *Environmental Valuation in South Asia* (Cambridge: Cambridge University Press).

invaluable for the review on the economics of biodiversity that I have had occasion to prepare recently (see Footnote 1).

In an earlier and equally remarkable collection of essays, SANDEE scholars reported a variety of ways in which communities in South Asia have managed their local ecosystems.[7] To be sure, there was in hand a well-known literature that had uncovered myriads of communitarian institutions that had served well to protect spatially localised ecosystems (woodlands, water sources, threshing grounds, coastal fisheries) from excessive use. What made the SANDEE collection striking was that it uncovered a variety of unexpected institutional channels through which good intentions can be thwarted. I do not know of any other publication that has so systematically uncovered 'systems failure' even in a world where no one intends to do harm. The moral is revealing: humanity's engagement with Nature can be best understood if institutions are modelled as systems with well-defined links among their components. The studies also cautioned us that our understanding can only be incremental, because every system will have links among its components no one has imagined are there or will arise if the system is perturbed. The studies in that collection have also proved invaluable for my review on the economics of biodiversity (see Footnote 1).

The present collection reflects SANDEE's move towards a new class of issues: humanity's propensity to discover ways to meet ecological problems by inventing new ways to do things. The context of the studies is global climate change, and the authors identify ways to mitigate an increased variability of rainfall and other ecological services as well as explore technological possibilities that move away from activities that are intensive in their emissions of carbon. As ever, the focus on South Asia has uncovered a remarkable range of findings, for 'technology' does not have to mean men in hard hats substituting produced capital for natural capital. It can mean rerouting natural capital in appropriate ways (e.g. water harvesting technologies). Modernisation of life has involved new technologies, but it has also meant forgetting old technologies that could come to our rescue. The possibility of reviving forgotten technologies that relied on Nature to avoid natural disasters—today they are called 'Nature-based solutions'—the shift from macro- to micro-focus and from high-income to low-income societies have together altered our common understanding of humanity's engagement with the natural world. SANDEE has been pivotal in that shift.

Partha Dasgupta
St John's College
Cambridge, UK

---

[7] R. Ghate, N. S. Jodha, and P. Mukhopadhyay, eds. (2008), *Promise, Trust and Evolution: Managing the Commons of South Asia* (Oxford: Oxford University Press).

The original version of the book was revised: Book Editor's Name "A. K. Enamul Haque" updated correctly in citations as "A. K. E. Haque" and Open Access license and logo has been changed in the whole book has been updated. The correction to the book is available at https://doi.org/10.1007/978-981-16-0680-9_30.

Foreword by Pema Gyamtsho

It is my privilege to introduce the edited book *Climate Change and Community Resilience: Insights from South Asia*. This is a timely publication that highlights local initiatives on climate change adaptation and resilience building in South Asia, which is one of the most vulnerable regions of the world to climate change impacts.

ICIMOD is committed to the process of knowledge generation with a focus on livelihood enhancement, sustainable resource management and use, and integrating indigenous knowledge and culture into solutions. We organise our work in regional programmes, which build on ICIMOD's deep history of engagement and are formulated to deliver strategic results; promote transboundary cooperation; meet capacity-building needs in the region; and support long-term testing, piloting, and monitoring of innovative approaches. Adaptation and resilience building at the grass-roots level are central to ICIMOD's mission and vision.

Using a narrative style, this book draws on stories and examples from seven South Asian countries—Bangladesh, Bhutan, India, Maldives, Nepal, Pakistan, and Sri Lanka—to highlight how communities in South Asia are building resilience to climate change. A total of 58 authors have contributed to this volume, and I am delighted that most of them are from the South Asian Network for Development and Environmental Economics (SANDEE) representing all the seven South Asian countries, five of which are from the Hindu Kush Himalayan region. The 29 chapters in the book are organised under six themes: concepts and models; traditional knowledge and sustainable agriculture; technology adoption; disaster risk reduction; urban sustainability; and alternative livelihoods. These chapters highlight stories of creativity, community engagement, and locally applicable solutions. They are powerful and instructive. They offer valuable lessons for researchers, practitioners, and policy-makers.

I applaud the editors—A. K. Enamul Haque, Pranab Mukhopadhyay, Mani Nepal, and Md Rumi Shammin—for dedicating this book to the memory of Prof. Karl-Göran Mäler, one of the co-founders of SANDEE, with a foreword by his friend and SANDEE Co-founder Sir Partha Dasgupta. The book also includes reflections on Prof. Mäler from his wife Sara Aniyar, along with a short recollection by Priya Shyamsundar, SANDEE's Founding Director, and Carl Folke, Director of the Beijer

Institute of Ecological Economics, which Prof. Mäler founded and where he served as Funding Director. Interestingly, the writing of the book was undertaken from start to finish during the COVID-19 pandemic and is indicative of the resilience of the network that Prof. Mäler and Prof. Dasgupta helped establish 20 years ago.

ICIMOD is honoured to sponsor open-access publication of the book to make it widely accessible and offer assistance with language and style editing. This collection is likely to have a broad impact on advancing knowledge and lessons on community-based climate adaptation initiatives in South Asia and other developing countries around the world. We expect it to be widely read and used as a reference for policy-making and programme development that will make a difference in people's lives—especially the marginalised population that are often left behind—by helping them embark on a path to resilience.

Pema Gyamtsho
Director General, International Centre
for Integrated Mountain Development
(ICIMOD)
Patan, Nepal

Acknowledgements

This book was conceived and given life during the COVID-19 pandemic. We editors met every week, mostly on Mondays, online from March 2020 to September 2021 often testing the patience of our families as these meetings lasted several hours at night or early morning since we were working from different time zones. We are grateful to them for their support without which this book could not have taken shape. We are thankful to all the contributors who agreed to participate in this project—including the family, friends, colleagues, and mentees of Karl-Göran Mäler and SANDEE associates.

Once the contributors came on board with the revised manuscripts after receiving comments from the reviewers, we organised a four-day online writeshop bringing together all contributing authors from seven South Asian countries and beyond with more than ten different time zones. The writeshop turned into an excellent peer review platform for all the manuscripts included in the volume, where participants provided critical and open feedback that helped improve the quality of the chapters. We thank Oberlin College for the use of their Zoom platform for the writeshop and all meetings of the editors and contributors throughout the process. We are grateful to the South Asian Network for Development and Environmental Economics (SANDEE) at the International Centre for Integrated Mountain Development (ICIMOD) for making it possible to publish this volume as an open access resource. Aunohita Mojumdar helped us with language editing, and Neesha Pradhan provided logistical support.

Contents

Editors and Contributors

About the Editors

Dr. A. K. Enamul Haque is a Professor of Economics and Dean, Faculty of Business and Economics at East West University (Dhaka, Bangladesh), Director of Economic Research Group, Executive Director of Asian Center for Development, a Member of the Steering and Advisory Committees of the South Asian Network for Development and Environmental Economics (SANDEE) and a member of the South Asian Network for Environmental Law and Policy (SANEL). He is an environmental economist with teaching, research and popular articles on climate change, agriculture and urban issues in developing countries with a focus on South Asia in general and Bangladesh in particular.

Dr. Pranab Mukhopadhyay is a Professor of Economics at the Goa Business School and the Program Director (Economics) at Goa University, India. He did his graduation from Presidency College, Calcutta and his master and doctoral work at the Jawaharlal Nehru University, New Delhi. He is a Fellow of SANDEE and INSEE and the former President of INSEE (2016–2018). Earlier he worked as Environmental Economist at IUCN, Nepal. He has also been a consultant to ICIMOD, Foundation for Ecological Security, and GIZ India. He works in the area of global change, managing commons, nature and society, sustainable development, ecosystem services, and economic growth.

Dr. Mani Nepal is SANDEE Program Coordinator and Lead Economist at ICIMOD. Earlier, he worked as Associate Professor of Economics, Tribhuvan University; Visiting Assistant Professor, University of New Mexico, USA; Visiting Professor, Kathmandu University; and Adjunct Professor, Agriculture and Forestry University. He also worked as Senior Economist at the Department of Finance and Administration, Santa Fe, New Mexico; and as an Economist at the Nepal Foundation for Advanced Studies. With research interest in the intersection of development and environmental issues, Dr. Nepal is a Fulbright scholar who received MS degree in

Policy Economics from the University of Illinois at Urbana-Champaign, and Ph.D. in Economics from the University of New Mexico, USA.

Dr. Md Rumi Shammin is a Professor of Environmental Studies at Oberlin College in Ohio, USA where he served as the Chair of the program between 2015 and 2019. He has a Ph.D. in Natural Resources and Environmental Sciences from the University of Illinois at Urbana-Champaign and a Master's in Natural Resources with a minor in Agricultural & Biological Engineering from Cornell University. Dr. Shammin's scholarship focuses on energy and climate change analysis; climate adaptation and resilience building in developing countries; behavioral and human dimensions of the environment; ecological economics; refugee camp environmental management; environmental justice; and urban sustainability.

Contributors

Rathnayake M. Abeyrathne Department of Sociology, University of Peradeniya, Kandy, Sri Lanka

Ajaz Ahmed Institute of Business Administration, Karachi, Pakistan

Raquibul Amin Bangladesh Country Office, International Union for Conservation of Nature (IUCN), Dhaka, Bangladesh

Chandra Sekhar Bahinipati Department of Humanities and Social Sciences, Indian Institute of Technology Tirupati, Tirupati, India

Raksha Balakrishna Foundation for Ecological Security, Anand, India

R. Balasubramanian Department of Agricultural Economics, Tamil Nadu Agricultural University, Coimbatore, India

Estiaque Bari Department of Economics, East West University, Dhaka, Bangladesh

Bishal Bharadwaj University of Queensland, St. Lucia, Queensland, Australia

Bijal Brahmbhatt Mahila Housing Sewa Trust, Ahmedabad, Gujarat, India

Rahul Chaturvedi Foundation for Ecological Security, Anand, India

Jamyang Choda Perth, Australia

Saudamini Das Institute of Economic Growth, Delhi, India

Kolpona De Costa Department of Economics, East West University, Dhaka, Bangladesh

Ngawang Dendup Waseda University, Tokyo, Japan

Murugaiah Devaraj JRD Tata Ecotechnology Centre, M. S. Swaminathan Research Foundation, Chennai, India

P. Indira Devi Kerala Agricultural University, Thrissur, Kerala, India

Priyanga M. Dunusinghe University of Colombo, Colombo, Sri Lanka

Islam M. Faisal Department of Conservation and Climate, Government of Manitoba, Winnipeg, MB, Canada

Remeen Firoz Dhaka, Bangladesh

Rucha Ghate Foundation for Ecological Security, Anand, India

Santadas Ghosh Department of Economics and Politics, Visva-Bharati University, Bolpur, West Bengal, India

R. P. Dayani Gunathilaka Department of Export Agriculture, Uva Wellassa University, Badulla, Sri Lanka

Rashadul Hasan Bangladesh Disaster Preparedness Centre, Dhaka, Bangladesh

A. K. Enamul Haque Department of Economics, East West University, Dhaka, Bangladesh

Nabila Hye Asian Center for Development, Sylhet, Bangladesh

Rishi Ram Kattel Department of Agricultural Economics and Agribusiness Management, Agriculture and Forestry University, Rampur, Chitwan, Nepal

Madan S. Khadayat Kathmandu, Nepal

Zakir Hossain Khan Fellow, Asian Center for Development, Dhaka, Bangladesh

Hemasiri B. Kotagama Sultan Qaboos University, Muscat, Oman

Heman Das Lohano Department of Economics, Institute of Business Administration, Karachi, Pakistan

Sakib Mahmud School of Business and Economics, University of Wisconsin-Superior, Superior, USA

Madhavan Manjula School of Development, Azim Premji University, Bengaluru, India

Pranab Mukhopadhyay Goa Business School, Goa University, Taleigao Plateau, Goa, India

Adnan Nazir Department of Economics, University College of Zhob (UCoZ), BUITEMS, Zhob, Pakistan

Apsara Karki Nepal International Centre for Integrated Mountain Development (ICIMOD), Kathmandu, Nepal

Mani Nepal South Asian Network for Development and Environmental Economics (SANDEE), International Centre for Integrated Mountain Development (ICIMOD), Kathmandu, Nepal

Harshkumar Nareshkumar Panchal School of Public Policy, IIT Delhi, New Delhi, India

Unmesh Patnaik Centre for Climate Change & Sustainability Studies, School of Habitat Studies, Tata Institute of Social Sciences, Mumbai, India

Ismat Ara Pervin Institute of Water Modelling, Dhaka, Bangladesh

Rajesh K. Rai School of Forestry and Natural Resource Management, Institute of Forestry, Tribhuvan University, Kathmandu, Nepal

Muntaha Rakib Department of Economics, Shahjalal University of Science and Technology, Sylhet, Bangladesh

Raj Rengalakshmi JRD Tata Ecotechnology Centre, M. S. Swaminathan Research Foundation, Chennai, India

Sreejit Roy Department of Economics and Politics, Visva-Bharati University, Bolpur, West Bengal, India

Anu Susan Sam Regional Agricultural Research Station, Kerala Agricultural University, Kumarakom, Kerala, India

P. S. M. Kalani J. Samarakoon Department of Export Agriculture, Uva Wellassa University, Badulla, Sri Lanka

V. Saravanakumar Department of Agricultural Economics, Tamil Nadu Agricultural University, Coimbatore, India

Archana Raghavan Sathyan College of Agriculture, Kerala Agricultural University, Vellayani, Kerala, India

Fathimath Shafeeqa Institute of Research and Development Pvt. Ltd., Malé, Maldives

Md Rumi Shammin Environmental Studies Program, Oberlin College, Oberlin, OH, USA

Upasna Sharma School of Public Policy, IIT Delhi, New Delhi, India

E. Somanathan Indian Statistical Institute, New Delhi, India

Maya Sosland Environmental Studies Program, Oberlin College, Oberlin, OH, USA

Liya Thomas Foundation for Ecological Security, Anand, India

Kuenzang Tshering Edith Cowan University, Perth, Australia

Tshotsho College of Natural Resources, Royal University of Bhutan, Lobesa, Bhutan

E. P. N. Udayakumara Department of Natural Resources, Faculty of Applied Sciences, Sabaragamuwa University of Sri Lanka, Belihiloya, Sri Lanka

Shamen P. Vidanage University of Kelaniya, Kelaniya, Sri Lanka

Amy Wang Environmental Studies Program, Oberlin College, Oberlin, OH, USA

Chapter 1
South Asian Stories of Climate Resilience

A. K. Enamul Haque, Pranab Mukhopadhyay, Mani Nepal, and Md Rumi Shammin

Key Messages

- South Asia is faced with both climate change threats and pre-existing development challenges such as poverty reduction, natural resource management, and social equity.
- Community-based adaptation initiatives in vulnerable communities of South Asia represent examples of multifaceted and holistic approaches.
- Lessons from grassroots responses in South Asia can be a source of knowledge and potential solutions for other developing countries.

1.1 Climate Resilience at the Community Level

Building resilience in communities against climate-induced disasters is a priority in many South Asian countries. These countries are also committed to attaining

A. K. E. Haque
Department of Economics, East West University, Dhaka, Bangladesh
e-mail: akehaque@gmail.com

P. Mukhopadhyay
Goa Business School, Goa University, Taleigao Plateau, Goa, India
e-mail: pm@unigoa.ac.in

M. Nepal
South Asian Network for Development and Environmental Economics (SANDEE), International Centre for Integrated Mountain Development (ICIMOD), Kathmandu, Nepal
e-mail: mani.nepal@icimod.org

M. R. Shammin (✉)
Environmental Studies Program, Oberlin College, Oberlin, OH, USA
e-mail: rumi.shammin@oberlin.edu

© The Author(s) 2022
Haque et al. (eds.), *Climate Change and Community Resilience*,
https://doi.org/10.1007/978-981-16-0680-9_1

the sustainable development goals (SDGs) by 2030. Community-based programmes offer promising opportunities for both these ends. When local communities become intimately engaged in adaptation initiatives, they can instill local and traditional knowledge, take advantage of existing networks, stimulate local capacity, and reduce dependence on long-term external support. Engaging with people at the grassroots level, empowering local institutions and building resilient communities will therefore be critical for humanity to navigate the journey into an uncertain climatic future.

This book documents how communities in South Asia are building climate resilience. At a time when climate change presents humanity with an uncertain and gloomy future, the stories of innovation, creativity, grassroots engagement, and locally applicable solutions in this book provide hope and pathways for sustainability. Grassroots initiatives already in place in South Asian countries suggest that locally engaged programmes often generate more effective solutions at lower costs. The narrative of writing makes the volume accessible to a diverse audience— from academics, researchers, and students to practitioners in various governmental, non-governmental, and international agencies.

1.2 Global Struggle with Climate Change

After many years of negotiations and failed attempts, the international community finally reached an agreement on a strategy to address climate change. The Paris Agreement was signed by 197 countries in December 2015 (UNFCCC, 2015). Nations around the world agreed on a plan of action to contain the global average temperature increase within 2 °C above pre-industrial levels by the end of the century. Although many of the vulnerable countries lobbied for keeping the temperature increase within 1.5 °C, they finally settled for a non-binding 'good faith' agreement where everyone would strive to reach this lower threshold.

While the Paris Agreement offered tangible hope for humanity to address climate change, it is still too little and too late to prevent significant human, ecological, and economic impacts. The risks posed by climate change are no longer distant phenomena and many countries are already facing the uproar of nature due to global warming. There has been an increase in the frequency of extreme events all over the world. The global North has experienced it through hurricanes and super storms in the USA, persistent heat waves in Europe, and unprecedented forest fires in Australia. The South has been subjected to multiple natural hazards such as severe cyclones, extended droughts, high-intensity rainfall, flooding, and landslides. These have caused widespread ecosystem degradation, loss of agricultural productivity, salinity intrusion, and erosion of soil in riparian zones. Between 1999 and 2018, about 495,000 lives were lost worldwide as a direct result of more than 12,000 extreme weather events, causing economic losses of US$ 3.54 trillion (in purchasing power parities) (Eckstein, Hutfils, & Winges, 2020). The overall impact on people in the South has been more severe due to direct loss of livelihoods amid pre-existing levels of poverty and the slow pace of development.

1.3   South Asia: A Climate Hotspot

Nowhere has the rage of climate change erupted with more vigour and multifaceted threats than in South Asia, a region enclosed by the Himalayas in the North and the Indian Ocean in the South. Here, a quarter of the global population resides in 3.5% of world's land area, making it the most populous and most densely populated region of the world. According to the Global Climate Risk Index, it is the most impacted region of the world in terms of fatalities and economic losses that occurred between 1998 and 2017 due to climate change. The region has been subjected to floods, landslides, cyclones, and heavy rainfall. Bangladesh, India, Nepal, and Pakistan are among the top 20 countries that have been most severely impacted by extreme weather events. India, Bangladesh, and the Maldives have been identified as the most vulnerable countries to rising sea level and increased river flooding, with Pakistan appearing in the top ten list (Eckstein et al., 2019).

A 2018 IPCC report provides a comprehensive account of the impacts expected even if the goal of the Paris Agreement to contain global warming at 2°C is achieved. The key findings of the report that apply to climate change impacts and responses in South Asia are summarised below (IPCC, 2018).

- Extreme temperatures and increase in frequency and intensity of heavy precipitation and drought.
- Higher levels of heavy precipitation associated with tropical cyclones and a larger fraction of the global land area affected by flood hazards.
- Amplified exposure of small islands, low-lying coastal areas and deltas to the risks associated with sea level rise for many human and ecological systems, including increased saltwater intrusion, flooding, and damage to infrastructure.
- Increasing risks to fisheries and aquaculture via impacts on the physiology, survival, habitat, reproduction, disease incidence, and risk of invasive species.
- Several hundred million people at disproportionately higher risk of adverse consequences and poverty by 2050, including disadvantaged and vulnerable populations, indigenous peoples and local communities dependent on agricultural or coastal livelihoods.
- Overlapping risks across energy, food, and water sectors spatially and temporally, creating new hazards and exacerbating current ones, exposures, and vulnerabilities for increasing numbers of people and regions.

While the above examples highlight the major shocks that South Asian countries have experienced in recent decades and are expected to confront in years to come, climate change also contributes to slow and persistent impacts on livelihoods and economies in the region through gradually declining flow of ecosystem services, reducing agricultural productivity, and increasing coastal and river erosion. Discourses of climate justice suggest that the worst impacts of climate change will be borne by some of the most vulnerable and poorest populations of the world. The livelihood impacts are already displacing a large number of people who have become internally or externally displaced climate refugees. As the world continues to warm, many more people are expected to be disenfranchised and displaced within the timeline of the Paris Agreement. Even if they are able to survive the wrath of violent wind or water, they will still be subject to the gradual loss of livelihood–gram by gram of grain and inch by inch of land–slowly, but surely.

1.4 South Asian Stories of Resilience Building

Given the geographic diversity of South Asia spreading from the Everest to the islands in the Indian Ocean, the heterogeneity of communities with diverse languages, ethnicities, lifestyles, and cultures, this region offers unique opportunities for studying community-based initiatives on building climate resilience. This book tells stories of climate change adaptation initiatives in seven South Asian countries highlighting grassroots level solutions, documenting lessons learned, and identifying gaps and opportunities. There are 27 studies selected from the island nations of Maldives and Sri Lanka, the floodplains of Bangladesh, India and Pakistan, and the mountainous countries of Bhutan and Nepal. These studies, which highlight how communities have been working to deal with climate change, are designed to guide others who are searching for examples to replicate in their own communities. These case studies highlight that win–win solutions may exist for communities battered by climate change, poverty, and environmental degradation.

The book is organised into six thematic areas—each constituting an important area for climate change and sustainability that is of concern in South Asia.

The book begins with a section on *Concepts and Models* to introduce issues related to building climate resilience at the community level. An integrated framework is developed that connects community-based climate adaptation (CBA) with the UN Sustainable Development Goals (SDGs) and principles of resilience and efforts at disaster risk reduction. A review of academic and grey literature offers insight into the concept, application, barriers, and opportunities for CBA along with a few case studies from outside South Asia to situate the stories in this book in the broader context of global grassroots resilience building initiatives. This section also develops the taxonomy related to resilience building efforts at the community level. Agriculture is an important sector which will be severely impacted by climate change and India is a large economy with a variety of agricultural products. Adaptation strategies are very important for agricultural communities. Therefore, a literature review of adaptation practices in Indian agriculture is also presented in this section. The section ends with an example of the application of a resilience analysis protocol in Bangladesh as a model for resilience building that can be applied elsewhere as a tool for community-based climate adaptation.

Communities across the regions in South Asia have used indigenous and traditional knowledge to combat many challenges of nature. Paddy growers in Bhutan have been growing traditional rice varieties to deal with water scarcity. It has given them a 'safety net' against shortage of irrigation water. Farmers in Pakistan facing floods have used local knowledge to develop resilience, using both individual wisdom as well as community knowledge to deal with flash floods. Similarly, farmers in Kerala formed collectives to manage the impact of massive flooding with a variety of post-flood measures in conjunction with the state. Farmers in Maldives use innovative local knowledge to grow vegetables. They learn from each other to build an agricultural practice to withstand the onslaught of sea waves. Similarly, stories from agriculture in India further suggest that farmers need to adapt to use water sustainably in order to protect their income. Hence, there is a genuine need to find water-saving technologies using traditional varieties of crops rather than using modern varieties which are water-intensive. The *Traditional Knowledge and Sustainable Agriculture* section highlights the value of local knowledge, autonomous adaptation, conservation initiatives, and water resource management.

Throughout human civilization, technology has been the foundation for change and progress. Facing harsh environmental conditions, human civilization has depended on technology adaptation. An array of modern technologies as well as indigenous know-how is explored in the section under *Technology Adoption*. Communities living in areas far from the influence of government institutions have used traditional knowledge to harvest rainwater for sustaining their agriculture-based livelihoods during the dry season and to protect riverbanks. This section also shows how markets often facilitate adoption of technologies that might build resilience in communities faced with the threat of cyclones. It shows how farmers in Sri Lanka used a cascading tank system to continue to irrigate their crop land for thousands of years. The cas studies show how markets and institutions facilitated adoption of modern technologies like LPG and Solar Home Systems in traditional and rural communities.

The section on *Disaster Risk Reduction* shows how communities have worked together to deal with natural disasters. It shows how communities in the coastal areas of Bangladesh used help from NGOs to build a resilient community in a cyclone-affected area. It also provides evidence on the value of mangroves on the coast of Odisha in India, in reducing damage caused by super-cyclones, and shows how in Bangladesh, villager's decision to adapt against cyclones depends on natural capital like the mangroves. It shows how farmers and the entire value chain in agriculture use seasonal weather forecasts to reduce risks and how farmers in Sri Lanka adapt locally to deal with soil erosion.

As more and more people migrate to urban areas, *Urban Sustainability* against climate change has become a major issue. In particular, it is more important for the people who are poor and living in slums and the low-lying flood prone areas. This section highlights creative approaches to water and waste management as part of climate adaptation in urban areas. This helps communities to reduce the risk of urban flooding and waterlogging due to events of high-intensity short duration rainfall. A case study also illustrates how local communities can come together and sustainably manage the supply of drinking water when natural springs are drying up in the mountains. Cases from Bangladesh, India, and Nepal show how waste management remains an important component for building resilience in urban settlements against climate extremes and how these can be addressed locally using market signals and by building awareness which not only reduces the risk of urban flooding and water logging but also provides a way of sustainable financing of municipal waste management.

Finally, the section on *Alternative Livelihood* documents how people in vulnerable communities are adjusting traditional practices and using innovation in new enterprises to cope with the inevitability of ongoing climate stresses. The section highlights win–win strategies for local communities to find alternative livelihood options using natural capital like mangroves and mountains to diversify income which is threatened by climate change. It also shows how agricultural diversification is used by farmers in Sri Lanka and Pakistan to build resilience.

1.5   The End of the Beginning

The Paris Agreement brought hope that bold initiatives through Nationally Determined Contributions (NDCs) would help combat climate change. While developed countries are better prepared to provide direct assistance and support services towards adaptation and mitigation activities, developing countries need financial and technical support to do so. This has been recognised in all international agreements and conventions related to climate change. Unfortunately, the focus has been on public interventions which often are biased towards major economic centres, and people and communities living away from the centre are left to find support on their own. Communities living in these areas often use their collective wisdom to address problems. The stories included in this book document the experiences to learn what worked and what did not in a variety of cultural, ecological, and geographic circumstances.

This book illustrates the promising opportunities for grassroots initiatives to become a significant part of climate adaptation in developing countries while also addressing other developmental needs in an integrated way. It provides locally grounded community solutions to climate change in different institutional settings, binding the geographical diversity of the region thematically. We anticipate that the case studies discussed here will become essential reading for anyone looking for examples from the grassroots in South Asia.

In the process of writing this volume, we were able to gather 58 researchers from the region to be part of a shared project with a shared vision. We envision the stories and experiences shared in this book as part of a broader movement to situate community-based climate adaptation as a mainstream approach to building climate resilient across communities in South Asia and beyond.

References

Eckstein, D., Hutfils, M., & Winges, M. (2019). *Global climate risk index 2019: Who suffers most from extreme weather events? Weather-related loss events in 2017 and 1998 to 2017.* Briefing Report, Germanwatche.V.

Eckstein, D., Künzel, V., Schäfer, L., & Winges, M. (2020). *Global climate risk index 2020: who suffers most from extreme weather events? Weather-related loss events in 2018 and 1999 to 2018.* Briefing Paper, Germanwatche.V.

Intergovernmental Panel on Climate Change. (2018). *Global warming of 1.5°C: An IPCC special report on the impacts of global warming of 1.5°C above pre-industrial levels and related global greenhouse gas emission pathways, in the context of strengthening the global response to the threat of climate change, sustainable development, and efforts to eradicate poverty,* IPCC.

United Nations Framework Convention on Climate Change. (2015). *The Paris Agreement.* https://unfccc.int/process-and-meetings/the-paris-agreement/the-paris-agreement. Last accessed 9 Feb 2020.

Part I
Concepts and Models

Chapter 2
A Framework for Climate Resilient Community-Based Adaptation

Md Rumi Shammin, A. K. Enamul Haque, and Islam M. Faisal

Key Messages

- Community-based approaches create innovative opportunities for building climate resilience.
- Local initiatives can also address other preexisting socioeconomic issues that are part of the UN Sustainable Development Goals.
- An integrated approach can foster the development of locally relevant, culturally appropriate, and resource-efficient solutions.

2.1 Introduction

Saving our planet, lifting people out of poverty, advancing economic growth… these are one and the same fight. We must connect the dots between climate change, water scarcity, energy shortages, global health, food security, and women's empowerment. Solutions to one problem must be solutions for all. Ban Ki-moon (United Nations, 2011).

Climate change is the defining environmental challenge of the twenty-first century—posing a global threat to the sustainability of environmental, social, and economic systems spanning from the North Pole to the South and all areas in between.

M. R. Shammin (✉)
Environmental Studies Program, Oberlin College, Oberlin, OH, USA
e-mail: rumi.shammin@oberlin.edu

A. K. E. Haque
Department of Economics, East West University, Dhaka, Bangladesh
e-mail: akehaque@gmail.com

I. M. Faisal
Department of Conservation and Climate, Government of Manitoba, Winnipeg, MB, Canada
e-mail: imfaisal@gmail.com

© The Author(s) 2022   11
Haque et al. (eds.), *Climate Change and Community Resilience*,
https://doi.org/10.1007/978-981-16-0680-9_2

Despite the urgency, progress in forging international climate change agreements and implementing effective mitigation programs has been slow. Political shifts in the United States and elsewhere have stalled comprehensive and coordinated efforts needed to implement the far-reaching, multilevel, and cross-sectoral mitigation and adaptation that the Intergovernmental Panel on Climate Change (IPCC) determined would be needed to deal with the projected impacts of climate change (Costanza, 2017; Yeo, 2019). In addition, the COVID-19 pandemic and its impact on the global economy further threaten timely disbursement of climate finances outlined in the 2015 Paris Agreement (UNFCCC, 2015). The latter particularly affects the implementation of top-down adaptation programs in developing countries that rely on large infrastructure investments.

The coastal communities in South Asia and other developing countries are particularly vulnerable to climate impacts due to their geographic location, demographics, and associated development challenges. The IPCC emphasizes that far-reaching, multilevel, and cross-sectoral climate mitigation needs to be upscaled and accelerated; both incremental and transformational adaptation strategies would be required to effectively address future climate-related impacts (IPCC, 2018).

Community-based initiatives are emerging as promising approaches to lessen the impacts of climate change while empowering people and bolstering community resilience (Kirkby, Williams, & Huq, 2017). These approaches have been applied to a wide range of climate adaptation programs in climate vulnerable communities—from disaster risk reduction (DRR), emergency preparedness, and flood/drought protection to sustainable agriculture, water resource management, food security, and resilient livelihood solutions (Shammin, Wang, & Sosland, 2022, Chap. 3 of this volume; UNDP, n.d.). Local innovation and agency are critical complements of these programs in fostering sustained community resilience. Therefore, community-based approaches with direct engagement of the vulnerable population that are adequately supported by international agencies, national and local government, academics, experts, and nonprofit organizations have the potential to develop locally relevant, culturally appropriate, and sustainable solutions (IPCC, 2014a).

Sustainable Development Goals (SDGs) and the principles of resilience offer additional tools to situate current climate adaptation initiatives and explore opportunities for enhancing existing adaptation models. There is documented evidence of successful outcomes of community-based initiatives in improving both the adaptive capacity and resilience. There is a need to learn from these initiatives to develop a more integrative and standardized framework that can be widely replicated (McNamara & Buggy, 2017).

This chapter adopts a holistic approach to designing future community-based adaptation programs that builds on past approaches while maximizing opportunities presented by recent developments in SDGs, resilience principles, and disaster risk reduction (DRR) initiatives. First, we deconstruct the landscape of climate change response and define key concepts and terms. Next, we contextualize the broader concepts of sustainability, sustainable development, and resilience. This is followed by a discussion on the opportunities and challenges of community-based adaptation to climate change. Finally, we present an integrative framework based on emerging concepts and lessons learned.

2.2 Key Concepts and Definitions

Climate change poses significant threats to both the abiotic (physical) and biotic (living) parts of the environment as well as economic growth and social well-being—especially in less developed countries. A report from the Asian Development Bank (ADB) predicts that by 2050, the collective economy of six South Asian countries—Bangladesh, Bhutan, India, the Maldives, Nepal, and Sri Lanka—will lose, on average 1.8% of its annual gross domestic product due to climate change impacts, rising to 8.8% by 2100 (ADB, 2014).

Climate change can also push millions of people to migrate when they reach the limits to adaptation, further intensifying intrastate and interstate competition for food, water, resources, and livelihood opportunities. There is documented evidence that climate change and weather variability are negatively impacting crop yields in Bangladesh, India, and Pakistan leading to an increase in the number of people migrating from the countryside into cities (Lohano, Iqbal, & Viswanathan, 2016). Mass migration of climate-displaced populations can have a "spillover" effect across national borders leading to heightened geopolitical tension and other global security concerns (Podesta, 2019). Against this backdrop, achieving the goals of the Paris Agreement by keeping the global average temperature rise in this century below 2 degrees Celsius over pre-industrial levels will require global collaboration and response on two fronts: *mitigation* and *adaptation* (UNFCCC, 2015).

IPCC defines mitigation of climate change as "A human intervention to reduce the sources or enhance the sinks of greenhouse gases" (IPCC, 2014c). This involves reducing the burning of fossil fuel by switching to cleaner renewable energy sources and enhancing carbon sinks via sustainable forest/land-use management practices.

Adaptation is defined by IPCC as "The process of adjustment to actual or expected climate and its effects. In human systems, adaptation seeks to moderate harm or exploit beneficial opportunities. In natural systems, human intervention may facilitate adjustment to expected climate and its effects" (IPCC, 2014c). Enhancing the adaptive capacity of a community will improve its ability to plan, respond to, and recover from external shocks, thereby making it more resilient.

2.3 The Impact Response Pathway

The process by which climate change impacts get translated into concrete societal response is complex and multifaceted. Figure 2.1 presents an overview of this process, where the societal response to the biophysical impacts of climate change is collectively determined by a range of mediating factors identified as "actors" and "enablers." The lists of mediating factors included in Fig. 2.1 are not exhaustive, but each of them play an important role in determining their collective ability to anticipate, prepare for, and respond to the present and future threats of climate change.

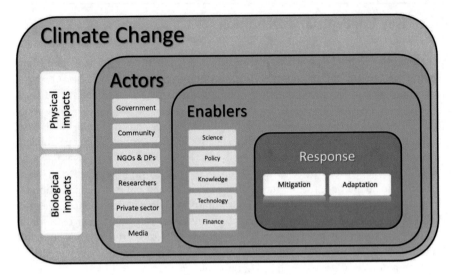

Fig. 2.1 Anatomy of climate change response. *Source* Author's conceptualization

2.3.1 *Climate Change Actors*

Both state and non-state actors play a dual role in determining the nature of response to climate change. On the one hand, they experience the biophysical impacts as institutions and individuals. On the other hand, they are also the drivers of change. These actors interact with each other within the given socioeconomic and political contexts, identify and prioritize key concerns, develop strategies, collaborate with internal and external stakeholders, mobilize resources, and ultimately design and implement the climate response programs. Coordination among the various levels of government, community-based organizations, domestic and international NGOs, international development partners (DPs), researchers (including academics), media, and the private sector is critical for achieving better outcomes.

Systemic issues can also hinder progress. For example, governments tend to suffer from "institutional inertia" and they are generally reluctant to antagonize influential private or communal interests who benefit from maintaining the status quo. As noted by Meadowcroft (2010), "Conflicts of power and interest are inevitable in relation to climate change policy" and "Climate change governance requires governments to take an active role in bringing about shifts in interest perceptions so that stable societal majorities in favour of deploying an active mitigation and adaptation policy regime can be maintained." This implies that groups aiming to bring about systemic change must focus on "building coalitions for change" at every level of decision making and "establishing new centers of economic power" so that systemic weaknesses can be reversed.

Similarly, multilateral and bilateral development partners play a delicate role in mediating climate change policies and actions. On the one hand, their stated mission is to align their programming with the priorities of national governments they partner with. On the other hand, they are accountable to their trustees and/or taxpayers in ensuring value for money from development aids, which is often clouded by other foreign policy agenda.

2.3.2   Climate Change Enablers

This complex and often contentious negotiation around social response is further mediated by the enabling factors. As depicted in Fig. 2.1, these factors influence the quality of response when the actors finally agree on the threat and are willing to take action. To begin with, climate science provides information on the cause and extent of climate change. This helps to generate future scenarios that inform evidence-based policy development including sector-specific goals for climate response.

It is important that climate policy is closely aligned, and preferably well integrated, with other key sectoral policies related to energy, infrastructure, industry, agriculture, natural resources, health, and the environment. One area where this policy alignment is crucial is disaster risk reduction. The Hyogo Framework for Action specifically identifies the need to "promote the integration of risk reduction associated with existing climate variability and future climate change into strategies for the reduction of disaster risk and adaptation to climate change" (UNISDR, 2005). Similar sentiment is echoed in the Sendai Framework, where one of the guiding principles is that "The development, strengthening and implementation of relevant policies, plans, practices and mechanisms need to aim at coherence, as appropriate, across sustainable development and growth, food security, health and safety, climate change and variability, environmental management and disaster risk reduction agendas" (UNDRR, 2015).

The integration of indigenous knowledge in policy formulation offers opportunities for developing community-based adaptation initiatives that are socially and culturally compatible, and consistent with the long-term sustainability requirements of the climate vulnerable communities. According to the IPCC Fifth Assessment Report, indigenous peoples' holistic view of community and the environment is a major resource for adapting to climate change (IPCC, 2014a).

Technological innovation is essential in climate proofing national and community-level infrastructure (e.g., roads, bridges, embankments, and cyclone shelters), protecting natural resources (e.g., forests fisheries), and creating new livelihoods options (e.g., by introducing salt-tolerant crops in coastal areas). It is important that various climate actors adopt an evidence-based and collaborative approach in developing and implementing technological solutions to achieve mitigation and adaptation goals.

Climate finance is another crucial enabler as access to adequate funding is essential for supporting adaptation initiatives in less developed countries. The polluter pays principle is now well established, and the 2015 Paris Agreement includes provisions for developed countries to mobilize financial support to assist developing country parties with climate change mitigation and adaptation efforts. It affirms a collective pledge by developed countries of providing $100 billion annually by 2020 and calls for continuing this collective mobilization through 2025 (Lattanzio, 2017)—the actual disbursement of which is complicated by political shifts in the USA and the COVID-19 pandemic, as mentioned earlier.

2.3.3 Adaptation Gap

"Adaption Gap" is the difference between the actual level of adaptation and the level required to achieve a societal goal, reflecting resource limitations and conflicting priorities (UNEP, 2014). Unfortunately, estimating the adaptation gap is far more challenging than estimating the emission gap, simply because there is no globally agreed goal or metrics for adaptation. Furthermore, adaptation is a response to specific climatic threats that may vary widely over geographic locations and time.

According to UNEP (2014), there will likely be a significant adaptation funding gap after 2020, which has been estimated to be in the order of US$70 billion to US$100 billion per year. Given the size of this deficit, multilateral and bilateral grants, domestic public finance, and contributions from the private sector—all are expected to play a role in closing this gap. In reality, a major gap in adaptation finance still persists. According to UNEP (2018), the global public finance flows in support of adaptation reached only about US$23 billion in 2016. Approximately 64 percent of this went to less developed countries via bilateral climate finance, multilateral climate funds, and multilateral development banks.

2.3.4 Limits to Adaptation

Adaptation limits are closely related to the notion of adaptation gap. A limit to adaptation is reached when adaptation efforts are unable to provide an acceptable level of security from risks to the existing objectives and values (IPCC, 2014d). Scaling down global GHG emissions and enhancing adaptive capacity may not, in the end, guarantee the desired level of risk reduction. This is because social and institutional actors are often constrained by opportunities, resources, and time needed to successfully implement and/or scale up adaptation efforts.

Dow et al. (2013) proposed a risk-based definition of adaptation limits, where adaptation efforts are considered as incremental efforts to keep the risk of adverse impacts within tolerable limits. For a given frequency and intensity of an adverse impact, the threshold for intolerable risks represents a point at which an actor must either live with the risk of escalating loss and damage, or drastically change behavior to avoid the risk, say by adopting transformational measures including relocation. According to the authors, "such a discontinuity in risk or behavior is symptomatic of an adaptation limit being reached."

2.3.5 Loss and Damage

The phrase *loss and damage* is used in the context of residual impacts that society is bound to experience beyond what can be avoided through adaptation. In addition, "loss refers to things that are lost forever and cannot be brought back, such as human lives or species loss, while damages refer to things that are damaged, but can be repaired or restored, such as roads or embankments" (Pidcock & Yeo, 2017). With the inclusion of Article 8 of the Paris Agreement, loss and damage is now a thematic pillar under the United Nations Framework Convention on Climate Change.

2.3.6 Sustainable Development Goals

IPCC defines sustainable development as development that meets the needs of the present without compromising the ability of future generations to meet their own needs (IPCC, 2014c). This definition was originally introduced by the World Commission on Environment and Development (WCED, 1987). Werners et al. (2013) argue that climate change shifts the sustainability challenge from conservation to adaptation. The authors contend that thresholds, and tipping and turning points are important focal points for sustainability under climate change that help bridge the gap between science and policy. Essentially, climate change has forced us to embrace the notion of a system that is likely changing at an accelerating pace. The scale of the issue has changed as well from local to global, which was not always the case previously. A new discipline, sustainability science, evolved using interdisciplinary research involving scientists and social actors to produce knowledge that supports and informs solutions, transformations, and transitions toward sustainability (Caniglia et al., 2017).

While these developments provide further conceptual clarity on sustainability and sustainable development, application of these concepts as guidelines and toolkits needed more detailed and comprehensive metrics. The initial work toward implementation of sustainable development began at the Earth Summit in 1992 through the preparation of Agenda 21—the Rio Declaration on Environment and Development. This was followed by the Millennium Development Goals at the turn of the century and further refined in the Sustainable Development Goals (SDGs) adopted by the United Nations in 2015 as part of the 2030 Agenda for Sustainable Development. The agenda provides a shared blueprint for peace and prosperity for people and the planet. The SDGs recognize that addressing poverty alleviation and other development challenges must go hand-in-hand with strategies that improve health and education, reduce inequality, and spur economic growth while tackling climate change and safeguarding the environment (United Nations, n.d.).

The SDGs shown in Fig. 2.2 build on more than two decades of global endeavors to operationalize sustainable development. These goals provide a detailed, practical,

Fig. 2.2 Sustainable development goals *Source* United Nations (n.d.)

and comprehensive deconstruction of the concepts of sustainability and sustainable development that captures the spirits of economic advancements, environmental responsibility, and social justice. The goals are supplemented by additional resources for implementation in a wide variety of contexts. According to an IPCC special report published in 2018, these Sustainable Development Goals provide a new framework to consider climate action within the multiple dimensions of sustainability (IPCC, 2018). The Government of Bangladesh, for example, has included SDGs in the 2016–20 National Plan for Disaster Management (Shammin, Firoz, & Hasan, 2021, Chap. 16 of this volume).

2.3.7   Resilience Principles

Ecologists have long used the concept of resilience for investigating why some ecological systems survive while others fail when faced with disturbance. It draws from the basic principles of physics of a spring that expands when pulled and retracts when released. It is the property of a system to bounce back or re-establish stability after being disturbed or perturbed. IPCC defines resilience more broadly as the capacity of social, economic, and environmental systems to cope with a hazardous event or trend or disturbance, responding or reorganizing in ways that maintain their essential function, identity, and structure, while also maintaining the capacity for adaptation, learning, and transformation (IPCC, 2014c).

Elmqvist et al. (2019) argue that the concept of resilience goes far beyond the mere recovery from disturbances and builds on the adaptive and transformative capacities of subsystems across time and scales. Tanner et al. (2014) introduced the concept of livelihood resilience as the capacity of all people across generations to sustain and improve their livelihood opportunities and well-being despite environmental, economic, social, and political disturbances. Such resilience is underpinned by human agency and empowerment, by individual and collective action, and by human rights, set within dynamic processes of social transformation. For example, Jordan (2015) analyzed case studies of specific communities in the southwest coastal region of Bangladesh and found a complex relationship between social capital and enhancing resilience to climate stress. Amin and Shammin (2022, Chap. 5 of this volume) detail the experience of the application of a resilience analysis protocol to facilitate community-led initiatives on livelihood and nature-based solutions to climate adaptation.

Elmqvist et al. (2019) also point out that resilience to climate change could mean social resilience or community resilience, or technological infrastructure resilience or ecological resilience if applied through a framing where social, ecological, and technological subsystems may differ in ways that challenge any kind of general system-level resilience. For example, United Nations Office for Disaster Risk Reduction (UNISDR) specifically defines resilience as: "The ability of a system, community or society exposed to hazards to resist, absorb, accommodate to and recover from the effects of a hazard in a timely and efficient manner, including through the preservation and restoration of its essential basic structures and functions" (UNISDR, 2009). In terms of development and climate change adaptation, the concept of resilience provides one of the most promising approaches to poverty reduction, development, growth, and sustainability (Ayeb-Karlsson et al., 2015).

Stockholm Resilience Centre has developed seven principles of resilience building that are designed to guide program development and implementation (Stockholm Resilience Centre, 2020).

1. Maintain Diversity and Redundancy.
2. Manage Connectivity.
3. Manage Slow Variables and Feedback.
4. Foster Complex Adaptive System Thinking.
5. Encourage Learning.
6. Broaden Participation.
7. Promote Polycentric Governance.

The Paris Agreement emphasizes the importance of fostering resilience as a key goal for addressing both adaptation, and loss and damage. The Sustainable Development Goals (SDGs) are closely linked to resilience building. The Global Climate Risk Index 2019 report concluded that carefully, locally, and inclusively designed adaptation measures can contribute to achieving the SDGs and increasing the resilience of communities (Eckstein et al. 2019). The Sendai Framework on Disaster Risk Reduction includes understanding of disaster risks, strengthening disaster management governance, investing in risk reduction, and resilience building. The SDGs and resilience principles thus offer opportunities for integrative community-based climate change adaptation programs at the grassroots level.

2.4 Community-Based Climate Adaptation

The word "community," as defined by the German Sociologist Ferdinand Tönnies, is meant to represent "a cohesive social entity" bound by a "unity of will" (New World Encyclopaedia, 2020). In this sense, a community-based adaptation activity could be led by an NGO, a cooperative, a closely bonded ethnic group, a village or even a "society" formed under a law with a specific set of objectives in mind. A community-based adaptation (CBA) program is closely associated with rural and other vulnerable people striving to improve their livelihood against the imminent and long-term threat of climate change. IPCC recognized that community-based adaptation activities for climate change in developing countries reveal a range of lessons as well as their limits (IPCC, 2014b). See Shammin, Wang, & Sosland (2022, Chap. 3 of this volume) for more on the concept, evolution, barriers, and opportunities of CBA.

Historically, richer nations have been able to adapt to the impact of climate change by making significant public investments in projects such as river training, embankments, and barrages. However, in many developing countries where government machinery is weak or even nonexistent, it is the communities that have protected life, used collective knowledge, and supported the weakest members against the wrath of nature. In fact, communities in Asia are often seen as a substitute for governments in terms of building strategies to navigate through difficult times. It is in this context that CBA plays a special role in reducing impacts of climate change in a community.

Community-led interventions can differ widely. Some interventions are indigenous in nature-built with the knowledge and strength within the community and its resources, while others require help from outside in terms of knowledge and/or resources to address the risks beyond the capacity of the community. Another variety of CBA that exists today has evolved over time through support from international partners and donors—who lent support to the vulnerable communities to build their ability to face the climate-induced disasters. In developing countries, adaptation constraints result from lack of access to credit, land, water, technology, markets, information, and perceptions of the need to change (IPCC, 2014b). However, several recent examples demonstrate that CBA initiatives can be developed at a low cost using domestic resources and existing international support (Shammin, Firoz, & Hasan, 2021, Chap. 16 of this volume).

After examining various adaptation strategies promoted as CBA activities, we classified them into six broad categories as shown in Fig. 2.3 and explained below.

2.4.1 Livelihood Diversification

Many of the livelihood activities are based on seasonality and other long-term climatic conditions. Climate change alters this status quo that supports the livelihood of millions of people. IPCC AR5 (2014b) reports clearly stated that due to climate change tropical agriculture will lose yield. This will have significant impact on the lives and livelihood of people across the world and particularly on people living in tropical zones (IPCC, 2014a). For example, farmers in Bangladesh are used to planting rice in July and harvest the crop in November. This has been the case for centuries resulting in the crop calendar being designed to emulate this pattern. Similar patterns exist in many other countries. However, if due to climate change, rainfall patterns change either nationwide or in the catchment areas of the rivers which flow into Bangladesh, the timing of riverine floods will shift. As a result, it may no longer be feasible for farmers to plant rice crops in July. As such, they may lose a seasonal crop and any associated income. Unless they are prepared to deal with untimely flooding with alternative strategies (like transplanting the crops in September), they might find it difficult to maintain their livelihood. Alternatively, they may switch to a rice variety which can withstand temporary flooding. For example, farmers in Bhutan used their traditional knowledge to use four indigenous breeds of rice to sustain their livelihoods. In many cases, CBA interventions supported vulnerable communities to diversify their sources of income so that if a crop is partly damaged due to climate variability, they can still manage to support their families.

2.4.2   Capacity Building

Vulnerable communities are often located in remote areas and do not have access to modern technology and knowledge base. These communities not only have low access to many public facilities but often do not have access to programs supporting formal or informal education and training. Many community-based adaptation programs, therefore, seek to leverage support from institutions, agencies, and donors. These programs introduce new technologies and products through building

Fig. 2.3 Classification of CBA activities. *Source Author's conceptualization*[1]

---

[1] Some graphic elements in this figure have been obtained from free templates at www.presentat iongo.com.

community-based awareness and/or sensitization programs. Rainwater harvesting in Nepal, desalination of water in coastal areas of Bangladesh, and introducing "floating agriculture" on a water hyacinth-made bed in water-logged areas are examples of such endeavors.

2.4.3 Ecosystem Integrity

Climate change is nature's response to human abuse on its integrity through excessive greenhouse gas emissions into the atmosphere. As such, we see nature's wrath on vulnerable communities in the Tarai lands of Nepal, on the coastal areas of Tamil Nadu, Odisha, and Bengal in India, and in coastal and riparian areas in Bangladesh, Pakistan, and so on. In many instances, communities work together with or without supports to restore the integrity of nature. Such endeavors include reforestation of degraded forest land, restoring tidal flooding of coastal rivers, protection of riverbanks, and building natural erosion control bunds. These actions eventually give the vulnerable communities a strategy to restore nature's ability to deal with extreme weather conditions like erosion, floods, water-logging, and similar events.

2.4.4 Infrastructure

Climate vulnerable communities in South Asia use local and indigenous techniques to build physical infrastructure to prevent flooding, reduce intrusion of saline water into the crop fields, control erosions, etc. For example, communities in Nepal have devised strategies to store rainwater for irrigation while communities in Bangladesh used bamboo-based structures to prevent riverbank erosions. Multi-purpose cyclone shelter in Bangladesh is another successful example of adaptation infrastructure in coastal communities.

2.4.5 Microfinance and Insurance

Microfinance has been a popular strategy in many South Asian nations used to liberate unbanked communities from the clutches of informal and often local moneylenders. These institutions give access to loans in local communities to groups of 5 to 20 members to help diversify their income sources. These loans often incorporate capacity building goals, whereby local communities develop the ability to understand the impact of climate change and find strategies to remove such risks. Many microfinance institutions in Bangladesh use microinsurance programs to ensure that borrowers are protected against different risks including flood, diseases, and so on. In addition, many loan packages come with provisions for microinsurance programs that protect them against climate-related risks. Besides registered NGOs[2] involved in lending, there are many other organizations like forest users' groups, fisheries' cooperatives, and farmers' organizations who are also engaged in lending to their members and supporting communities in dealing with climate risks.

2.4.6 Resource Management

While many community-based organizations are working on restoring degraded ecosystems, some community-based actions include protection of lakes, rivers, and forests in order to ensure that these natural resources provide natural insurance against climate risks. Similarly, in coastal lands, ensuring conservation of mangroves supports coastal communities and provides protection against cyclones and coastal flooding. Farming communities in the Western Ghats of Tamil Nadu and Kerala have restored their forest lands to improve groundwater recharge and ensure that enough water is available in their storage tanks for irrigation.

2.5 CBA Framework for Resilient Communities

Analysis of the successes and failures of CBA initiatives indicate the need to re-conceptualize the process and scope of CBA (Kirkby et al., 2016; Reed et al., 2015; Wright et al., 2014). Programs that build on indigenous knowledge and local resources where community members are active participants in both program development and implementation have the potential for advancing despite the global uncertainties mentioned earlier. Academic literature has emphasized that building adaptation processes from the bottom-up would enhance the adaptive capacity of communities (McNamara & Buggy, 2017). It is important to develop flexible models where programs may be initiated by the community, an external government, or a non-governmental agency. The process would include participatory solutions and iterative learning at the local level complemented by transformative action at national, regional, and international scales. The scope would consider the impacts of climate change alongside poverty, ecological integrity, gender equality, and other development priorities.

CBA programs designed with the explicit purposes of climate adaptation, SDG attainment, and disaster risk reduction can create opportunities for optimizing operational efficiencies and maximizing program outcomes by pooling financial, institutional, and human resources (Eckstein et al. 2019; Ayers & Forsyth, 2009; Ensor & Berger, 2009). Additionally, resilience principles can be integrated in the process

---

[2] NGOs are often registered with governments to receive donations from home and abroad in support of their activities, whereas there are other organizations which are financed through memberships and are engaged in money lending.

of program development to develop systems that increase community resilience. We present a framework that integrates all these considerations into the life cycle of climate adaptation programs with three primary goals:

- **Maximize outcome** (climate adaptation, community development, and community resilience): Reduce vulnerability and exposure for disaster risk reduction while increasing co-benefits in terms of maximizing SDG attainment and resilience building.
- **Maximize efficiency** (financial, resource, logistics): Enhance coordination to avoid duplication of efforts, reduce reporting burden, improve financial oversight, and enhance cost-effectiveness of measures that contribute to multiple agendas.
- **Maximize equity** (equity and justice): Empower socially and economically marginalized groups by ensuring their active participation in the design and implementation of CBA initiatives.

The framework consists of three levels: resources and strategies that situate CBA programs within the *context* of community knowledge, development needs and adaptive capacities; factors that need to be considered during the *process* of developing CBA programs; and enhancements that would lead to the desired *outcome* of a climate resilient community. The schematics of the CBA Framework is shown in Fig. 2.4 and explained below.

2.5.1 *Context*

CBA programs include enabling factors that can offer a better understanding of local impacts and potential solutions.

- Knowledge integration: Develop solutions by taking a holistic perspective informed by indigenous knowledge, scientific information, and global experiences (McNamara & Buggy, 2017). Maximize the use of local materials and resources and combine valuable information distilled from the community with transferable lessons from other relevant case studies to design projects.
- Livelihood solutions: Seek solutions that enhance sustainable livelihood opportunities. Adaptive strategies such as modifying agricultural practices or switching to alternative livelihoods that are designed to strengthen people's livelihood resilience (Ayeb-Karlsson et al., 2015).
- Dynamic governance: Establish a system of community-level governance with clear mandates and supporting rules to foster ownership, agency, and self-organization. Build on the existing social capital of the community and optimize connectivity across multiple programs (Stott & Huq, 2014).

2.5.2 Process

The process of developing CBA programs includes building blocks that embody the principles of resilience and lessons learned from case studies of CBA applications in developing countries. These can be applied toward the development of a wide range of CBA solutions across contexts and vulnerabilities.

- *Root cause analysis*: Identify and address specific properties of communities, social or environmental, that contribute to or exacerbate their climate risks. For example, if women are disproportionately impacted, then their participation and empowerment should be integrated in the solutions.
- *Capacity building*: Foster capacity building at the local levels to ensure independence, sustainability, and resilience. Integrate training programs for local implementation teams including staff from government and non-government organizations and volunteers. Everyone should be trained on relevant technical information, action steps, and workflow. Communication modes, methods, and chains should be clear and well understood by all.

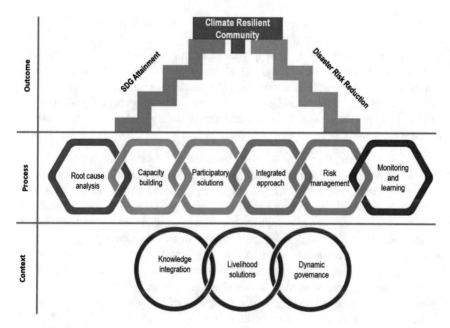

Fig. 2.4 CBA framework for climate adaptation and community resilience. *Source* Authors' conceptualization[3]

---

[3] Some graphic elements in this figure have been obtained from free templates at www.presentat iongo.com.

- *Participatory solutions*: Develop community-based initiatives with representation of key community stakeholders including vulnerable and underrepresented groups over the life cycle of the programs. Prioritize community-led solutions. Establish and maintain trust and local authority.
- *Integrated approach*: Develop multiple interventions using a diverse set of solutions with built-in redundancies instead of large infrastructure solutions that rely on singular solutions.
- *Risk management*: Ensure that risk perceptions are aligned and mutually understood between outside actors and community members. Protect the long-term integrity or stability of the local environment with careful attention to socioeconomic and ecological systems. Both slow and fast variables are considered.
- *Monitoring and learning*: Implement monitoring, feedback, and learning systems to navigate through the complexity of adaptation actions across scales and contexts (IPCC, 2014b).

2.5.3 Outcome

CBA programs are designed to reduce vulnerability and exposure to risks posed by climate change. The elements included under process and context in the CBA framework presented above can help attain a certain level of community resilience and climate adaptation. Well-designed CBA programs constitute the foundations for achieving both sustainable development and climate resilience in communities. They can narrow the gap between adaptation programs and developmental goals. Progress can be made on two parallel fronts to pursue the ultimate goal of developing climate resilient communities.

- *SDG Attainment*: While the steps outlined under process and context may help accomplish advancements toward some closely related SDGs, they are not comprehensive enough to fully harness the potential co-benefits of adaptation programs toward the attainment of all possible SDGs. The aspirational goal of community resilience would require proactive planning for maximizing the attainment of SDGs alongside planning for climate adaptation.
- *Disaster Risk Reduction*: Risk of natural disasters exacerbated by climate change remains the most widespread and consequential threat to climate vulnerable communities. Community-based adaptation programs can significantly improve emergency preparedness, reduce loss of life and property, and deliver well-coordinated responses at the local level.

2.6 Conclusions

Climate change is an evolving, intensifying, and multidimensional global environmental threat that is endangering vulnerable communities in South Asia and many other parts of the world. Many of these communities are already faced with a host of development challenges such as poverty, poor access to education and health services, and inadequate infrastructure. Progress to improve socioeconomic well-being in these communities is stymied by climate risks—often resulting in loss of livelihood and climate-induced migration. The global community has been slow to come up with the necessary international agreements, institutions, and financing to drive aggressive mitigation measures in the developed countries and support comprehensive adaptation programs in developing countries. The adaptation gap including the financing gap persists at the global level preventing more substantial investments in adaptation to climate change.

CBA initiatives that focus on the climate vulnerable population and bottom-up approaches offer a beacon of hope in this uncertain world. They present opportunities for innovation in developing place-based, community-engaged, resource-efficient, cost-effective, and sustainable responses to climate risks through better coordination of the actors and enablers (see Fig. 2.1).

CBA in the past has been ad hoc and often driven by necessity at the community level. There are many isolated examples of creative initiatives on livelihood diversification, resource management, and capacity building as well as innovative financing and infrastructure schemes to improve adaptive capacity in communities (see Fig 2.3).

There are also unique examples of adaptation initiatives that focus on building community resilience that are yet to be mainstreamed. While progress has been made toward achieving the SDGs and DRR, these initiatives have not been widely linked with climate adaptation and the results have not been uniform across different geographic areas and vulnerable groups.

A framework based on a holistic approach to community-based adaptation that embodies the resilience principles with explicit links to SDG attainment and DRR, as presented in Fig. 2.4, offers a new and more effective way for building resilience in climate vulnerable communities.

References

Amin, R., & Shammin, M. R. (2022). A resilience framework for climate adaptation: The Shyamnagar experience. In A. K. E. Haque, P. Mukhopadhyay, M. Nepal, & M. R. Shammin (Eds.), *Climate Change and community resilience: Insights from South Asia.* Singapore: Springer Nature.

Asian Development Bank. (2014). *Assessing the cost of climate change and adaptation in South Asia.* June 2014.

Ayeb-Karlsson, S., Tanner, T., van der Geest, K., & Warner, K. (2015). *Livelihood resilience in a changing world-6 global policy recommendations for a more sustainable future.* UNU-EHS Publication Series No. 22.

Ayers, J., & Forsyth, T. (2009). Community-based adaptation to climate change. *Environment: Science and Policy for Sustainable Development, 51*(4), 22–31.

Caniglia, G., Schapke, N., Lang, D., Abson, D., Luederitz, C., Wiek, A., Laubichler, M., Gralla, F., & von Wehrden, H. (2017). Experiments and evidence in sustainability science: A typology. *Journal of Clean Production., 169*, 39–47.

Costanza, R. (2017). Trump: A confluence of tipping points? *Nature, 542*, 295.

Dow, K., Berkhout, F., Preston, B., Klein, R. J. T., Midgley, G., & Shaw, M. R. (2013). Limits to adaptation. *Nature Climate Change, 3*, 305–307.

Elmqvist, T., Andersson, E., Frantzeskaki, N., McPhearson, T., Olsson, P., & Gaffney, O., et al. (2019). Sustainability and resilience for transformation in the urban century. *Nature Sustainability, 2*, 267–273.

Eckstein, D., Hutfils, M., & Winges, M. (2019). Global climate risk index 2019. *Who Suffers Most From Extreme Weather Events? Weather-related Loss Events in 2017 and 1998 to 2017. Briefing Report.* Germanwatche.

Ensor, J., & Berger, J. (2009). *Understanding climate change adaptation.* Rugby: Practical Action.

Intergovernmental Panel on Climate Change. (2014a). Technical summary. In *Climate Change 2014: Impacts, Adaptation, and Vulnerability. Part A: Global and Sectoral Aspects. Contribution of Working Group II to the Fifth Assessment Report of the Intergovernmental Panel on Climate Change.* Cambridge, UK: Cambridge University Press.

Intergovernmental Panel on Climate Change. (2014b). Summary for policymakers. In *Climate Change 2014: Impacts, Adaptation, and Vulnerability. Part A: Global and Sectoral Aspects. Contribution of Working Group II to the Fifth Assessment Report of the Intergovernmental Panel on Climate Change* (pp. 1–32). Cambridge, UK.

Intergovernmental Panel on Climate Change. (2014c). Annex II: Glossary. In *Climate Change 2014: Synthesis report. Contribution of Working Groups I, II and III to the Fifth Assessment Report of the Intergovernmental Panel on Climate Change, Geneva, Switzerland* (pp. 117–130)

Intergovernmental Panel on Climate Change. (2014d). Adaptation opportunities, constraints, and limits. In *Climate Change 2014: Impacts, Adaptation, and Vulnerability. Part A: Global and Sectoral Aspects. Contribution of Working Group II to the Fifth Assessment Report of the Intergovernmental Panel on Climate Change* (pp. 899–943). Cambridge, UK: Cambridge University Press.

Intergovernmental Panel on Climate Change. (2018). *Global warming of 1.5 °C: An IPCC special report on the impacts of global warming of 1.5 °C above pre-industrial levels and related global greenhouse gas emission pathways, in the context of strengthening the global response to the threat of climate change, sustainable development, and efforts to eradicate poverty.* IPCC.

Jordan, J. (2015). Swimming alone? The role of social capital in enhancing local resilience to climate stress: A case study from Bangladesh. *Climate and Development, 7*(2), 110–123

Kirkby, P., Williams, C., & Huq, S. (2017). Community-based adaptation (CBA): Adding conceptual clarity to the approach, and establishing its principles and challenges. *Climate and Development, 10*(7).

Lattanzio, R. K. (2017). *Paris Agreement: U.S. Climate Finance Commitments.* Congressional Research Service. 7–5700, R44870.

Lohano, H., Iqbal, K., & Viswanathan, B. (2016). *The impact of climate change on internal migration through the agriculture channel: Evidence from Bangladesh, India and Pakistan.* Policy Brief. The South Asian Network for Development and Environmental Economics (SANDEE).

McNamara, K. E., & Buggy, L. (2017). Community-based climate change adaptation: A review of academic literature. *Local Environment, 22*(4), 443–460.

Meadowcroft, J. (2010). *Climate change governance.* World Bank.

New World Encyclopaedia. (2020). https://www.newworldencyclopedia.org/entry/Community#Sociology. Last accessed 9 June 2020.

Pidcock, R., & Yeo, S. (2017). *Dealing with the "loss and damage" caused by climate change.* Carbonbrief. Explainer. Retrieved from: https://www.carbonbrief.org/explainer-dealing-with-the-loss-and-damage-caused-by-climate-change. Last accessed November 8, 2020.

Podesta, J. (2019). *The climate crisis, migration, and refugees.* Brookings. July 25, 2019. Retrieved from https://www.brookings.edu/research/the-climate-crisis-migration-and-refugees/. Last accessed November 8, 2020.

Reed, S. O., Friend, R., Jarvie, J., Henceroth, J., Thinphanga, P., Singh, D., Tran, P., & Sutarto, R. (2015). Resilience projects as experiments: Implementing climate change resilience in Asian cities. *Climate and Development, 7*(5), 469–480.

Shammin, M. R., Firoz, R., & Hasan, R. (2021). Frameworks, stories and lessons from disaster management in Bangladesh. In A. K. E. Haque, P. Mukhopadhyay, M. Nepal, & M. R. Shammin (Eds.), *Climate change and community resilience: Insights from South Asia.* Springer Nature.

Shammin, M. R., Wang, A., & Sosland, M. (2022). A survey of community-based adaptation in developing countries. In A. K. E. Haque, P. Mukhopadhyay, M. Nepal, & M. R. Shammin (Eds.), *Climate change and community resilience: Insights from South Asia.* Springer Nature.

Stockholm Resilience Centre. (2020). https://www.stockholmresilience.org/. Last accessed October 25, 2020.

Stott, C., & Huq, S. (2014). Knowledge flows in climate change adaptation: exploring friction between scales. *Climate and Development, 6*(4), 382–387.

Tanner, T., Lewis, D., Wrathall, D., Bronen, R., Cradock-Henry, N., Huq, S. & Lawless, C. (2014). Livelihood resilience in the face of climate change. Perspective. Nature Climate Change. December 18, 2014.

United Nations. (1992). United Nations framework convention on climate change.

United Nations. (n.d.). https://sdgs.un.org/goals. Last accessed April 9, 2020.

United Nations. (2011). *We the peoples.* Address to the 66th General Assembly by Ban Ki-moon. September 21, 2011. https://www.un.org/sg/en/content/sg/speeches/2011-09-21/address-66th-general-assembly-we-peoples. Last accessed September 2, 2020.

United Nations. (2015). *Transforming our world: the 2030 Agenda for Sustainable Development.* https://sdgs.un.org/2030agenda. Last accessed on 25 Oct 2020.

United Nations. (2019). *unprecedented impacts of climate change disproportionately burdening developing countries.* October 8, 2019. Meetings coverage: GA/EF/3516.

United Nations Development Programme. (n.d.). *Climate change adaptation.* https://www.adaptation-undp.org/. Last accessed August 29, 2020.

United Nations Environment Programme. (2014). *The adaptation gap report 2014.* Nairobi, Kenya: UNEP.

United Nations Environment Programme. (2018). *The adaptation gap report 2018.* Nairobi, Kenya: UNEP.

United Nations Framework Convention on Climate Change. (2015). *The Paris agreement.* United Nations.

United Nations International Strategy for Disaster Reduction. (2005). *Hyogo Framework for Action 2005–2015: Building the resilience of nations and communities to disasters* (HFA). UNISDR.

United Nations International Strategy for Disaster Reduction. (2009). *UNISDR terminology on disaster risk reduction.* Geneva: UNISDR, May 2009.

United Nations Office for Disaster Risk Reduction. (2015). *The Sendai framework for disaster risk reduction 2015–2030.* United Nations.

Werners, S., Pfenninger, S., Slobbe, E., Haasnoot, M., Kwakkel, J., & Swart, R. (2013). Thresholds, tipping and turning points for sustainability under climate change. *Current Opinion in Environmental Sustainability, 5*(3–4), 334–340.

World Commission on Environment and Development. (1987). *Our common future.* Oxford University Press.

Wright, H., Vermeulen, S., Laganda, G., Olupot, M., Ampaire, E., & Jat, M. L. (2014). Farmers, food and climate change: Ensuring community-based adaptation is mainstreamed into agricultural programmes. *Climate and Development, 6*(4), 318–328.

Yeo, S. (2019). Where climate cash is flowing and why it's not enough. News Feature. *Nature, 573*(7774)

Chapter 3
A Survey of Community-Based Adaptation in Developing Countries

Md Rumi Shammin, Amy Wang, and Maya Sosland

Key Messages

- Community-based climate change adaptation (CBA) programs are now widely used in developing countries to address climate change impacts in vulnerable communities.
- Growth in academic and gray literature provides insights into the concepts, evolution, and barriers of CBA.
- Case studies from the Philippines, Thailand, and Ethiopia illustrate the diversity of ways national and international organizations are working with local partners to implement community-based solutions in developing countries outside of South Asia.
- Lessons from literature review and case studies offer opportunities for the development and evolution of CBA beyond 2020.

3.1 Introduction

Without necessary reflection on lessons learnt from the recent history of CBA academic literature, as well as more substantial lessons from similar areas of practice, including

M. R. Shammin (✉) · A. Wang · M. Sosland
Environmental Studies Program, Oberlin College, Oberlin, OH, USA
e-mail: rumi.shammin@oberlin.edu

A. Wang
e-mail: amy.wang@oberlin.edu

M. Sosland
e-mail: maya.sosland@oberlin.edu

Haque et al. (eds.), *Climate Change and Community Resilience*,
https://doi.org/10.1007/978-981-16-0680-9_3

community-based natural resource management and participatory approaches to development, CBA initiatives risk continuing as isolated interventions, limited in both scale and impact. (McNamara & Buggy, 2017)

The term "community-based adaptation" or CBA was promoted by Huq and Reid (2007), but the principles associated with "community-based" or "bottom-up" approaches are not new in climate change discourse (Piggott-McKellar et al., 2019). A widely cited definition of CBA was articulated by Reid et al (2009) as "a community-led process, based on community's priorities, needs, knowledge, and capacities, which should empower people to plan for and cope with the impacts of climate change." Ayers and Forsyth (2009) provided further deconstruction of the concept by emphasizing the need for CBA to focus on vulnerability at the local level, identify ways to augment the adaptive capacity of communities, foster active engagement of local stakeholders and practitioners, integrate existing cultural norms, and address the root causes of vulnerability.

Early climate change adaptation (CCA) initiatives were dominated by "command and control" approaches that resulted in techno-centric, engineered, and infrastructure-type solutions (McNamara & Buggy, 2017). These initiatives were mostly driven by external entities—national or international—with limited direct engagement of the local community in the planning and decision-making process. Grassroots level initiatives—whether self-directed or externally facilitated—have become more widespread in the global south in the new millennium (Ayers & Forsyth, 2009; Schipper et al., 2014). There is increasing awareness that vulnerability, resilience, and adaptive capacity are best understood at the local level—contextualized within exogenous macro-level factors.

We conduct a survey on CBA using literature review and case studies. First, we draw key observations from a review paper on the conceptual clarity of CBA. Second, we summarize the lessons learned from review studies of both scientific literature and gray literature. Third, we distill lessons and observations from selected papers on the barriers to developing effective CBA programs. Fourth, we present three case studies from the Philippines, Thailand, and Ethiopia. These examples situate the South Asian CBA initiatives highlighted in this book in the broader context of other developing countries of the Global South. Finally, we summarize the opportunities and pathways for CBA beyond 2020—during and in the aftermath of the COVID-19 crisis.

3.2 Core Properties of CBA

Kirkby, Williams, and Huq (2017) have conducted a comprehensive study to explore the conceptual clarity of CBA. They trace the origin of the usage of the term to the early 2000s and note that there is a lack of unified definition and understanding of the term. They draw from discourses in the literature and experiences from CBA in practice. Here, we present four key takeaways from their paper that provide insights into the core properties of CBA programs.

CBA is bottom-up, participatory, and place-based

CBA requires direct engagement of communities over the life cycle of the projects and by all relevant local stakeholders. Vulnerabilities are identified based on both scientific information and indigenous knowledge. It is a process that manifests in the sociopolitical landscape of the community where local people are involved through formal and voluntary arrangements and their participation is often institutionalized. There is an emphasis on training and capacity building at the grassroots level. Solutions to short-term crises are situated within the goal of developing long-term resilience. CBA initiatives build on a community's strengths, indigenous knowledge, and local resources to maximize sustainable and resilient solutions (Shammin, Firoz, & Hasan, 2021, Chap. 2 of this volume).

CBA directly addresses the needs of the poor and the vulnerable

Top-down initiatives that dominated adaptation efforts globally were focused on hard infrastructure, resulting in bureaucratic and costly measures that failed to improve the long-term adaptive capacity of the poor and the vulnerable. On the other hand, CBA focuses on building local capacity, fostering community participation, integrating indigenous knowledge, prioritizing community empowerment, and investing in long-term well-being and resilience that can deliver adaptation solutions that are cost-efficient, representative, better managed, transparent, and effective as observed in the Shyamnagar experience (Amin and Shammin, 2021, Chap. 5 of this volume). It is a vulnerability-led approach which necessitates a broader recognition of both climate and other preexisting socioeconomic and ecological factors affecting poor and marginalized communities. Since these communities do not always have the resources to develop adaptation solutions autonomously, there is a need for extensive and transformative adaptation initiatives that directly target their needs.

CBA initiatives are multilateral, multi-level and cross-sectoral

CBA is a movement that involves financial, technical, and project development support from international agencies (e.g., the World Bank), global regimes (e.g., the Paris Agreement), the academic and research community, national and local governments, and international and local NGOs. In practice, CBA is often implemented as small-scale, localized, stand-alone, short-term, NGO-led projects that need to be scaled up, scaled out, and mainstreamed to be broadly effective. Grassroots engagement in disaster management in Bangladesh is an example of successful integration of community-level interventions into a national framework (Shammin, Haque, & Faisal, 2022, Chap. 16 of this volume). Even though people and organizations in the local communities remain the central focus in implementing CBAs, external involvement is still necessary to support the development of appropriate policies and frameworks for defining the rules of engagement. However, their role must be facilitative and supportive through decentralization and devolution of decision-making authority and administrative control. Finally, CBA programs cannot deliver sustainable and resilient solutions without cross-coordination with other related sectors. Climate change response interacts closely with initiatives for economic development, livelihood solutions, reduction of poverty and hunger, and the pursuits of justice and equality.

CBA has a growing community of practice

Unlike top-down adaptation programs, CBA relies on the experiences of a growing community of practitioners. Learning and sharing of successes, failures, best practices, and persistent challenges take place through publications, reports, conferences, workshops, and online forums. Since 2005, 14 international CBA conferences have been held through partnership of organizations such as CARE, International Institute for Environment and Development (IIED), and other international and local organizations with the most recent iteration taking place online in September 2020 (IIED, n.d.). One of the most comprehensive online platforms for CBA knowledge sharing is *weADAPT* that allows practitioners, researchers, and policymakers to access credible, high-quality information and connect with one another (weADAPT, n.d.).

3.3   CBA in Academic Literature

Academic literature on CBA has evolved significantly in the past two decades with contributions from scholars across disciplines. In this section, we summarize selected findings of McNamara and Buggy (2017) who reviewed the evolution of academic literature on CBA starting from the early 2000s based on 128 publications identified through a systematic database search. They accomplished it through keyword search of the Scopus database starting with the broad topic of climate change adaptation and subsequently narrowing down to community-based studies in the social sciences.

CBA discourses during 2000–2010

In the early years, CBA literature recognized a trend away from top-down, technological approaches to community-based initiatives that build on local knowledge and resources. Communities have embedded natural resource management strategies and coping mechanisms for dealing with natural disasters that evolved over generations and can inform and strengthen adaptation efforts. There was increasing focus on deconstructing climate change impacts and responses at different scales leading to a better understanding of how climate vulnerability, resilience, and adaptive capacity manifest at the local level. CBA offered a different paradigm for adaptation solutions that are grounded in the sociopolitical, economic, and ecological landscape of affected communities. This led to the exploration of cross-scalar approaches that bridge the separation between top-down and bottom-up initiatives. Local knowledge paired with scientific knowledge created opportunities for the development of integrated, multi-level solutions. There was growing emphasis on preexisting social drivers of vulnerability such as poverty, economic inequality, social injustice, and marginalization that exacerbate vulnerability and exposure to climate change impacts. Subsequently, ideas about conflating CBA initiatives with other development agendas such as poverty reduction, food security, and economic mobility emerged. Instead of reactive projects to address the impacts of climate change, CBA began to emerge as a pro-poor, pro-development project that advances the broader goals of sustainable development while increasing the resilience and adaptive capacity of communities.

CBA discourses beyond 2010

Since 2010, there has been significant growth in CBA literature on enabling factors including an emphasis on the use of participatory approaches to actively engage communities and facilitate the use of local knowledge over the life cycle of adaptation programs. What emerges from this are place-based initiatives that build on existing social and ecological capital and coordinate with existing efforts. These programs seek to ensure equitable and just solutions with local priorities and ownership as central to the process. The outcomes were community empowerment, enhanced adaptive capacity, and resilient solutions inspired by indigenous knowledge and grounded in local leadership. As a result, CBA was increasingly considered as a social process that manifests within the existing sociopolitical landscape of the community leading to a positive environment for collective problem-solving and decision-making. This laid the foundation for transformative structural changes that increased community resilience. Since risks and vulnerabilities were approached broadly to include a range of other preexisting development challenges, a more nuanced identification of vulnerability was possible. Increasingly, the social and economic inequalities faced by women and other marginalized groups were considered and made part of the solutions to climate vulnerabilities. The relationships between existing heterogeneity of communities and issues of power and governance were considered to reveal the root causes of climate vulnerabilities.

Institutions, communication, and finances for CBA

A large body of work is called for developing support for CBA across multiple scales to foster synergies between infrastructure and community-based solutions and strengthen the necessary institutions, policies, and rules of engagement. One key component of these multi-scalar interventions was ensuring mechanisms for effective communication and information sharing—from the upper echelons of the government or international agencies to the grassroots microcosm of the community. Social movements, advocacy, and the work of local NGOs made considerable progress in this regard to preserve the voice of the vulnerable population from being overshadowed by vested interests within and beyond the communities. Another key component was ensuring the availability of necessary funds by leveraging international financing sources. The Global Environment Facility, Green Climate Fund, United Nations Development Program (UNDP), and NGOs such as CARE International created grants and programs to stimulate the proliferation of CBAs across the developing world.

Critical discourses on CBA

In recent years, there have been a few critical discourses on CBA in academic literature. There is a need to *upscale* CBA from isolated, unique projects to a coordinated, standardized framework that can be replicated for it to be broadly effective. CBA needs to be *mainstreamed* to integrate local climate change adaptation concerns and development priorities into national development planning objectives. There are untapped opportunities for CBA initiatives to also address preexisting development issues in a community. There is a push to *re-conceptualize* CBA to become a flexible, learning process as part of a comprehensive toolbox of approaches for tackling climate change impacts and pursuing sustainable development. Mainstreaming CBA to address multi-scalar and multifaceted challenges will also require calibrating local initiatives with transformative action at the national, regional, and international scales. Finally, there is a need to develop *monitoring and evaluation* tools and outcomes to facilitate the learning process. For example, Bahinipati and Patnaik (2021, Chap. 4 of this volume) have found through a systematic literature review that there is a gap in farm-level adaptation research in India.

3.4 CBA in Gray Literature

Detailed information on the design, delivery, and experience of CBA often exists in the domain of gray literature. These include reports and evaluative studies by international development agencies, national and local governments, and non-government organizations. Since CBAs are often small scale, community driven and local, many initiatives do not get attention from scientific communities and hence are never scrutinized using scientific techniques. Many of these initiatives are reported and evaluated in the gray literature.

Piggott-McKellar et al. (2019) conducted an evaluative study on the barriers to successful CBA through a comprehensive review of gray literature available online and published in English between 2006 and 2016. They searched 21 Web sites from multilateral agencies and donors, 16 from CCA funding bodies, 71 from international NGOs, and 13 from research institutes and networks. The study found that the predominant issues addressed by these projects involved food security and agriculture followed by water security and coastal protection. Some projects also focused on broader issues such as conservation of natural resources, general livelihood solutions, and disaster risk reduction. The types of interventions used by these projects are summarized in Table 3.1.

There are a few challenges with gray literature on CBA. It is difficult to find publicly accessible online reports on CBA projects that offer detailed evaluation of outcomes and include adequate supplementary documents on project details. More comprehensive reporting and a systematic process of knowledge sharing would allow for useful learning experiences to be distilled from individual CBA projects across

Table 3.1 Types of CBA project activities

Activity type	Percent of projects (%)
Capacity building and training Natural resource management New agricultural techniques	> 50
Awareness raising Infrastructure Technology Targeting marginalized groups	30–40
Planning or policy development Establish community groups	20–30
Livelihood diversification Early warning systems Financial schemes	< 20

Note There were no entries in the 40–50% range
Source Created by the authors based on Table 3 of Piggott-McKellar et al (2019)

geographies and implementing agencies. The *weADAPT* platform mentioned earlier is a move in the right direction in this regard.

3.5 Barriers and Challenges to CBA

Piggott-McKellar et al. (2019) identified three overarching categories of barriers through their review of gray literature: sociopolitical, resource, and physical systems and processes. The sociopolitical barriers were divided into four thematic categories: (1) cognitive and behavioral (such as reluctance to adopt new technology or misalignment with cultural and religious values); (2) government structures and governance (such as poor resource and financing flows within the government); (3) communication and language (such as incompatible communication or unclear stakeholder roles and responsibilities); and (4) inequity, power, and marginalization (such as domination by the elite and powerful or gendered barriers for women). The resource-related barriers involved five categories: (1) financial (such as financial deficiencies or high implementation costs); (2) human resources (such as lack of oversight or high staff turnover); (3) time (such as limited time for relationship building, or long-term uncertainties); (4) access to information and technology (such as lack of access to information or communication hardware); and (5) infrastructural (such as poor quality of infrastructure or unintended consequences). Finally, the physical systems and processes constitute ongoing and potentially intensifying natural hazards induced by climate change that supersede the capacities of CBA interventions.[1]

---

[1] The information in this paragraph has been adapted from Table 4 of Piggott-McKellar et al (2019).

Kirkby, Williams, and Huq (2015) suggest that the aspirational goals of CBA to be inclusive, place-specific, and empowering, to support adaptation needs of the most vulnerable, are confronted with the following challenges:

- Inadequate understanding of heterogeneity, internal differences, and uneven distribution of power in communities.
- Difficulty in achieving meaningful participation of the poorest and the most marginalized.
- Top-down perspectives embedded in the international scientific knowledge systems dominating local perspectives.
- Excessive focus on the "local" downplaying important factors from outside the communities.
- Insufficient and uncertain financing.
- Lack of distinction between CBA and other development initiatives.
- Inadequate integration of CBA into government policies and programs—sometimes due to corruption, political uncertainties, and lack of interagency coordination.
- Deficiencies in the required technical expertise, funds, resources, and labor capacities to integrate CBA into government policies.
- Improper sensitivity to local cultures inhibiting planned or autonomous adaptation.

McNamara and Buggy (2017) noted a few persistent questions regarding CBA that have been identified by scholars in the academic literature. While most CBA programs address local-level vulnerabilities of the present, they need to be transformed to be able to address long-term, unforeseen climate change scenarios. Unique local lessons need to be upscaled and mainstreamed for broader application. CBA needs to become multi-level and cross-sectoral to leverage greater long-term policy and financial support. CBA needs to adequately acknowledge and address the prevailing power dynamics and inequalities in communities. Short-term CBA goals need to be situated within longer-term projects and linked with other development priorities.

3.6   CBA in Practice

3.6.1   Urban Resilience in the Philippines

The World Risk Report 2018 ranks the Philippines as the nation at third-highest risk to climate change impacts worldwide (Kirch, 2018). In 2017, the country experienced 22 named typhoons and storms followed by 21 in 2018 as well as 21 in 2019 (PAGASA, n.d.). According to a recent IPCC report (IPCC, 2019), "Extreme El Niño and La Niña events are likely to occur more frequently with global warming and are likely to intensify existing impacts … even at relatively low levels of future global warming." The urban poor in informal settlements is one of the groups most vulnerable to climate-related impacts, due in part to the additional pressures on urban systems created by rapidly increasing population growth (The World Bank, 2013). Figure 3.1 shows the progressively increasing occurrence of storms in the Philippines since 1900.

In September 2009, Typhoon *Ketsana* traversed aggressively through Metro Manila area, the national capital region and the most densely populated urban area of the Philippines with more than 12 million people living in its 636 square-kilometer land area near the coastline, causing unprecedented damages (GMMA READY Project Document, 2013). The typhoon caused more than 200 deaths and damaged more than 2 million homes in the Philippines (Scarano & Dunbar, 2009). The severity of impacts revealed the vulnerability of local systems and prompted Metro Manila to improve its disaster risk management. In addition to natural hazards, the vulnerability of the area is intensified by prevailing internal problems such as rapid urbanization and increased informal settlements (GMMA READY Project Document, 2013).

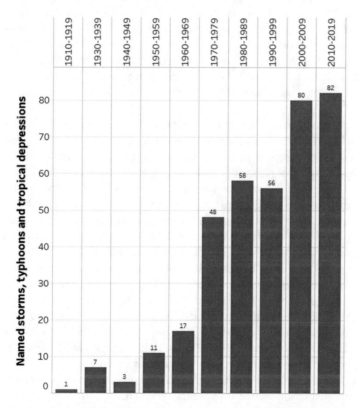

Fig. 3.1 Number of named storms, typhoons, and tropical depressions in the Philippines by decade since 1900. *Source* Author's calculations based on data from the Emergency Events Database (EM-DAT), Centre for Research on the Epidemiology of Disasters (CRED), Université catholique de Louvain (UCLouvain)

In 2009, Philippines passed the Climate Change Act that initiated the formulation of a National Framework Strategy on Climate Change which defined the overall parameters for the National Climate Change Action Plan (NCCAP)—the lead policy document guiding climate agenda at all levels of government from 2011 to 2028. The Climate Change Act and the NCCAP represent a clear evolution of priorities from mitigation to adaptation in the Philippines. The multi-agency READY project, a community-based adaptation program initiated before the Climate Change Act, was further expanded and integrated into emerging institutional frameworks. Supported by the United Nations Development Program (UNDP) and Government of Australia's Australian Aid (AusAID), READY was implemented to target disaster risk management at the local level in 27 high-risk provinces in the Philippines. The project prioritized three main approaches (Sassa & Canuti, 2008): assessment of disaster/climate risk vulnerabilities; priority of local mitigation actions; and mainstreaming climate risk management (CRM) and disaster risk management (DRM) into local planning and regulatory processes.

In the aftermath of *Ketsana*, it made sense for the READY project to be implemented in the Greater Metro Manila Area (GMMA). The GMMA READY project conducted a risk-based vulnerability and adaptation (V&A) assessment of the GMMA based on short- and long-term climate scenarios to aid in the re-examination of land use (GMMA READY Project Document, 2013). Through analyzing multi-hazard maps, the project enabled local government units (LGUs) to re-examine land use development plans to determine the safest location of future public facilities and housing.

As part of local mitigation action to reduce climate risks, the GMMA READY project focused on educating and training local communities and established a community-based warning system. The early warning systems are neither cost-intensive nor highly sophisticated, functioning at the household level with simple and easily accessible instruments, such as rain gauges. Community members learn to install early warning systems, increase awareness, and allow LGUs to better implement priority mitigation measures. For example, pilot campaigns were implemented to perform drills and rapid assessments of the building structures under earthquake risk scenarios (GMMA READY Project Document, 2013).

Under mainstreaming CRM and DRM into local planning, software called Rapid Earthquake Damage Assessment System (REDAS) was developed to simulate and estimate the impacts of multiple hazards including earthquakes, floods, and tsunami. The REDAS was developed by Philippines Institute of Volcanology and Seismology (PHIVOLCS) supported by local expertise and applied to the GMMA area. As an integrated risk assessment tool, REDAS allowed immediate mainstreaming of hazard data into resource planning. With free access and training, REDAS is used by 71 provinces, 30 state universities and colleges, private companies, and nonprofit and government institutions across the country (PHIVOLCS, 2018).

The READY project adopts a top-down approach to CBA. At the national level, the project institutionalizes and standardizes DRM measures, while at the local level, it prepares vulnerable communities in the Philippines to integrate DRM into their development plans (Agustin, 2016). However, community participation is the core accomplishment of this project. All three approaches used in the READY project are developed through pilot campaigns, public engagement, and leveraging local expertise. The READY project demonstrates a balance between local knowledge and technological innovation that offers a replicable model for other communities around the world.

3.6.2 Coastal Resilience in Thailand

Thailand, located in the southeastern region of Asia, has long coastlines mostly facing the Gulf of Thailand and partly exposed to the Andaman Sea and the Malacca Strait. The country has experienced changing temperatures and precipitation patterns for several decades. A period of recurrent and prolonged droughts between 2015 and 2016 led to critically low levels of water in reservoirs nationwide causing significant impact on agricultural productivity (Anuchitworawong, 2016). Thailand's suscepti-bility to extreme weather events such as tropical storms, floods, and drought makes it vulnerable to climate change. In addition to natural hazards, saltwater intrusion due to sea level rise is affecting rice fields, mangrove ecosystems, and coral reefs. Loss of ecosystem services from these natural resources and decline in fisheries across the coastal region of Thailand are endangering the livelihood of coastal communities. In recent decades, the local mean sea level has been increasing by 5 mm/year and can be attributed to land subsidence at the river mouth posing the impending threat of severe coastal recession in the Upper Gulf (Sojisuporn, 2013).

Recognizing the importance of indigenous knowledge in developing effective CBA programs, CARE Climate and Resilience Academy developed a Climate Vulnerability and Capacity Analysis (CVCA) method. This method facilitates the systematic collection of local knowledge on changes in living environments, resource scarcity in livelihood, local strategies against natural hazards, and past or even repet-itive hazards. This is used to identify the most vulnerable resources and the crit-ical institutions in the community, accessibility to essential services, and other key properties. The CVCA method was implemented in 20 coastal areas in Indonesia and 16 coastal areas in Thailand under the Building Coastal Resilience to Reduce Climate Change Impact in Thailand and Indonesia (BCR-CC) program. The BCR-CC program was a three-year project funded by the European Commission. This project focused on strengthening the capacity of coastal authorities and civil society organizations to deal with the challenges of climate change (Ketsomboon & Dellen, 2013). The BCR-CC project targeted four major categories in southern Thailand.

The first category was the promotion of climate-resilient livelihoods. Coastal communities in Thailand heavily rely on natural resources, such as fishing and agri-culture, as their main source of income. Hence, it is necessary to provide resources and a knowledge-sharing platform for the local communities to understand the impact of climate change on their livelihoods. Climate data from national agencies should be converted into local languages and socioeconomic contexts to improve local community's accessibility (Ketsomboon & Dellen, 2013). The process prioritized a bottom-up information flow since it can better address the specific needs and vulnerabilities of a community.

The second category was disaster risk reduction. Since mangrove forests play an important role in reducing the impacts of storms and floods in coastal areas, preventive measures were developed to protect the mangrove ecosystems. The measures included awareness of ecosystem services and restoration of degraded forest lands. The community in the village of Tambon Koh Klong Yang, one of the sixteen coastal areas, went through a series of educational programs about mangrove forests. This created incentives for the community members to safeguard the forests in an informed way. The restoration of mangrove forest in a deserted prawn farming location was initiated, and plans were put in place to expand the planting of additional mangrove forests (King, 2014).

The third category was capacity building for local civil society and government institutions. Prior to the intervention, the ministries on the provincial level in Thailand operated with a narrow focus on climate adaptation. This was most likely due to conflict of interest among different government agencies (Ketsomboon & Dellen, 2013). This created an institutional culture that resulted in inconsistencies and redundancies in adapting policies and programs. The BCR-CC program developed collaborative training on the big picture on climate change that educated people about the interconnectedness of climate adaptation initiatives with other development priorities. Climate and disaster-related information was put into context to make it accessible to local communities. Government agencies were encouraged to improve access to funding and opportunities for direct participation by community members.

The fourth category addressed the underlying causes of vulnerability. CBA programs often prioritize resource conservation and enforcement of laws and permits in the decision-making process. However, local communities who execute the national development strategies are most directly affected by the programs and outcomes. Their feedback, knowledge, questions, and concerns can better inform the national decision-making process. The BCR-CC project facilitated knowledge sharing by engaging community members in various stages of program development and implementation and integrating climate change education as part of those processes (Ketsomboon & Dellen, 2013). For example, in the village of Tambon Koh Klong Yang, students, community members, and tourists were engaged in planting mangroves during community events while learning about climate change in local communities. Moreover, a local regulation was drafted in a village meeting and shared with neighboring villages. The adopted regulation was written on a billboard, making it easily accessible to the community (King, 2014).

The BCR-CC project illustrates the links between ecosystem conservation and climate resilience. It also exemplifies the essential role of community working groups in knowledge sharing and implementing CBA programs.

3.6.3 Livelihood Resilience in Ethiopia

Ethiopia is a tropical landlocked country on the Horn of Africa. It is one of the most drought-prone countries in the world. Climate change has increased the incidence of drought due to rising temperatures, erratic rainfall, unpredictability of seasonal rain, and increased incidence of extreme events. The country also has high levels of food insecurity and ongoing conflicts over natural resources that exacerbate its vulnerability to climate change. These are projected to have significant impact on agriculture, livestock, water, and human health in Ethiopia (USAID, 2016).

The Graduation with Resilience to Achieve Sustainable Development (GRAD) project was designed to strengthen community resilience through livelihood solutions in Ethiopia. The focus was on building a system that facilitates financial management and loan access for poor households and engaging community members in analyzing key hazard maps. The main areas of focus of the GRAD project were to: enhance livelihood options, improve household and community resilience, and strengthen enabling environment to increase impact and sustainability. Between December 2011 and December 2016, the project helped build a more secure food network in 16 districts of Ethiopia through empowering women and developing sustainable farming practices.

GRAD was a USAID-funded project led and implemented by CARE, collaborating with local organizations. The essential tool established in the GRAD project is the Village Economic and Social Associations (VESAs) built on local traditions. VESAs provided community members, both men and women, with access to economic skills such as savings and credit, and financial literacy. Members of VESAs hold each other accountable to save money regularly. Those savings qualified them for small loans. According to a CARE report, 57,175 members had joined VESA by 2014 and 65% of VESA members have formal microfinance credit. Some other VESAs training sessions included good nutrition practices and climate change adaptations such as early maturing crop varieties and drought-tolerant crop types. By the end of the project, women made up 39% of the VESA participants (USAID, 2014).

GRAD prioritized livelihood solutions to climate-induced droughts. Although not related to the acronym, GRAD literally graduated participants from dependence on cash and food entitlements. It shifted the focus from aid and emergency assistance to building self-reliance and adaptive capacity—thus improving short-term well-being while building long-term resilience. Many of the GRAD interventions were designed to provide financial assistance to foster economic opportunities and enterprises. This was accomplished through provisioning of savings and credit, creation of production and marketing groups, and introduction of women-accessible economic activities. These economic interventions were accompanied by education initiatives on improving health and nutrition, gender relations, and harmful norms and practices.

GRAD instilled inner strength, ambition, and motivation among the participants. It promoted a culture of "aspiration" as opposed to "dependency." "Before GRAD, I was alone in struggling for gender equality in my community," and Kassa, a project participant, shared her experience with the GRAD project. Kassa started farming for her family when her father was getting old but did not receive much support from her family or the community because women did not traditionally work in the agricultural fields. GRAD triggered a shift in that tradition resulting in the engagement of both men and women to engage in and share household work and farming activities (CARE, 2016).

The GRAD project is an example of CBA initiatives delivering co-benefits of broader development challenges. In this case, co-benefits of building climate resilience were alternative livelihood opportunities, advancement in gender equality in the community, improved food security, development of community groups, and reduced dependency on external aid.

3.7 CBA Beyond 2020

At the time of writing this chapter, the entire world has been turned upside down by the COVID-19 pandemic. More than 85 million people worldwide have been known to be infected by December 31, 2020, with over 1.8 million confirmed deaths. The social order has been completely changed to combat this crisis—masks, social distancing, travel restrictions, and limits to the size of gatherings are among the most common measures taken globally. However, climate change impacts continue to take place in this transformed world and are making adaptation initiatives even more challenging. Developing community-engaged solutions is an intimate process. It involves staff from government departments, international agencies, and NGOs traveling to remote areas and interacting closely with many individuals and groups. Cyclone response involves directing people from vulnerable communities to congregate in the limited confines of cyclone shelters where any social distancing measure would reduce the capacity of the shelters to provide refuge. At least in the short run—assuming that this pandemic is a transient crisis—CBA initiatives are expected to face challenges that were not anticipated in the literature reviewed here. It does, however, pose the question of how future CBA programs can be designed to be resilient against unforeseen situations that may alter the social, political, or ecological order.

Building community resilience is at the heart of successful CBA. At the end of the day, vulnerable communities need to be able to withstand current and future threats of climate change—ideally on their own—and still be able to pursue their quest for improved well-being and a better quality of life. The literature review and case studies presented in this chapter illustrate that there have been exciting developments in the design, delivery, and outcomes of CBA initiatives in the last two decades. The importance of "community" in CBA has been established and fleshed out. For CBA to be successful, the community must be meaningfully engaged over the life cycle of the projects and local knowledge must be appropriately combined with scientific knowledge. Programs must be situated in the socioeconomic, cultural, and ecological realities of the communities including the nuances of preexisting inequities in power and privilege. This is true for programs administered by national or local governments and international agencies or NGOs—irrespective of whether the program is top-down, bottom-up, or some combination of the two. However, the extent and nature of community engagement may vary in these different circumstances. CBA initiatives also need to be multi-sectoral. The role of ecosystem conservation and natural resource management needs to be recognized and internalized through nature-based solutions. This is important because marginalized vulnerable populations are often disproportionately dependent on natural ecosystems for their survival.

Inclusive processes are likely to lead to building trust and ownership in the community which in turn would contribute toward long-term sustainability of the CBA solutions. The voice of the most vulnerable needs to be heard and meeting their needs should be the common denominator. There needs to be vertical and horizontal integration of institutions, information flows, and governance structures. The roles and responsibilities of all stakeholders need to be clearly defined. Improvements need to be made in documenting the details of CBA programs, instituting monitoring and evaluation processes, and sharing of lessons learned so that individual CBA initiatives can be upscaled and mainstreamed.

The GMMA READY, BCR-CC, and GRAD projects highlighted in this chapter illustrate a few of the myriad ways CBA programs are addressing some of these goals. Building on the successes of past initiatives and addressing the known barriers, CBA offers both immediate and lasting relief for climate-vulnerable communities in developing countries of the Global South. The scope of CBA solutions needs to be broadened to address other preexisting development issues as potential co-benefits. The Sustainable Development Goals (SDGs) developed by the United Nations offer a well-established framework to integrate a broad range of social, economic, and ecological solutions with CBA initiatives. Shammin, Haque, & Faisal (2022, Chap. 2 of this volume) incorporates resilience principles and the SDGs into an integrated CBA framework for the development of climate-resilient communities. While the international community struggles to reduce greenhouse gas emissions to mitigate climate change, CBA solutions liberate people in these communities from relying on the mercy of developed countries who are responsible for the bulk of the emissions. CBA not only empowers vulnerable communities with climate resilience, but also provides them with the freedom of agency and hope for a better future—against all odds.

Acknowledgements Funding for student research assistants was supported by the Arthur M. Blank Fund through the Environmental Studies Program at Oberlin College. We thank Ramsha Babar and Leo Lasdun, student research assistants at Oberlin College, for assistance with literature review and data collection.

References

Agustin, N. (2016). *GMMA READY Project terminal evaluation report.* United Nations Development Programme (UNDP).

Amin, R., & Shammin, M. R. (2021). A resilience framework for climate adaptation: the Shyamnagar experience. In A. K. E. Haque, P. Mukhopadhyay, M. Nepal, & M. R. Shammin (Eds.), *Climate change and community resilience: Insights from South Asia.* Singapore: Springer Nature.

Anuchitworawong, C. (2016). *Analysis of water availability and water productivity in irrigated agriculture.* Research Report. Thailand Development Research Institute Foundation.

Ayers, J., & T. Forsyth. (2009). Community-based adaptation to climate change. *Environment: Science and Policy for Sustainable Development, 51*(4), 22–31.

Bahinipati, C. S., & Patnaik, U. (2021). What motivates farm level adaptation in India? A systematic review. In A. K. E. Haque, P. Mukhopadhyay, M. Nepal, & M. R. Shammin (Eds.), *Climate change and community resilience: Insights from South Asia.* Singapore: Springer Nature.

CARE (2016). *Aspire: Building resilience in the Ethiopian Highlands.* CARE Ethiopia.

GMMA Ready Project Document. (2013). *Philippines disaster and climate risks management.* Australian Government Department of Foreign Affairs and Trade. https://www.dfat.gov.au/sites/default/files/dcrm-gmma-ready-project-pd.pdf. Last accessed September 17, 2020.

Huq, S., & Reid, H. (2007). *Community-based adaptation: A vital approach to the threat climate change poses to the poor.* International Institute for Environment and Development.

Intergovernmental Panel on Climate Change. (2019). Summary for policymakers. In H.-O. Pörtner, D.C. Roberts, V. Masson-Delmotte, P. Zhai, M. Tignor, E. Poloczanska, K. Mintenbeck, M. Nicolai, A. Okem, J. Petzold, B. Rama, N. Weyer (Eds.), *IPCC Special Report on the Ocean and Cryosphere in a Changing Climate.* IPCC.

International Institute for Development Studies. (n.d.). https://www.iied.org/cba14-how-you-can-help-design-programme. Last accessed on August 14, 2020.

Ketsomboon, B., & Dellen, K. (2013). *Climate vulnerability and capacity analysis report: building coastal resilience to reduce climate change impact in Thailand and Indonesia (BCR-CC).* CARE Deutschland-Luxemburg E.V. and Raks Thai Foundation.

King, S. (2014). *Community-based adaptation in practice: A global overview of CARE international's practice of community-based adaptation (CBA) to climate change.* CARE International.

Kirch, L. (Ed.) (2018). *World Risk Report 2018.* BündnisEntwicklungHilft and Ruhr University Bochum—Institute for International Law of Peace and Armed Conflict (IFHV).

Kirkby, P., Williams, C., & Huq, S. (2015). A brief overview of Community-Based Adaptation. The International Centre for Climate Change and Development (ICCCAD) at the Independent University, Bangladesh (IUB). Last accessed on July 29, 2019.

Kirkby, P., Williams, C., & Huq, S. (2017). Community-based adaptation (CBA): Adding conceptual clarity to the approach and establishing its principles and challenges. *Climate and Development, 10*(295), 1–13.

McNamara, K. E., & Buggy, L. (2017). Community-based climate change adaptation: A review of academic literature. *Local Environment, 22*(4), 443–460.

Philippine Atmospheric, Geophysical and Astronomical Services Administration. (n.d). Department of Science and Technology, Government of Philippines. http://bagong.pagasa.dost.gov.ph/. Last accessed September 12, 2020.

Philippine Institute of Volcanology and Seismology. (2018). REDAS (Rapid Earthquake Damage Assessment System). Department of Science and Technology. https://www.phivolcs.dost.gov.ph/index.php/redas. Last accessed on September 20, 2020.

Piggott-McKellar, A. E., McNamara, K. E., Nunn, P. D., & Watson, J. E. M. (2019). What are the barriers to successful community-based climate change adaptation? A review of grey literature. *Local Environment, 24*(4), 374–390.

Reid, H., Alam, M., Berger, R., Cannon, T., Huq, S., & Milligan, A. (2009). *Community-based adaptation to climate change: an overview.* Special Edition on Community-Based Adaptation to Climate Change. Participatory Learning and Action. Nottingham: Russell Press.

Sassa, K., & Canuti, P. (2008). *Landslides-disaster risk reduction.* Springer.

Scarano, J., & Dunbar, B. (2009). *NASA-Hurricane Season 2009: Typhoon Ketsana (Western Pacific).* NASA. https://www.nasa.gov/mission_pages/hurricanes/archives/2009/h2009_Ketsana.html. Last accessed on September 20, 2020.

Schipper, E. L. F., Ayers, J., Reid, H., Huq, S., & Rahman, A. (Eds.). (2014). *Community based adaptation to climate change: scaling it up.* Routledge.

Shammin, M. R., Haque, A. K. E., & Faisal, I. M. (2022). A framework for climate resilient community-based adaptation. In A. K. E. Haque, P. Mukhopadhyay, M. Nepal, & M. R. Shammin (Eds.). *Climate change and community resilience: Insights from South Asia.* Singapore: Springer Nature.

Shammin, M. R., Firoz, R., & Hasan, R. (2021). Frameworks, stories and lessons from disaster management in Bangladesh. In A. K. E. Haque, P. Mukhopadhyay, M. Nepal, & M. R. Shammin (Eds.), *Climate change and community resilience: Insights from South Asia.* Singapore: Springer Nature

Sojisuporn, P., Sangmanee, C., & Wattayakorn, G. (2013). Recent estimate of sea-level rise in the Gulf of Thailand. *Maejo International Journal of Science and Technology, 7*(Special Issue), 106–113.

The World Bank. (2013). *Getting a grip on climate change in the Philippines: overview (English).* Washington, D.C.

United States Agency for International Aid. (2014). *Graduation with resilience to achieve sustainable development (GRAD).* https://www.careevaluations.org/wp-content/uploads/GRAD-MTE-Report-Vol-1-Main-Text-FINAL.pdf. Last accessed September 12, 2020.

United States Agency for International Development. (2016). *Climate risk profile: Ethiopia.* Climate links: A global knowledge portal for climate and development practitioners. USAID. https://www.climatelinks.org/resources/climate-change-risk-profile-ethiopia. Last accessed September 17, 2020.

weADAPT (n.d.). https://www.weadapt.org/. Last accessed on August 14, 2020.

Chapter 4
What Motivates Farm-Level Adaptation in India? A Systematic Review

Chandra Sekhar Bahinipati and Unmesh Patnaik

Highlights

- A systematic review of determinants of farm-level adaptation measures in India.
- Lack of research with respect to climatic factors, perception and risk attitude behaviour and government policies.
- Although behavioural economics discourse has a lot to do with human behaviour, there is little research on this in the climate change adaptation research in India.
- We identify the gaps in adaptation research in India, so it provides a scope for scholarly communities to expand the domains of future research.

4.1 Introduction

Several studies report the negative impact of climate change on agriculture in India, now and in the foreseeable future (Pingali, 2019). Half of the labour force in India is engaged in agriculture and has been facing climate-related stress. In view of this, agricultural policies in India always aim to enhance uptake of adaptation mechanisms and to reduce variability in farmer's income. Previous studies acknowledge farm-level adaptation options undertaken by Indian farmers to reduce potential damage to crops from climate change and extreme events (Aryal et al., 2019). This includes

C. S. Bahinipati (✉)
Department of Humanities and Social Sciences, Indian Institute of Technology Tirupati, Yerpedu, Tirupati 517619, India
e-mail: csbahinipati@iittp.ac.in

U. Patnaik
Centre for Climate Change & Sustainability Studies, School of Habitat Studies, Tata Institute of Social Sciences, Mumbai 400088, India
e-mail: unmesh.patnaik@tiss.edu

© The Author(s) 2022   49
Haque et al. (eds.), *Climate Change and Community Resilience*,
https://doi.org/10.1007/978-981-16-0680-9_4

crop diversification, drought/flood tolerant seeds, altering crop-calendar, crop choice; water management options, soil conservation techniques, etc. (see Below et al., 2010). Although there is a paucity of micro-level studies to estimate pecuniary benefits in India, Patnaik, Das and Bahinipati (2019) find that adopters are getting more output and revenue in the drought-prone regions. Further, Cariappa et al. (2020) observe that insured agricultural households have more crop income and less outstanding debt. There are numerous qualitative papers that point out the positive benefits of adaptation (Singh, 2020). Several papers from South Asia also document farmer adaptation in context of flood in India and Pakistan (Ahmed, 2022, Chap. 7 of this volume and Devi, Sam and Sathyan, 2022, Chap. 8 of this volume), in Sri Lanka drought (Vidanage, Kotagama and Dunusinghe, 2022, Chap. 15 of this volume) and extreme events in India (Ghosh & Roy, 2022, Chap. 26 of this volume).

Even though several policy initiatives are in place to promote farm-level adaptation measures, a low adoption rate is prevalent across states (see Kharumnuid et al., 2018). Hence, identifying factors that influence farmer's adaptive behaviour is the major research inquiry for several papers. However, there is limited research on a systematic review of the literature on determinants of farm-level adaptation in India, and this chapter attempts to address this gap. The papers by Shaffril, Krauss and Samsuddin (2018) and Dang et al. (2019) are noteworthy exceptions and this chapter builds on Dang et al. (2019) by focussing specifically on Indian literature. This chapter is organised as follows: the next section explains the methodology for selection of papers. The third and fourth sections reflect the factors that influence farmer's adaptive behaviour based on quantitative and qualitative studies. The fifth section provides concluding remarks and highlights the avenues for various research domains in the climate change adaptation research in India.

4.2 Methods and Materials

This chapter aims to do a systematic review of determinants of farm-level adaptation options in India. In doing so, it has followed the four strategies of Preferred Reporting Items for Systematic Reviews and Meta-analysis method (PRISMA) method to select articles, i.e. identification, screening, eligibility and the inclusion (Shaffril et al., 2018). With regard to the identification or literature search, several strings/keywords related to farm-level adaptation were used to search articles in the Web of Science database (see Table 4.1). The search was conducted during January 2019, and

Table 4.1 The search string employed to identify studies for systematic review

Database	Keywords used
Web of science	("Climate smart agricultural practices" or "Coping strategies" or "Climate change adaptation" or "Farm-level adaptation measures") AND ("Farming Communities" or "Agriculture" or "Farmer") AND ("India")

Source Author's table

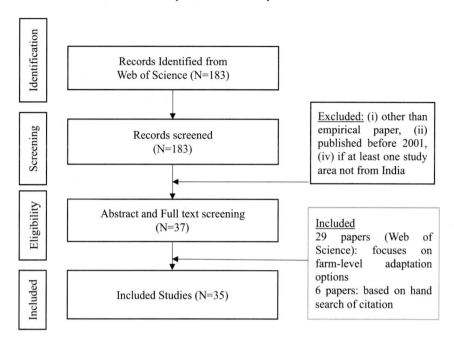

Fig. 4.1 PRISMA flow chart for the selection process. *Source* Author's figure

therefore, the papers obtained through this search must have been published before December 2018. In doing so, we have identified a total of 183 papers. After this inclusion and exclusion criteria were adopted. Inclusion was based on the literature type (empirical journal paper), time period (between 2001 and 2018),[1] and country (India). Following this, around 37 articles were selected for the eligibility stage (see Fig. 4.1), and the final selection of 29 papers was based on whether it looks into determinants of farm-level adaptation (See Fig. 4.1). Following Bird et al. (2019), we have also hand-searched citations of articles, and in doing so, we have selected an additional six papers. In sum, we arrived at a basket of 35 papers for systematic review, of which, 12 are quantitative and 23 are qualitative papers (see Fig. 4.2).

---

[1] For the first time, the notion adaptation was discussed in the COP 13 (conference of the parties) held in Bali 2007, and after this, the scholarly communities have given emphasis to different dimensions of adaptation research (Pielke et al., 2007); this is the main reason for choosing the time period between 2001 and 2018.

Fig. 4.2 Description of
study design and regional
focus. *Source* Author's figure

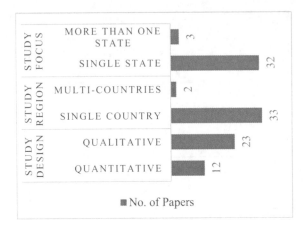

4.3 Factors Influencing Farmer's Adaptive Behaviour: Quantitative Papers

The review of quantitative papers reveals that the most frequently used determinants can be classified under seven broad categories, viz. (i) experience/perception on climate variability and extreme events, (ii) other risk and shocks (iii) socio-economic characteristics, (iv) farm characteristics, (v) access to institutions that include formal, informal, development policies and information, (vi) risk attitude behaviour and behavioural anomalies and (vii) others. Seven studies especially focussed on extreme events like cyclonic storms, floods and droughts are Bahinipati and Venkatachalam (2015), Bahinipati (2015), Mehar, Mittal and Prasad, (2016), Panda, (2013), Panda (2013), Patnaik et al. (2019) and Sahu and Mishra (2013). The remaining five papers concentrate on climate variability, monsoon onset, etc., e.g. Jain et al. (2015), Kakumanu et al. (2016), Khatri-chhetri et al. (2017), Kumar et al. (2011) and Patil et al. (2018). Table 4.2 lists out the independent variables taken by the studies in the regression models.

4.3.1 Climate Change and Extreme Events

Variables related to climatic aberrations are used as prime explanatory variables to identify whether farmers are adapting to climate change and climate sensitive options. Within these, two variants are found: (i) experienced climate variability and extreme events and (ii) perception. Out of the 12 papers, six focussed on the former while three papers consider variables related to the latter (see Fig. 4.3). Although it is anticipated all the papers should consider variables related to either climate variability and extreme events or perception to find out event specific farm-level adaptation mechanisms, only eight papers have done so. Adaptation takes place in

Table 4.2 List of independent variables considered in the papers

Author	Bahinipati and Venkatachalam (2015)	Bahinipati (2015)	Jain et al. (2015)	Kakumanu et al. (2016)	Khatri-chhetri et al. (2017)	Kumar et al. (2011)	Mehar et al. (2016)	Panda et al., (2013)	Panda (2013)	Patil et al. (2018)	Patnaik et al. (2019)[a]	Sahu and Mishra (2013)
Independent variables												
Climatic change and shocks												
Experienced and/or affected	Y	Y	–	Y	–	Y	–	Y	–	–	Y	–
Perception	–	–	Y	–	–	–	Y	Y	Y	–	–	–
Other risks and shocks	–	–	–	–	–	–	Y	–	–	–	Y	-
Socio-economic characteristics of household and household head												
Demographic characteristics	Y	Y	–	Y	Y	Y	Y	Y	Y	–	Y	Y
Caste	–	–	–	–	Y	–	–	–	–	Y	–	–
Education	Y	Y	–	Y	–	Y	Y	Y	Y	Y	Y	Y
Asset and income	Y	Y	Y	Y	Y	Y	Y	Y	Y	Y	Y	Y
Migration and remittance	Y	Y	–	Y	–	–	–	Y	Y	–	Y	–
Agricultural livelihood	Y	Y	–	Y	Y	Y	Y	Y	Y	Y	–	Y
Farm characteristics	–	–	Y	Y	Y	Y	–	Y	Y	Y	Y	Y
Access to institutions												
Formal	Y	Y	–	Y	–	Y	Y	Y	Y	–	Y	Y

(continued)

Table 4.2 (continued)

Author	Bahinipati and Venkatachalam (2015)	Bahinipati (2015)	Jain et al. (2015)	Kakumanu et al. (2016)	Khatri-chhetri et al. (2017)	Kumar et al. (2011)	Mehar et al. (2016)	Panda et al., (2013)	Panda (2013)	Patil et al. (2018)	Patnaik et al. (2019)[a]	Sahu and Mishra (2013)
Informal	Y	Y	–	Y	–	–	–	Y	Y	–	Y	–
Development policy	Y	Y	–	–	–	–	–	–	–	–	Y	–
Information	–	–	Y	–	–	Y	–	Y	Y	–	–	–
Other Characteristics												
Risk attitude behaviour	–	–	Y	Y	–	Y	–	–	–	–	–	–
Others	–	–	–	Y	–	Y	–	–	–	Y	–	–

Y denotes Yes and "–" represents no; [a]This paper was published online in the year 2017, and thus, it has appeared during search through web of science. *Source* Author's table

Fig. 4.3 Frequency of determinants appeared in the quantitative papers. *Source* Author's figure

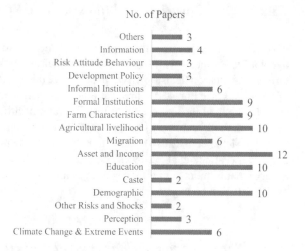

two steps, i.e. first is the farmer's perception about climate change, and then adoption of several adaptation measures (Bahinipati & Venkatachalam, 2015). Six papers consider extreme events as explanatory variables, and these are Panda (2013), Panda et al. (2013), Bahinipati (2015), Bahinipati and Venkatachalam (2015) and Kakumanu et al. (2016), Patnaik et al. (2019). Bahinipati and Venkatachalam (2015) and Bahinipati (2015) consider three variables to account for the impact of cyclones and floods, e.g. high, moderate and low. Kakumanu et al. (2016) report that variables related to climatic shocks such as drought, untimely rain and irregular weather influence the adoption of farm mechanisation and supplementary irrigation. Likewise, Panda et al. (2013) and Panda (2013) find that loss incurred due to drought has significant positive association with adaptation options, but not statistically significant. Whereas, Patnaik et al. (2019) observe that drought-affected farmers are more likely to undertake agricultural adaptation options.

Further, three papers considered perception as an independent variable, e.g. Jain et al. (2015), Panda et al., (2013) and Panda (2013). Two perception variables were considered by Jain et al. (2015), e.g. total rainfall amount and timing of monsoon onset[2], and in fact, these confounders are reported as the strongest determinant for farmer's adoption behaviour. As per Panda et al., (2013) and Panda (2013), perceived drought has significant relationship with water conservation, and likewise, perceived change in time of first rain affects farmers' decision on planting timing.

---

[2] In particular, farmers were asked to report whether they have observed any change in the total amount of rainfall and the monsoon onset date (Jain et al., 2015).

4.3.2   Other Covariate and Idiosyncratic Risks

Apart from climatic aberrations, other covariate and idiosyncratic risks also negatively affect farmer's well-being, and as a result, it could have a spillover effect on farmer's adaptation decisions. Two papers have taken variables to capture the impact of other risks (see Fig. 4.3). Three variables, for instance, were considered by Patnaik et al. (2019) and those are infant mortality rate, health expenditure and number of months of food scarcity. Similarly, one variable such as food security status was taken by Mehar et al. (2016). The infant mortality rate positively influences farmer's behaviour (Patnaik et al., 2019). With reference to food security, Mehar et al. (2016) observe that farmers facing food shortages are more likely to adapt, whereas Patnaik et al. (2019) find a negative association.

4.3.3   Socio-economic Characteristics of Household and Household Head

Numerous studies assert that the combination of social, economic and political factors reflects the adaptive capacity of an entity (Wisner, 2003), and these factors could have influenced farmer's behaviour. We have considered six sub-categories such as demographic characteristics, social category, education, asset and income, migration and remittances and agricultural livelihood. It is observed that most of the papers consider at least one variable from the demographic characteristics (see Fig. 4.3). The indicators mostly used are household size, gender and age of the household head. Except Kakumanu et al. (2016), the remaining papers have taken at most three variables under demographic characteristics. Mixed results are observed across these set of studies. It is expected that literate households are able to access information on various adaptation options, and thus, they are more likely to adapt (Bahinipati & Venkatachalam, 2015). Hence, a few studies suggested organising farmer-to-farmer meetings for spillover of information (Bahinipati & Venkatachalam, 2015). Around 10 papers have identified education as one of the determinants, and it is mostly observed as significant across the studies.

The social category of the household as an explanatory variable appeared in two papers; i.e. Khatri-chhetri et al. (2017) and Patil et al. (2018). The latter outlines significant result for the caste variable (Patil et al., 2018). Richer households opt for several adaptation mechanisms because of access to resources, asset and income (Wisner et al., 2003). Besides, Kakumanu et al. (2016), the rest of the papers have taken at most four variables, and the frequently used ones are income, agricultural income, non-agricultural income and asset. Several papers have taken income as one of the variables (Bahinipati, 2015; Bahinipati & Venkatachalam, 2015), whereas Patnaik et al. (2019) have included asset variable, and Jain et al. (2015) have constructed an asset index.

Income diversification is considered as a major coping option to smoothen both income and consumption. It is largely adopted by households to withstand all types of risks and shocks, as reported in several papers within the development economics discourse (see Morduch, 1995). In particular, either the household head or other family members migrate for better income opportunities and the remittances enhance adaptive capacity. As a result, the families with at least one migrated member are better equipped to undertake several adaptation measures. It appeared as an explanatory variable in six papers (see Fig. 4.3). If the major share of income comes from agriculture, then those households are more likely to adopt various farm-level adaptation options to reduce potential damage to agricultural crops, and in turn, reduce variability in agricultural income. Around 10 papers have considered agriculture dependency livelihood as one of the variables.

4.3.4   Farm Characteristics

Farm characteristics are mostly viewed as a major determinant for taking up adaptation options, because, farmers could undertake measures on the basis of location of land, rainfed/ irrigation, soil quality, etc. Across the studies, the frequently adopted variables are soil characteristics, location of land, irrigation, input use, etc. Around 9 papers have taken up it as one of the independent variables (see Fig. 4.3).

4.3.5   Access to Institutions

Adoption of farm-level mechanisms requires capital and liquidity, for which, farmers depend on various formal and informal sources. Around 9 papers consider the former and 7 papers included variables related to the latter (see Fig. 4.3). The papers by Kumar et al. (2011), Bahinipati and Venkatachalam (2015) and Bahinipati (2015) have considered access to formal credit, and crop loss compensation. Panda (2013) and Panda et al. (2013) included access to credit and crop insurance in the model. The access to agricultural extension is taken by Mehar et al. (2016). In regard to informal sources, most of the studies considered borrowing from money lender and self-help groups (Bahinipati, 2015; Bahinipati & Venkatachalam, 2015; Panda, 2013; Panda et al., 2013; Patnaik et al., 2019). In summary, these variables show a positive association with farmer's adaptation decision.

Around three papers have engaged with development policy-related variables. Termed as generic adaptation measures, they are expected to enhance adaptive capacity of vulnerable households (Sharma & Patwardhan, 2008). Access to Mahatma Gandhi National Rural Employment Guarantee Scheme (MGNREGS) is used as a confounder by three studies; Bahinipati and Venkatachalam (2015), Bahinipati (2015) and Patnaik et al. (2019). Four papers considered information (agronomic and agro-climatic) as variable such as Jain et al. (2015), Kumar et al. (2011) Panda et al. (2013) and Panda (2013).

4.3.6 Risk Attitude Behaviour and Other Variables

Farmer's risk preference played a major role in influencing their decision on adoption of agricultural technologies and farm-level measures (Kakumanu et al., 2016). Two papers considered risk attitude behaviour such as Jain et al. (2015) and Kakumanu et al. (2016). Following the Arrow-Pratt absolute risk aversion method, Kakumanu et al. (2016) calculate risk premium and how it is influencing adoption of technology in agriculture. However, none of these studies have examined behavioural anomalies. In fact, behavioural economics discourse in recent past years has significantly contributed to how different behavioural anomalies influence people's choice architecture, and the possible anomalies are status quo bias, uncertainty and discounting future, social norms, nudge-like interventions, heuristics, etc. Besides these variables, a few other variables are also considered in the econometric models by three papers such as Kakumanu et al. (2016), Kumar et al. (2011) and Patil et al. (2018).

4.4 Factors Influencing Farmer's Adaptive Behaviour: Qualitative Papers

A contribution of the present chapter is the inclusion of findings from qualitative papers in the review. In total, we include around 23 papers for this review with the study areas spread over entire India. Appendix outlines different determinants of farm-level adaptation options extracted from these papers. The stressor across all these papers is climate variability and climate-related extreme events from droughts to cyclonic storms and floods. Around 11 papers focus on climate variability, and 3 papers consider both climate variability and extreme events. Drought is a major stressor for 8 papers, while one paper is related to cyclonic storm and flood, e.g. Duncan et al. (2017).

Across studies related to climate variability, the reported major constraints are lack of knowledge about adaptation options (e.g. new varieties of seeds, soil conservation techniques, water management options, etc.), credit, wealth, climate information, shortage of inputs, access to agricultural extension, access to market, etc. (Kharumnuid et al., 2018; Singh et al., 2018b; Singh, 2017, 2018a; Tripathi & Mishra, 2017a, 2017b). Kharumnuid et al. (2018) and Tripathi and Mishra (2017a, 2017b) outline that although the farmers perceive a change in climate over a temporal scale, they are not adopting effective mechanisms. Singh, (2018a) finds that farmers mostly rely on short-term coping options. Rao (2017) suggest creation of farm-ponds in the semi-arid regions of Telangana to arrest rapid decline of groundwater level. Only one paper discusses the gender dimension of adaptation (Singh, 2017). Around three papers focus on climate information. These papers indicate that climate information influences farmer's adaptation decision in Maharashtra (Lobo, Chattopadhyay and Rao, 2017), West Bengal (Mishra, 2013) and Tamil Nadu (Rengalakshmi, Manjula and Devaraj, 2018), and dissemination of information through mobile phone was found to be most cost-effective (Lobo et al., 2017). In Maharashtra and Tamil Nadu, social network, farmer-to-farmer interaction, agricultural extension service, training, technology know how and interaction between farmer, financial institutions and local government should be initiated for large scale dissemination of climate information (Lobo et al., 2017; Rengalakshmi et al., 2018). Looking at both climate variability and extremes, Bhatta et al. (2016a, 2016b) conclude that adaptation options are driven by rainfall patterns, and more emphasis should be given for social protection, participatory action research to identify climate smart agricultural options, and widen the focus area of agricultural research, policy and development agencies. Also, market-related drivers are observed as major determinants for changing agricultural practices (Bhatta et al., 2016a).

Studies related to drought have taken case studies from Andhra Pradesh (Balaji, Ganapuram, and Devakumar2015; Banerjee, 2015; Banerjee et al., 2013), South India (Venot,), Odisha (Mishra & Mishra, 2010; Panda, 2016), Gujarat (Mwinjaka, Gupta, and Bresser, 2010) and Maharashtra (Banerjee, 2015; Banerjee et al., 2013; Udmale, 2014). Balaji et al. (2015) infer that the likelihood of adoption of options is high for smallholder farmers if they are part of the learning process. Banerjee et al. (2013) observe the major constraints are poor governance, lack of collective action, inequality, political influence and social exclusion. Venot et al. (2010) observe that farmers in South India are opting for short-term coping measures. The intensity of droughts is seen as key for adoption of adaptation mechanisms in Odisha (Mishra & Mishra, 2010). Panda (2016) mentions that at the community level, the constraints are lack of government intervention, no knowledge about drought resistant crops, lack of renovation of water bodies and irrigation systems. Overall, integration of indigenous knowledge and traditional practices with scientific knowledge and development planning is supported by Banerjee et al. (2015) and Singh et al. (2018a), Singh et al. (2018b).

4.5   Concluding Remarks

Adopting farm-level options is imperative for the farmers in India to mitigate potential damages to agricultural output. Over the years, low uptake was cited across the studies, and therefore, identifying determinants is a major research question for several papers. There is a gap in the literature with regard to comprehensive reviews of existing studies examining the factors that facilitate farmers to adopt farm-level adaptation measures. This chapter attempts to address this by reviewing 12 quantitative and 23 qualitative papers published between 2001 and 2018. We have come up with four major conclusions. *First*, non-climatic factors like demographic character-istics, asset and income, education, dependency on agriculture, farm characteristics and access to financial institutions are observed as major motivating factors. *Second*, this chapter advocates future studies to establish the causal relationship of farm-level adaptation options with climate change and extreme events and to identify climate and extreme event specific adaptation measures. *Third*, there is a paucity of studies with reference to perception, risk attitude behaviour, climate information and development-related policies. *Fourth*, although behavioural economics discourse has lot to do with human behaviour, little research has been taken up on climate change adaptation research in India. These findings provide important avenues for scholarly communities and policy makers for expanding the domains of future research while realigning existing plans to address the observed gaps and develop evidence-based policies to undertake adaptation mechanisms.

Although research on issues related to agriculture is extensive in India, focus on the aspects related to the impact of climatic aberrations and extremes started only during the past one and half decades. We have attempted to include papers in this chapter based on seemingly robust inclusion criteria, but the non-inclusion of many research articles related to agricultural interventions has been a limitation. Further, we have excluded several papers in grey literature. Another limitation of the review is that we have not examined the farm-level adaptation decisions across different categories of farmers, for instance, large farmers compared to small and marginal ones. A final weakness concerns the exclusion of non-farm adaptation measures. Addressing some of these could be a starting point for future research.

Acknowledgements The first author has received financial support from the Indian Institute of Technology Tirupati, Yerpedu, India through the New Faculty Initiative Grant (NFIG) to carry out this study. The authors would like to acknowledge the research support received from Vijay Kumar and Dinamani Biswal. The usual disclaimers apply.

Appendix: Summary of Major Findings from the Qualitative Studies

S. No.	Author (Yyear)	Study location	Major findings
	Climate variability		

(continued)

(continued)

S. No.	Author (Yyear)	Study location	Major findings
1	Kharumnuid et al. (2018)	Meghalaya	• Large percentage of households undertake medium level adaptation, i.e. persistence of adaptation deficit • Major constraints: market price distortion, lack of technical knowhow, communication and market, shortage of inputs, particularly quality seeds and non-availability of climate information
2	Lobo et al. (2017)	Maharashtra	• Dissemination of agro-climatic information through mobile is cost-effective • Training and awareness campaign, technology knowhow and interaction among the farmers, civil society organisations, financial institutions and local government bodies are required for further scaling up
3	Mishra et al. (2013)	West Bengal	• Access to rainfall forecast information reduces the use of water for irrigation
4	Rao et al. (2017)	Telangana	• Farm pond could reduce water footprint, in turn, increase groundwater level, and hence, increase cropped area and productivity
5	Rengalakshmi et al. (2018)	Tamil Nadu	• Climate information should be context specific and gender sensitive, and integration of traditional knowledge with scientific information is essential • Social network, farmer-to-farmer interaction and agricultural extension services should be promoted for large scale dissemination of climate information

(continued)

(continued)

S. No.	Author (Yyear)	Study location	Major findings
6	Singh et al. (2016)	Rajasthan	• Apart from resource endowment, social cognitive factors like perceived adaptive capacity and efficacy to carry out adaptation options influence farmers' decision
7	Singh et al. (2017)	Arunachal Pradesh	• Gender and wealth are found as major determinants
8	Singh et al., (2018a)	Karnataka	• Responses are multi-dimensional; vary with geography, identity, social capital and economic status • Farmers rely more on short-term coping strategies and less long-term measures • Interventions may not be climate specific, therefore, integration of climate component is necessary, i.e. comprehensive risk-response framework
9	Singh et al. (2018b)	Punjab and Telangana	• Major constraints are lack of knowledge, credit, climate information, land holding and shortage of input and agricultural labour • Integration of local level perception and adaptation strategies in development planning
10	Tripathi and Mishra (2017a, 2017b)	Uttar Pradesh	• Farmers perceive climate variability, but not undertaking effective response • Advises to provide climate information, extension services and to organise capacity building programme
	Climate variability and Extreme Events		

(continued)

(continued)

S. No.	Author (Yyear)	Study location	Major findings
11	Bhatta et al. (2016a)	Bihar	• Adaptation practices are associated with rainfall patterns, e.g. farmers in medium rainfall areas (900–1500 mm) are taking several adaptation practices • Large farmers undertake on-farm livelihood options, and also more innovative • More emphasis for social protection, participatory action research to identify efficient climate smart agricultural options; widen the focus area of agricultural research, policy and development agencies
12	Bhatta et al. (2016b)	Bihar	• Farmers in the high rainfall region (1500–2100 mm) are likely to undertake several on-farm livelihood diversification activities • Market-related drivers are major determinants • Evokes for developing rainfall and farmers' resource endowment strategies
13	Pandey (2018)	Uttarakhand	• Communities are largely depending on traditional knowledge and historical climate information, and thus, integration of these with scientific knowledge could lead to cost-effective options • Major barriers are shortage of cash, lack of information and awareness
	Drought		
14	Balaji et al. (2015)	Andhra Pradesh	• High likelihood of undertaking adaptation by smallholder farmers when they are part of a learning context, e.g. Mobi-MOOC • Advocates for requirement of effective channels of information sharing and capacity development

(continued)

(continued)

S. No.	Author (Yyear)	Study location	Major findings
15	Banerjee et al. (2013)	Andhra Pradesh and Maharashtra	• Minimum of impact of information dissemination on the poor farmers • Poor governance, collective action, inequality, local level political power and social exclusion are the major constraints • Advocates for undertaking adaptation at institutional and community level, so it minimises the additional burden on the marginalised communities. Further, support to be provided with respect to finance, information, technology and infrastructure
16	Banerjee (2015)	Andhra Pradesh and Maharashtra	• Integration of indigenous knowledge and traditional practices while designing climate action plan • At policy level, financial inclusion and land right to women should be promoted
17	Venot et al. (2010)	South India	• Farmers are changing the cropping patterns followed by a drought and back to normal crops in other years, i.e. mostly adopting short-term coping strategies
18	Mishra and Mishra (2010)	Odisha	•Adaptation options are undertaken based on the drought intensity
19	Mwinjaka et al. (2010)	Gujarat	• Additional government interventions to assist poor farmers are essential • Greater coordination between research, monitoring and technological institutes are required • Though several policies are in place to withstand drought, these are not benefiting poor farmers (e.g. food for work programme, credit, insurance, etc.)

(continued)

(continued)

S. No.	Author (Yyear)	Study location	Major findings
20	Panda (2016)	Odisha	• Major barriers: • Household level: shortage of water and irrigation facility, knowledge about several adaptation options and early warning system • Community level: lack of government intervention, no knowledge about drought resistant crops and pro-active renovation of water bodies and irrigation systems
21	Udmale et al. (2014)	Maharashtra	• Farmers perceive climate change, however, not adopt sufficient options • Low satisfaction for government supported drought mitigation measures
	Other hazards (Cyclonic Storms and Floods)		
22	Duncan et al. (2017)	Odisha	• Factors related to social, economic, institutional and environmental are affecting farmers' resilience capacity

Source Author's compilation from different studies

References

Ahmed, A. (2022). Autonomous adaptation to flooding by farmers in Pakistan. In A. K. E. Haque, P. Mukhopadhyay, M. Nepal & M. R. Shammin (Eds.), *Climate change and community resilience: Insights from South Asia* (pp. 101–112). Springer Nature.

Aryal, J. P., Sapkota, T. B., Khurana, R., et al. (2019). Climate change and agriculture in South Asia: Adaptation options in smallholder production systems. *Environment, Development and Sustainability, 22*, 5045–5075.

Bahinipati, C. S. (2015). Determinants of farm-level adaptation diversity to cyclone and flood: Insights from a farm household-level survey in Eastern India. *Water Policy, 17*, 742–761.

Bahinipati, C. S., & Venkatachalam, L. (2015). What drives farmers to adopt farm-level adaptation practices to climate extremes: Empirical evidence from Odisha, India. *International Journal of Disaster Risk Reduction, 14*(4), 347–356.

Balaji, V., Ganapuram, S., & Devakumar, C. (2015). Communication and capacity building to advance adaptation strategies in agriculture in the context of climate change in India. *Decision, 42*(2), 147–158.

Banerjee, R. (2015). Farmer's perception of climate change, impact and adaptation strategies: A case study of four villages in the semi-arid regions of India. *Natural Hazards, 75*, 2829–2845.

Banerjee, R., Kamanda, J., Bantilan, C., et al. (2013). Exploring the relationship between local institutions in SAT India and adaptation to climate variability. *Natural Hazards, 65*, 1443–1464.

Below, T., Artner, A., Siebert, R., & Sieber, S. (2010). Micro-level practices to adapt to climate change for African small-scale farmers: A review of selected literature. *IFPRI Discussion Paper 00953.*

Bhatta, G. D., Aggarwal, P. K., Shrivastava, A. K., et al. (2016a). Is rainfall gradient a factor of livelihood diversification? Empirical evidence from around climatic hotspots in Indo-Gangetic Plains. *Environment, Development and Sustainability, 18*(6), 1657–1678.

Bhatta, G. D., Aggarwal, P. K., Kristjanson, P., et al. (2016b). Climatic and non-climatic factors influencing changing agricultural practices across different rainfall regimes in South Asia. *Current Science, 110*(7), 1272–1281.

Bird, F. A., Pradhan, A., Bhavani, R. V., & Dangour, A. D. (2019). Interventions in agriculture for nutrition outcomes: A systematic review focused on South Asia. *Food Policy, 82*, 39–49.

Cariappa, A. G. A., Mahida, D. P., Lal, P., & Chandel, B. S. (2020). Correlates and impact of crop insurance in India: Evidence from a nationally representative survey. *Agricultural Finance Review.* https://doi.org/10.1108/AFR-03-2020-0034

Dang, H. L., et al. (2019). Factors influencing the adaptation of farmers in response to climate change: A review. *Climate and Development.* https://doi.org/10.1080/17565529.2018.1562866

Devi, P. I., Sam, A. S., & Sathyan, A. R. (2022). Resilience to climate stresses in South India: Conservation responses and exploitative reactions. In A. K. E. Haque, P. Mukhopadhyay, M. Nepal, & M. R. Shammin (Eds.), *Climate change and Community resilience: Insights from South Asia* (pp. 113–127). Springer Nature.

Duncan, J. M., Tompkins, E. L., Dash, J., et al. (2017). Resilience to hazards: Rice farmers in the Mahanadi Delta India. *Ecology and Society, 22*(4), 3. https://doi.org/10.5751/ES-09559-220403

Ghosh, S., & Roy, S. (2022). Climate change, ecological stress and livelihood choices in Indian Sundarban. In A. K. E. Haque, P. Mukhopadhyay, M. Nepal, & M. R. Shammin (Eds.), *Climate change and community resilience: Insights from South Asia* (pp. 399–413). Springer Nature

Jain, M., Naeem, S., Orlove, B., et al. (2015). Understanding the causes and consequences of differential decision-making in adaptation research: Adapting to a delayed monsoon onset in Gujarat, India. *Global Environmental Change, 31*, 98–109.

Kakumanu, K. R., Kuppanan, P., Shalander, K., et al. (2016). Assessment of risk premium in farm technology adoption as a climate change adaptation strategy in the dryland systems of India. *International Journal of Climate Change Strategies and Management, 8*(5), 689–717.

Kharumnuid, P., Rao, I., Sudharani, V., et al. (2018). Farm level adaptation practices of potato growing farmers in East Khasi Hills district of Meghalaya India. *Journal of Environmental Biology, 39*(5), 575–580.

Khatri-Chhetri, A., Aggarwal, P., Joshi, P. et al., (2017). Farmer's prioritization of climate-smart agriculture (CSA) technologies. *Agricultural Systems, 151*, 184–191

Kumar, D. S., et al. (2011). An analysis of farmer's perception and awareness towards crop insurance as a tool for risk management in Tamil Nadu. *Agricultural Economics Research Review, 24*, 37–46.

Lobo, C., Chattopadhyay, N., & Rao, K. V. (2017). Making smallholder farming climate-smart: Integrated agrometeorological services. *Economic and Political Weekly, 52*(1), 53–58.

Mehar, M., Mittal, S., & Prasad, N. (2016). Farmers coping strategies for climate shock: Is it differentiated by gender? *Journal of Rural Studies, 44*, 123–131.

Mishra, S. K., & Mishra, N. (2010). Vulnerability and adaptation analysis in flood affected areas of Orissa. *Social Change, 40*(2), 175–193.

Mishra, A., et al. (2013). Short-term rainfall forecasts as a soft adaptation to climate change in irrigation management in North-East India. *Agricultural Water Management, 127*, 97–106.

Morduch, J. (1995). Income smoothing and consumption smoothing. *Journal of Economic Perspectives, 9*(3), 103–114.

Mwinjaka, O., Gupta, J., & Bresser, T. (2010). Adaptation strategies of the poorest farmers in drought-prone Gujarat. *Climate and Development, 2*(4), 346–363.

Panda, A. (2013). Climate variability and the role of access to crop insurance as a social-protection measure: Insights from India. *Development Policy Review, 31*(2), 57–73.

Panda, A. (2016). Exploring climate change perceptions, rainfall trends and perceived barriers to adaptation in a drought affected region in India. *Natural Hazards, 84*(2), 777–796.

Panda, A., et al. (2013). Adaptive capacity contributing to improved agricultural productivity at the household level: Empirical findings highlighting the importance of crop insurance. *Global Environmental Change, 23*, 782–790.

Pandey, R., et al. (2018). Climate change adaptation in the Western-Himalayas: Household level perspectives on impacts and barriers. *Ecological Indicators, 84*, 27–37.

Patil, V. S., et al. (2018). Adapting or changing water? Crop choice and farmer's responses to water stress in peri-urban Bangalore, India. *Irrigation and Drainage, 68*, 140–151.

Patnaik, U., Das, P. K., & Bahinipati, C. S. (2019). Development interventions, adaptation decisions and farmer's well-being: Evidence from drought prone households in rural India. *Climate and Development, 11*(4), 302–318.

Pielke, R. J., et al. (2007). Lifting the taboo on adaptation. *Nature, 455*(8), 597–598.

Pingali, P., et al. (2019). *Transforming food systems for a rising India.* Palgrave Macmillan.

Rao, C. S. R., et al. (2017). Farm-pond for climate resilient rain-fed agriculture. *Current Science, 112*(3), 471–477.

Rengalakshmi, R., Manjula, M., & Devaraj, M. (2018). Making climate information communication gender sensitive: Lessons from Tamil Nadu. *Economic and Political Weekly, 53*(17), 87–95.

Sahu, N. C., & Mishra, D. (2013). Analysis of perception and adaptability strategies of the farmers to climate change in Odisha, India. *APCBEE Procedia, 5*, 123–127.

Shaffril, H. A. M., Krauss, S. E., & Samsuddin, S. F. (2018). A systematic review on Asian farmer's adaptation practices towards climate change. *Science of the Total Environment, 644*, 683–695.

Sharma, U., & Patwardhan, A. (2008). An empirical approach to assessing generic adaptive capacity to tropical cyclone risk in coastal districts of India. *Mitigation and Adaptation Strategies for Global Change, 13*(8), 819–831.

Singh, C., Dorwarda, P., & Osbahra, H. (2016). Developing a holistic approach to the analysis of farmer decision-making: Implications for adaptation policy and practice in developing countries. *Land Use Policy, 59*, 329–343.

Singh, R. K., et al. (2017). Perceptions of climate variability and livelihood adaptations relating to gender and wealth among the Adi community of the Eastern Indian Himalayas. *Applied Geography, 85*, 41–52.

Singh, C., et al. (2018a). Risks and responses in rural India: Implications for local climate change adaptation action. *Climate Risk Management, 21*, 52–68.

Singh, N. P., Anand, B., & Khan, M. A. (2018b). Micro-level perception to climate change and adaptation issues: A prelude to mainstreaming climate adaptation into developmental landscape in India. *Natural Hazards, 92*, 1287–1304.

Singh, C., et al. (2020). Assessing the feasibility of adaptation options: Methodological advancements and directions for climate adaptation research and practice. *Climatic Change, 162*, 255–277.

Tripathi, A., & Mishra, A. K. (2017a). Farmers need more help to adapt to climate change. *Economic and Political Weekly, 52*(24), 53–59.

Tripathi, A., & Mishra, A. K. (2017b). Knowledge and passive adaptation to climate change: An example from Indian farmers. *Climate Risk Management, 16*, 195–207.

Udmale, P., et al. (2014). Farmer's perception of drought impacts, local adaptation and administrative mitigation measures in Maharashtra state, India. *International Journal of Disaster Risk Reduction, 10*, 250–269.

Venot, J. P., et al. (2010). Farmer's adaptation and regional land-use changes in irrigation systems under fluctuating water supply, south India. *Journal of Irrigation and Drainage Engineering, 136*(9), 595–609.

Vidanage, S. P., Kotagama, H. B., & Dunusinghe, P. M. (2022). Sri Lanka's small tank cascade systems: Building agricultural resilience in the dry zone. In A. K. E. Haque, P. Mukhopadhyay,

M. Nepal, & M. R. Shammin (Eds.), *Climate change and community resilience: Insights from South Asia* (pp. 225–235). Springer Nature.
Wisner, B., et al. (2003). *At risk: Natural hazards, people's vulnerability and disasters* (2nd ed.). Routledge.

Chapter 5
A Resilience Framework for Climate Adaptation: The Shyamnagar Experience

Raquibul Amin and Md Rumi Shammin

Key Messages

- Scenario analysis for resilience building delivers community-led, locally relevant programme development and nature-based solutions.
- Concentration of multiple integrated interventions at the community level produces effective and efficient outcomes.
- Inclusive and collaborative community engagement produces a shared vision for pathways towards a sustainable future.

5.1   The Old Guard

> But here, in the tide country, transformation is the rule of life: rivers stray from week to week, and islands are made and unmade in days. In other places, forests take centuries, even millennia, to regenerate, but mangroves can recolonize a denuded island in ten to fifteen years. Could it be the very rhythms of the earth were quickened here so that they unfolded at an accelerated pace?—Amitav Ghosh

---

R. Amin
Bangladesh Country Office, International Union for Conservation of Nature (IUCN), Dhaka, Bangladesh
e-mail: Raquibul.amin@iucn.org

M. R. Shammin (✉)
Environmental Studies Program, Oberlin College, Oberlin, OH, USA
e-mail: rumi.shammin@oberlin.edu

© The Author(s) 2022
Haque et al. (eds.), *Climate Change and Community Resilience*,
https://doi.org/10.1007/978-981-16-0680-9_5

In his tantalizing novel *The Hungry Tide*, award-winning writer Amitav Ghosh depicted the social and political struggles of the unique social–ecological setting of the Indian Sundarban near Kolkata (Ghosh, 2005). Nature's influence in shaping the past, present and future of the Sundarban area where land meets the sea was brought alive through his narrative. The Sundarban, a World Heritage Site, is the largest contiguous mangrove forest in the world spanning across the border between Bangladesh and India and home of the majestic Bengal tiger—a globally endangered species as per the Red List (IUCN, n.d.). Port Canning, a port city inaugurated in 1864 on the fringes of the Sundarban in South 24 Parganas district in the province of formerly undivided Bengal, was built on the banks of the *Matla* River but washed away by consecutive cyclones within a few years (Ghosh, 2016). The power the natural world has over land and people in this part of the world is simply overwhelming.

Shyamnagar, a small *upazila* (sub-district) in Bangladesh, not very far from Port Canning, represents a similar narrative of people's lives shaped by the tides of *Kholpetuya* and *Arpangasia* rivers. The oscillating rise and fall in prosperity in Shyamnagar is shaped by the social, economic and political landscape of the region, but more importantly by the tropical cyclones emerging from the Bay of Bengal. Unlike Port Canning, Shyamnagar sits behind the Sundarban which tames the fury of the cyclones to a large extent by the time they arrive here. Lives are still lost, and rice fields, shrimp *ghers (ponds)* and houses ravaged, but the damage could be a lot more without the natural protection of the Sundarban—the old guard shielding Shyamnagar from the spinning skies and the hungry tides of the Bay of Bengal for centuries.

Cyclone *Amphan*, which made landfall in May 2020, headed towards Port Canning and Kolkata with similar ferocity as in 1864 and 1867—this time narrowly missing Shyamnagar in its direct path. The area, however, has not always been so lucky. Cyclone *Sidr* in 2007 and Aila in 2009 left lives and livelihoods in disarray in their wake. When the International Union for Conservation of Nature (IUCN) decided to implement the Mangroves for the Future (MFF) initiative in 2014 with the goal of building resilience in vulnerable communities, Shyamnagar was still suffering from the aftermath of these disasters. This chapter tells the story of the MFF initiative in Shyamnagar and how it applied a resilience framework that deviated from the typical linear project development and implementation models by adopting an approach based on integrating the existing dynamics of the complex social and ecological systems. Lessons learned from this example can inform the development of climate resilient community-based adaptation initiatives and nature-based solutions in other vulnerable communities in South Asia and elsewhere in the world.

5.2 Shyamnagar: Life on the Edge

Shyamnagar *upazila* is located in south-western Bangladesh, near the border between Bangladesh and India (as shown in Fig. 5.1), with a population of more than 300,000

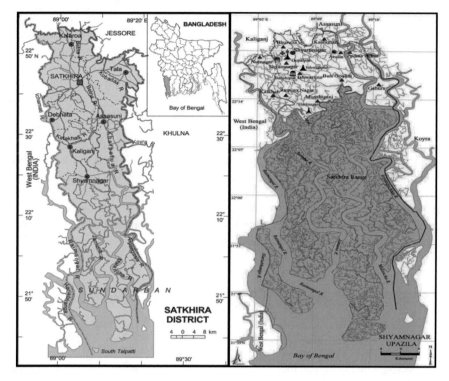

Fig. 5.1 Map of Shyamnagar *upazila* in the Satkhira district of south-western Bangladesh. *Source* Banglapedia (n.d.)

inhabiting an area of about 2000 km^2 (Bangladesh Bureau of Statistics, 2012). Large parts of the upazila comprise of rivers, canals, agriculture lands and shrimp *ghers*. This limits the areas available for human settlement, and the communities are concentrated in densely populated enclaves. The area is highly vulnerable to natural hazards such as cyclones and storm surges—compounded by drainage congestion due to the construction of massive polders (circular embankments protecting a land areas) and expansion of saline zones due to increasing shrimp cultivation (IUCN, 2013a).

Shyamnagar lies in a 10 km wide buffer area beyond the Sundarban Reserve Forest (SRF) boundary that is designated as the Ecologically Critical Area (ECA)[1] of the SRF. A significant part of the population depends on the Sundarban for its livelihood. For these people, survival is an everyday battle. The declining health of the mangrove ecosystem on the edges of the forest forces them to travel deeper inside the forest to harvest subsistence and economic resources such as honey, crab and other non-timber forest products. Along with storm surges, salinity intrusion and more intense cyclones, this results in further degradation of the ecosystem of the forest. When people venture deeper into the forest, they also face the threat of

---

[1] Ecologically defined areas or ecosystems affected adversely by the changes brought through human activities designated under Bangladesh Environment Conservation (Amendment) Act 2010.

being attacked by the Bengal tiger. A falling population growth rate from 1.7% in 2001 to only 0.14% in 2011 (BBS, 2012) highlights the precarious balance of the social and ecological systems that exists in Shyamnagar and indicates the possibility of increasing climate-induced displacement.

With prevailing poverty, proximity to the Sundarban, complexity of the socio-economic and ecological systems and changing demography with seasonal and permanent outmigration, Shyamnagar presented a challenging environment and creative opportunity for developing and testing a resilience-based approach to climate change adaptation. This required a clear understanding of how the dynamics between coastal ecosystems and human communities has evolved over time and factors that may shape their co-existence in the future. This prompted IUCN to develop a new diagnostic tool to distil local knowledge on human–environment interactions and the issues that impact the tipping points. The centrepiece of this approach was a concerted attempt to create the solutions with the local community to engender local ownership of the resilience building process.

5.3   RAP: A Systems Approach to Community Resilience

One of the cornerstones of the MFF initiative was a grant programme with the explicit goal of developing solutions grounded in the realities and needs of the local communities. The programme sought to implement projects with an area-based approach that would identify and incorporate the differing yet connected and inter-acting perspectives that drive the dynamics of complex social–ecological systems (SES). Systematic evaluation to stimulate a cycle of learning, improvements and generate policy-relevant lessons was also incorporated. One overarching goal was to address the two core questions in the process of resilience building posed by Walker et al. (2002): 'resilience of what?' and 'resilience to what?'. In order to achieve these goals, IUCN developed the Resilience Analysis Protocol (RAP)—a context-specific tool that could be applied to a wide variety of communities across the 11 countries where MFF was operating. The protocol was inspired by the work of the Stockholm Resilience Centre on resilience thinking. Global Climate Adaptation Partnership (GCAP) also provided technical input in drafting of the RAP.

RAP builds on the Drivers–Pressures–States–Impacts–Responses (DPSIR) framework used by the United Nations (UNEP, 2002). The DPSIR is based on a linear causality between drivers, pressures and impacts; a measurable and desirable state of ecosystems and communities; and accurate predictability of future impacts and responses. RAP, however, considers people and nature as interdependent subsystems within a larger dynamic, non-linear social–ecological system (Walker et al., 2010). This approach is the result of decades of research on resilience science that seeks to understand the co-evolving relationship between people and places. It provides insights for providing better support to communities to enable them to use and manage their resources in a sustainable way and respond to shocks at the local scale while

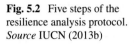

Fig. 5.2 Five steps of the resilience analysis protocol. *Source* IUCN (2013b)

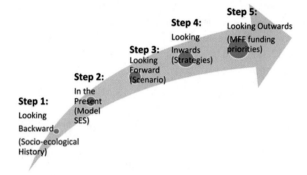

maintaining a broader long-term focus. The RAP consists of five steps illustrated in Fig. 5.2 and described below.

5.3.1 Step 1: Looking Backward

This step aims to gather baseline information on the socio-ecological history of the target area and cover issues related to ecosystems and ecological processes; institutional and governance frameworks and actor-network dynamics; characteristics of key drivers of change; and 'disturbances' and pressures including natural hazards and disasters. This process documents a historical narrative using a range of 'tools' or exercises such as unstructured qualitative interviews with individual informants; focused group discussions with stakeholder groups; and application of well-developed tools for participatory research such as resource maps, seasonal calendars, historical timelines, institutional Venn diagrams and actor-network mapping.

5.3.2 Step 2: In the Present

This step formulates a conceptual model for the coupled social and ecological system for the project site through focused group discussion using the information and knowledge generated in Step 1. The model is preferably represented as a graphic diagram with system boundaries, subsystems, interactions between various subsystems, narrative on the emerging system properties and explanations of the potential impact of unpredictable and uncontrollable drivers (operating at multiple scales) on the system. One of the important issues for developing the conceptual model is determining the possible disturbances that cause disruptions to the system. The disturbances can be episodic/discrete events like cyclones or gradual changes like sedimentation and sea level rise. The timing and magnitude of disturbances can be uncertain like many climate change impacts typically are. They can also be changes where causality can be understood by science or direct experience (e.g. land-use change). The other important aspect is to understand the existing governance system and power dynamics among different stakeholder groups. Such analysis is essential to identify the barriers and opportunities for developing and implementing strategies for improving resilience.

5.3.3 Step 3: Looking Forward

This step involves scenario building exercises to understand how the unpredictable, exogenous, and uncontrollable drivers might interact with stakeholder behaviour and policy processes to change community access to ecosystem services. Alternative scenarios and future visions are developed to determine the uncertainties associated with drivers, policies and resource management decisions. Stakeholder dialogues involve systematic consideration of a diverse set of information and perspectives to facilitate shared understanding of threats and responses to foster more informed decision-making.

5.3.4 Step 4: Looking Inwards

This step focuses on thinking through strategies or factors that might lead to building community resilience. Four filters are used to determine the appropriate context-specific strategies for implementation:

- Good practice: Tried and tested strategies to cope with predictable changes that would produce measurable results within a couple of years.
- Best practice: Strategies for replication drawn from case studies from elsewhere with similar context.
- Innovative ideas: New or updated place-specific strategies that are likely to lead to enhanced adaptive capacity and local resilience.
- Transformative ideas: Strategies that transform practice into a sustained effort, cross-sectoral solutions and long-term resilience.

Strategies passed through these four filters stand out for two important aspects of resilience: adaptability and transformability. Adaptability refers to the capacity to deal with changes within a regime, and transformability refers to the social–ecological capacities that enable shifts from one regime to another (Folke et al., 2010; Olsson et al., 2014; Walker et al., 2002).

5.3.5   Step 5: Looking Outwards

This final step entails leveraging the necessary financial and institutional support for the implementation of the chosen strategies. Funding priorities are documented, possible sources of funding are identified and pursued, and relevant institutional and policy procedures are established.

5.4   Implementation of RAP in Shyamnagar

To implement the RAP process in Shyamnagar, considerable time and effort was expended to collect primary and secondary data on the geography, ecology and demographics of the area. This was followed by the establishment of the stakeholder group that would participate in the five steps of the RAP. This group consisting of about 20 people included representatives of community leaders, local government agencies, local NGOs, resource user groups, women and youths. Experience of the RAP facilitators was a key factor in their ability to moderate discussions using the conceptual model. The local social–ecological history, current challenges and future scenarios are summarized based on this participatory process.

5.4.1   Socio-ecological History

During the implementation of Steps 1 and 2, four major changes were identified as important parts of the history of Shyamnagar:

- Large-scale ecological and social change: Clearing of the mangroves to reclaim the current settlement area and establishment of the Sundarban Reserve Forest (SRF) in 1876.
- Major hydrological system changes in the aftermath of the Green Revolution: National policy to expand *boro* rice (a hybrid rice variety) cultivation achieved through the construction of polders or ring dykes that converted the tidal floodplain into stagnant freshwater areas in the 1960s.
- Major shift in the agricultural system and ensuing social conflict: Sharp rise in global demand of shrimp prompted the development of a national policy that supported an exponential growth of shrimp aquaculture in the 1980s. It led to conflict between the freshwater-dependent rice farmers and shrimp farmers as rice farmlands were converted into shrimp aquaculture ponds.
- Climate change, intensifying natural disasters and new stimulants for economic development: Two consecutive cyclones, *Sidr* in 2007 and *Aila* in 2009, devastated Shyamnagar, but also brought it into the spotlight in national and international discourses on climate change and disaster risk reduction (DRR). This resulted in considerable external investment into development programmes in the area.

Shyamnagar became the test site for climate change adaptation and DRR projects for local and international donors and NGOs. Nationwide social and economic changes in terms of gender empowerment, female education and connectivity through mobile communication also penetrated these communities.

These changes triggered a series of social–political transformations in Shyamnagar:

- Embankments and polders built on riverbanks in the late 1960s made the community dependent on the government agency replacing the voluntary community service to build and maintain seasonal bunds/embankments for rice cultivation. The polders slowly weakened over the years till the super cyclones *Sidr* and *Aila* washed off major parts of it. An unintended consequence was the siltation of riverbeds due to sedimentation of silt load resulting in drying of rivers and canals and drainage congestion inside the polders.
- The area has traditionally relied on rain-fed agriculture and local varieties of rice. The construction of embankments triggered a shift to commercial agriculture and plantation of hybrid high-yielding varieties of rice. While yield per unit area increased, there was continued loss of agricultural land due to saline intrusion— especially after embankment failures during *Sidr* and *Aila*. With the disappearance of traditional rice varieties, farmers have experienced seasonal crop loss and lack of availability of straw and fodder for raising animals in recent years.
- Saline intrusion created a favourable condition for shrimp farming starting in the 1980s and continuing to expand through the 1990s. Many of these shrimp hatcheries were washed away during cyclone *Sidr* in 2007, following embankment failure. Farmers adapted to grow brackish-white fish. The shrimp farming revived after a while before facing another major shock due to cyclone *Aila* in 2009. The cycle of boom and bust continues till now. The export-oriented crab farming is getting popular and slowly replacing the shrimp. However, the COVID-19 crisis severely disrupted the supply chain.
- Forest management practices were introduced in the late nineteenth century and continued throughout the twentieth century. Management interventions in Shyamnagar historically followed a revenue-oriented model that propagated embankments, hybrid rice varieties, forest resource management and shrimp cultivation.
- While there were extensive damages to the area forests during *Sidr*, the forest ecosystem recovered rather quickly. A paradigm shift towards climate change adaptation and community-based disaster risk reduction began in the aftermath of *Sidr* and *Aila*.

5.4.2 Prevailing Challenges

The dilemma between rice and shrimp cultivation appears to be continuously shaping the fate of Shyamnagar and its people. The linear chain of hazard, vulnerability and response is compounded by regional and national policy changes outside Shyamnagar. The large and wealthy landholders have been the main beneficiary of the Green Revolution in the 1960s and were also the first to switch to shrimp cultivation by turning their rice fields into saline aquaculture ponds in the 1980s. They had the required capital and capacity to take risks by investing in a new enterprise. While the polder system was established in the 1960s to protect the rice fields by preventing the entry of saline water into the tidal floodplain, in the 1980s, the large landholders started pumping saline water from rivers to flood the rice field—thereby transforming the landscape with brackish water aquaculture systems.

The small landholders and landless seasonal lease-holder farmers were directly impacted by this shift that had sudden as well as long-term consequences. There were instances where powerful shrimp farmers inundated agricultural land with saline water—quickly transforming the hydrology and ecology of the area. Parallelly, gradual build-up of salinity forced farmers to abandon rice cultivation over time with accompanying loss of livestock and poultry. It also changed the role of women in these communities who used to lead the post-harvest activities of rice farming. Especially after *Aila*, trees started dying fast and the ponds remained saline. This caused scarcity of drinking water forcing women to travel long distances daily to collect water.

Shrimp postlarvae (PL.) collection and seasonal demands for day labourers in the shrimp farms offered some alternative sources of income. Rice cultivation continued in land was not suitable for shrimp farming. With monsoon rain washing out the soil salinity, a reduced level of rice cultivation is still possible in the area during the wet season. Seasonal outmigration has become a coping strategy for the landless agricultural labourers and small farmers. Another impact has been increased pressure on the Sundarban ecosystem for subsistence and economic resource extraction.

Both rice farmers and shrimp farmers need durable embankments for protection against regular tidal waves and recurring storm surges. However, there is conflict between the groups in terms of the operation of the sluice gates; shrimp cultivators want saline water inside the embankment, whereas farmers want to keep the saline water out. Some sluice gates were illegally constructed, and water pipes were placed through the polders to draw saline water from rivers. It weakened the polders and potentially contributed to their breach by *Aila*. Even though the government repaired the embankments quickly, people could not make productive use of the affected land for almost two years in many parts of Shyamnagar. Shrimp farmers faced litigation due to the damage they caused to the embankments which were eventually dropped due to strong political lobbying by the shrimp farmers association. However, the litigation triggered a change in practice, at least temporarily. The shrimp farmers agreed to use low-lift pumps to collect saline water from the rivers without incurring any physical damage to the polders.

The embankment was breached again by cyclone *Amphan* in May 2020. It appears that the same cycle of incidences was repeated leading to the embankment breach. The vulnerability of people and environment of Shyamnagar, thus, is the product of three factors: climate change increasing the frequency and intensity of cyclones and sea level rise; national policy decisions to pursue structural solutions (e.g. polders) and propagate hybrid rice varieties and shrimp farming; and the collective or individual actions that undermined the integrity of the embankments. Climate change is going to make some of these decisions and actions worse in the coming years. A thorough understanding of these interconnected endogenous and exogenous drivers is a crucial prerequisite of effective resilience planning.

5.4.3 Future Opportunities

During scenario planning, it was interesting to note that people did not mention the word 'climate change' despite the fact that they had been exposed to the concept as part of a number of ongoing climate change-related projects in Shyamnagar. People were more concerned with the immediate problems and issues that they face on a day-to-day basis such as recovery in the aftermath of Aila, the conflict between shrimp cultivation and agriculture, increasing tide height, embankment malfunction, local governance, salinity intrusion and freshwater crisis. However, participants demonstrated good understanding of the cause-and-effect linkages between environmental degradation and human suffering—which is a core concept of complex socio-ecological systems.

Two extreme scenarios set the boundaries of the scenario planning process: an optimistic vision that current trends and issues would reach a tipping point that would trigger a series of integrated planning initiatives fostering a positive growth trajectory for the community;and a pessimistic vision of further worsening of current socio-economic and ecological degradation leading to collapse. In between the extremes were scenarios with further consolidation of individual wealth and higher investment on natural resource-based enterprises such as shrimp farms and crab farms as well as progressive ideas such as women's empowerment, spread of information and communication technology, building social networks and local capacity building resulting in various social and environmental outcomes.

The Shyamnagar RAP participants ended up choosing a more optimistic future that envisioned Shyamnagar as the agent of change in writing the history of a prosperous future. The community envisaged a community-based disaster risk reduction approach to bounce back quickly after any disaster. An ecosystem protected by embankments and a mangrove forest established through co-management systems that would provide both protection from storm surges and a steady flow of ecosystem services. The community also predicted export of value-added products from Shyamnagar and a steady growth in ecotourism to generate the dual benefit of employment creation and conservation of the Sundarban. New technologies would increase production of rice, shrimp, crab and fish substantially. A community-based local water management system would create integrated water management systems, and a piped water distribution system would end the drinking water crisis and ensure safe drinking water for all. All of these would be supported by a strong local governance system that would transform Shyamnagar into a model community.

This narrative was very inspiring for the participants to begin to consider actions that need to be taken today and between now and the future to achieve the desired vision. The timeline was arbitrarily set at 2050. The extended time horizon helped participants to plan short, medium and long-term strategies. A total of 24 strategies were proposed which falls under the four key features of general resilience building: learning to live with change and uncertainty; nurturing diversity for reorganization and renewal; combining different types of knowledge for learning; and creating opportunity for self-organization towards social–ecological sustainability (Folke et al, 2003).

5.5 RAP Outcomes in Shyamnagar

The RAP process in Shyamnagar generated ideas for a diverse set of strategies ranging from local actions to national policy. These were mapped against the priorities of the MFF grant—which was intended to pilot short- to medium-term strategies, take feedback, and facilitate local actions while pursuing policy advocacy at the national level for issues which cannot be solved through local initiatives. This resulted in the following priorities that intersected the visions of the community with the scope of the MFF grant.

- Livelihoods support for forest-dependent communities, especially targeting women for eco-friendly enterprise development and market linkage
- Freshwater conservation and innovations in freshwater utilization
- Community initiatives for responsible ecotourism
- Community-based climate change adaptation to reduce climate change vulnerabilities
- Better environmental performance of the shrimp *ghers* and introducing innovations like mangrove shrimp culture
- Community initiatives for participatory mangrove restoration.

The priority issues formed the narrative for a call for proposals published in newspapers for local NGOs to apply for the MFF grants, and 16 grant projects were funded between 2013 and 2017. The stakeholder group that participated in the RAP process formed a strong network of community advocates during the implementation of these projects. They continued to liaise with the local government agencies, local government and the non-governmental organizations (NGOs). Some NGOs which were MFF grantees continued to work on the projects beyond the project timeline. The grant projects, co-designed with the local community and implemented by local NGOs, helped set a new trajectory for the socio-ecological systems in Shyamnagar to move towards a shared vision of an optimistic future. The process augmented local adaptive capacity to manage the uncertainty of living in a dynamic environment.

Restoration of freshwater supplies has made safe drinking water available to 750 families year-round. A total of 133 rainwater harvesting systems have been established. All reservoirs and water systems are managed by community members. More than 1200 marginal farmers cultivated their second crop with the water stored in the re-excavated canal and ponds and adopted climate-smart agriculture including alternative crops such as wheat, salt-tolerant rice and sunflower. Alleviation of household water stress has freed up women to explore alternative incomes such as mat weaving and ecotourism. It has also reduced their 'invisible labour' of having to travel long distances in search of potable water. Interventions at the farm and household level have improved household well-being through diversification of agricultural practices and income sources. Families who were involved in various MFF-funded projects formed community groups and engaged in communal entrepreneurship. Nearly, 500 entrepreneurs engaged in 12 enterprises, and their supplementary income increased by 12–15%. Development of a decentralized and representative governance system has resulted in improved local management of resources. About 200 shrimp farmers were trained in organic shrimp cultivation, and they planted mangroves in their shrimp farms. About 100 farmers received organic certification in 2017 (Karim et al., 2019).

Mangrove conservation is a socially and financially attractive approach for building resilience (Das, 2021, Chap. 17 of this volume). Targeted policies to ensure access to external finance for access to natural capital through extensive mangrove forest coverage can improve climate resiliency in coastal communities (Mahmud et al., 2021, chapter 20 of this volume). Under the MFF grant, about 150 ha of mangrove forest have been rehabilitated, and higher fish abundance has been observed in the vicinity of newly planted mangrove forest. The model was replicated by other development partners resulting in the protection of more than 400 ha of mangrove areas. The local government (Union Parishad) works with a 35-member forest management committee as stewards of the mangroves. The revenue from the forest (fruits, honey and nursery) is distributed between the community (40%), Union Parishad (40%) and a fund to cover forest management expenses (20%). A community-led ecotourism operation has been installed to diversify livelihoods while engaging the community in forest conservation. Local people have been trained to manage the ecotourism programme, and women have been offered culinary lessons. The ecotourism enterprise includes eco-cottages, eco-boats and motorboats and tours for tourists by local guides (MFF, 2018).

5.6 Lessons from Shyamnagar

5.6.1 Scenario Analysis is Useful

The scenario analysis process implemented in Shyamnagar was found to be a valuable tool for expanding thinking about how unpredictable and uncontrollable endogenous and exogenous drivers interact with stakeholder behaviour and policy processes. Scenarios can typically be developed either for a short timeline (5–10 years) or for much longer ones (20–30 years). For Shyamnagar, stakeholders found looking far into the future to be more appropriate due to the slow onset changes like climate change impacts, sea level rise, salinity ingress, ecological changes and social–cultural transformations like women's empowerment, outmigration and digitalization. The longer timeline allowed participants to think beyond the usual short-term 'project' timeframe and develop a shared vision.

5.6.2   Visioning with the Community is Valuable

The RAP process engaged a wide range of stakeholders from the local policymakers, local leaders, practitioners, different user groups, women and youths in collaborative visioning and planning. The 'Golden Shyamnagar 2050' vision that resulted from this process reflects the diverse perspectives of this representative group. The process also exposed the historical inequities in the transition from rice farming to shrimp cultivation in the 1980s that favoured the wealthy in the community. If the trends in shrimp cultivation shift towards crab farming in the future, the design and delivery of that transition can be informed by past experiences to develop systems that ensure a fair distribution of benefits in the community. Developing both freshwater and brackish water-based enterprises promises a more robust and resilient solution.

5.6.3   Program Density Matters

Karim et al. (2019) conducted a review of MFF projects in Shyamnagar and concluded that the grants positively contributed towards resilience building through increased income, diversification of income sources, improved health, time savings due to less sick days and better access to safe drinking water. The study also found that the programme density or concentration in a community, not the number of interventions, played a more important role in the diversification of income sources—leading to greater community resilience against climate change. The provision for 16 small and medium sized grants under the MFF facilitated a flexible planning and implementation process that stimulated the development of a set of synergistic and complementary initiatives. This fostered the creation and strengthening of decentralized, multi-dimensional communitynetworks involving broad participation and ownership at the local level (e.g. the community water association).

5.6.4   Focus on Future Opportunities Fosters Creative Solutions

An interesting lesson learned from the Shyamnagar example is the departure from discussions framed around climate change—which is more common in vulnerability analysis and adaptation project planning. The RAP process allowed people to expand the scope of stakeholder deliberations beyond hazard-specific discussions on problem identification and adherence to known solutions. It helped people to untangle the dynamics of place-based social and ecological issues to create a shared understanding of where and how to intervene in the journey towards a more sustainable future. While doing so, it allowed people to think of strategies or pathways for that journey; what to avoid to keep the system in a desired trajectory; how to empower local people to self-organize; mechanisms to foster greater local ownership and control; and ways to build communities' capacity to manage the future evolution of Shyamnagar.

5.7 The Way Forward

Ever since IUCN deployed the RAP, several new tools for resilience planning have emerged. The Stockholm Resilience Centre has developed the seven principles of resilience to guide programme development and implementation in the pursuit of community resilience (Stockholm Resilience Centre, n.d.). A new software platform, Wayfinder, has been developed for executing efficient and systematic community engagement by the Guidance for Resilience in the Anthropocene: Investments for Development programme (GRAID, n.d.)—hosted by the Stockholm Resilience Centre and funded by the Swedish International Development Cooperation Agency (Sida). The lessons learned in Shyamnagar have directly informed the development of these tools. Other tools, such as Nexus Assessment (Brunner et al., 2019), offer a multi-stakeholder engagement simulation on a given scale such as a river basin or watershed. Greater use of these types of tools and processes is likely to mainstream community-based approaches to climate change adaptation and resilience building. This will help establish greater legitimacy of the knowledge, agency and engagement of local people and communities as owners, implementers and shareholders in climate adaptation, natural resource management and community development solutions. Best practices from Shyamnagar fit well with IUCN's current initiatives on nature-based solutions (NBS) which involve 'actions to protect, sustainably use, manage and restore natural or modified ecosystems, which address societal challenges, effectively and adaptively, providing human well-being and biodiversity benefits' (IUCN, 2020). Shammin et al. (2022, Chap. 2 of this volume) offers an integrative framework for building climate-resilient communities that embodies the lessons from Shyamnagar.

The Shyamnagar story demonstrates the value of community-based approaches to resilience building. This is not an idealistic story of a perfect solution; rather, it is the story of using local knowledge and building local capacity to empower communities to navigate through the nuances of socio-economic realities, threats of climate change and the complexities of social–ecological systems. As the skies continue to spin faster and the hungry tides become stronger, programmes like the MFF are helping the community to find resilience through conviction, collective endeavours, human ingenuity and conservation of the mangrove forests of the Sundarban—the old guard that is poised to watch over Shyamnagar for centuries to come.

Acknowledgements Partial support for this research was provided by an implementation grant from the Henry Luce Foundation's Luce Initiative on Asian Studies and the Environment (LIASE) for Faculty Research Support at Oberlin College. The authors acknowledge the MFF donors— Swedish International Development Cooperation Agency, Norwegian Agency for Development Cooperation and Danish International Development Agency—as well as the MFF institutional partners and the MFF Secretariat staff.

References

Bangladesh Bureau of Statistics. (2012). *Statistical yearbook of Bangladesh*. Ministry of Planning, Government of Bangladesh.

Banglapedia. (n.d.). *The national encyclopaedia of Bangladesh*. http://en.banglapedia.org/index.php?title=Shyamnagar_Upazila and http://en.banglapedia.org/index.php?title=Satkhira_District. Accessed October 18, 2020

Brunner, J., Carew-Reid, J., Glemet, R., McCartney, M., & Riddell, P. (2019). *Measuring, understanding and adapting to nexus trade-offs in the Sekong, Sesan and Srepok Transboundary River Basins* (Viii + 70pp). IUCN: VietNam Country Office.

Das, S. (2021). Valuing the role of mangroves in storm damage reduction in coastal areas of Odisha. In A. K. E. Haque, P. Mukhopadhyay, M. Nepal, & M. R. Shammin (Eds.), *Climate change and community resilience: Insights from South Asia*. Springer.

Folke, C., Colding, J., & Berkes, F. (2003). Synthesis: Building resilience and adaptive capacity in social-ecological systems. In F. Berkes, J. Colding, & C. Folke (Eds.), *Navigating social-ecological systems: Building resilience for complexity and change* (pp. 352–387). Cambridge University Press.

Folke, C., Carpenter, R. R., Walker, B., Scheffer, M., Chapin, T., & Rockström, J. (2010). Resilience thinking: Integrating resilience, adaptability and transformability. *Ecology and Society, 15*(4), 20. http://www.ecologyandsociety.org/vol15/iss4/art20/.

Ghosh, A. (2005). *The hungry tide*. Houghton Mifflin.

Ghosh, A. (2016). *The great derangement: Climate change and the unthinkable*. The University of Chicago Press.

Guidance for Resilience in the Anthropocene: Investments for Development. (n.d.). Stockholm Resilience Centre and the Swedish International Development Cooperation Agency (Sida). https://graid.earth/projects/wayfinder/. Accessed September 11, 2020

International Union for Conservation of Nature. (n.d.). *Red list of threatened species*. https://www.iucnredlist.org/species/136899/4348945. Accessed September 11, 2020

International Union for Conservation of Nature. (2013a). *Past, present and future of Shyamnagar: Citizen's analysis*. Report prepared for the Mangrove for the Future (MFF) initiatives.

International Union for Conservation of Nature. (2013b). *Resilience Analysis Protocol (RAP) for MFF project sites: Scoping challenges and setting strategies* (unpublished report).

International Union for Conservation of Nature. (2020). *Global standard for nature-based solutions. A user-friendly framework for the verification, design and scaling up of NbS* (1st ed.). IUCN.

Karim, S., Ahmed, M. U., & Tulon, K. M. N. I. (2019). *Analyzing the effectiveness of MFF-SGF program on household and community resilience to climate change in Bangladesh*. Asian Center for Development.

Mangroves for the Future. (2018). Annual report to the Regional Steering Committee.

Mahmud, S., Haque, A. K. E., & De Costa, K. (2021). Climate resiliency and location specific learnings from coastal Bangladesh. In A. K. E. Haque, P. Mukhopadhyay, M. Nepal, & M. R. Shammin (Eds.), *Climate change and community resilience: Insights from South Asia.* Springer.

Olsson, P., Galaz, V., & Boonstra, W. J. (2014). Sustainability transformations: A resilience perspective. *Ecology and Society, 19*(4), 1.

Shammin, M. R., Haque, A. K. E., & Faisal, I. M. (2022). A framework for climate resilient community-based adaptation. In A. K. E. Haque, P. Mukhopadhyay, M. Nepal, & M. R. Shammin (Eds.), *Climate change and community resilience: Insights from South Asia.* Springer.

Stockholm Resilience Centre. (n.d.). Stockholm University. https://www.stockholmresilience.org/. Accessed September 11, 2020

United Nations Environment Programme. (2002). *Global Environment Outlook 3: Past, present and future perspectives.* Earthscan.

Walker, B., Carpenter, S., Anderies, J., Abel, N., Cumming, G., Janssen, M., Lebel, L., Norberg, J., Petersen, G. D., & Pritchard, R. (2002). Resilience management in social-ecological systems: A working hypothesis for a participatory approach. *Conservation Ecology, 6*(1), 14.

Walker, B., Gunderson, L., Quinlan, A., Kinzig, A., Cundill, G., Beier, C., Crona, B., &Bodin, Ö. (2010). *Assessing resilience in social-ecological systems: Workbook for Practitioners.* Version 2. Resilience Alliance.

Part II
Traditional Knowledge and Sustainable Agriculture

Chapter 6
Indigenous Practices of Paddy Growers in Bhutan: A Safety Net Against Climate Change

Tshotsho

Key Messages

- This chapter provides a community resilience story of rice growers in Bhutan.
- Traditional knowledge usage can provide a way forward for building and securing livelihood in the fight against climate change.
- The traditional knowledge needs to be streamlined into the common policy debate.

6.1 Introduction

Climate change has profound impacts on agriculture (Cline, 2007). One of the most prominent forms of this impact is water scarcity (Balasubramanian & Saravanakumar, 2022, Chap. 10 of this volume). The IPCC (2014) predicts as much as 50% loss in crop yield in rainfed agriculture that does not use any adaptation strategies. The drying-up of springs is a cause of concern in the hills and mountains of the Himalayas (Bharti et al., 2020; Rai & Nepal, 2022, Chap. 23 of this volume), and the agricultural community faces a risk to its livelihood due to the short window of rainy season which is the only source of water for irrigation (Gurung & Bhandari, 2009; Kattel & Nepal, 2021, Chap. 11 of this volume). Agriculture in Bhutan is carried out on small

---

Disclaimer: The presentation of material and details in maps used in this book does not imply the expression of any opinion whatsoever on the part of the Publisher or Author concerning the legal status of any country, area or territory or of its authorities, or concerning the delimitation of its borders. The depiction and use of boundaries, geographic names and related data shown on maps and included in lists, tables, documents, and databases in this book are not warranted to be error-free nor do they necessarily imply official endorsement or acceptance by the Publisher, Editor(s), or Author(s).

Tshotsho (✉)
College of Natural Resources, Royal University of Bhutan, Lobesa, Bhutan
e-mail: tshotsho1993@gmail.com

© The Author(s) 2022   87
Haque et al. (eds.), *Climate Change and Community Resilience*,
https://doi.org/10.1007/978-981-16-0680-9_6

scale and mostly rain fed and dry land and wetland farming, and provides livelihood to over 57% of the population (ICTA & World Bank, 2017). Since, the country is part of the Himalayas, the agriculture sector has been facing climate change impacts in the form of: reduction in agricultural water availability (increasing fallow land in rice cultivation due to lack of irrigation water); reduction in crop yield due to inadequate rainfall during the growing season; and erratic and excessive rainfall patterns leading to extreme events like flash floods, and reduced availability of arable land (ICTA & World Bank, 2017).

A recent strategy by the Government of Bhutan is the adoption of Climate Smart Agriculture (CSA) (CIAT & World Bank, 2017). CSA technologies are a set of agricultural technologies that include use of improved plants such as drought-tolerant, pest- and disease-resistant early-maturing seed varieties for cereals and vegetables; crop intensification such as maize intercropping with legumes; soil conservation and nutrient management such as manure; improved water and irrigation management such as drip irrigation; and alternate wetting and drying for paddy and upland rice cultivation. These adaptation technologies should be able to secure food security, increase food production and promote rural development (CIAT & World Bank, 2017). However, funding for CSA technologies is limited and adversely impacts the safety nets without an alternative source of livelihood for farmers (CIAT & World Bank, 2017).

In this situation, there is evidence of community adoption and mitigation through utilization of traditional agriculture knowledge (Galloway-McLean, 2017). Traditional knowledge can re-emerge, and traditional agriculture can become part of the resilience effort. The re-emergence and dependence on traditional agricultural knowledge will be driven by vulnerability originating from changes in climatic and environmental problems, further exacerbated by the lack of agricultural input and support (Shava et al., 2009). Among the many crop choices, farmers perceive rice to be more resistant to climate change (Bojang et al., 2020). Galloway-McLean (2017) present case studies from around the world that have cultivated paddy to build resilience to climate change, and the Mphunga community in Malawi, for example, have built resilience against climate change by switching from maize to rice. Farmers, for example, in Gambia, understand that water supply will be affected by climate change and rice production will reduce without adaptation measures and think that traditional rice varieties can withstand extreme weathers because they have survived for a long period (Bojang et al., 2020). The Nwadjahane community in Mozambique have sought to cultivate drought-resistant rice; and the Khagrachari community in Bangladesh relies on different varieties of rice to suit the recession of flood waters and duration of droughts, Dar et al. (2017) argue that farmers prefer traditional rice varieties over modern varieties because they give 30–42% more yield and are more flood-tolerant. Traditional local rice varieties have also been adopted as a climate change adaptation measure by the local community, because they have found it to be socially acceptable, economically beneficial, and environmentally sound in many parts of Bangladesh (Kabir & Baten, 2019).

This study looks at the adoption of traditional rice variety[1] as a strategy in building resilience to climate change for households in the five gewogs[2] of Punakha valley in west-central Bhutan. Specifically, the study proposes to: (a) identify the determinants which influence household's adoption of traditional local rice varieties with a focus on irrigation constraint; and (b) understand the extent of contribution of its adoption towards households' resilience. Estimates show that availability of water for irrigation has significant effect on farmer's decision to use traditional versus high-yielding variety of rice. In the upper hill area, where water is scarce, farmers choose traditional rice variety of rice whereas in the valley area where irrigation water is available farmers choose high-yielding variety.

6.2   Study Area and Sampling

The study uses data from the Renewable Natural Resources Survey carried out in 2019. The survey provides recent information on demography, land holding and ownership, land use and irrigation, crops, livestock, farm mechanisation, credit, labour and forestry. Specifically, the study uses 1088 households from the five gewogs spread over the district of Punakha, which are highly vulnerable to climate change in terms of water stress. The selection of sample gewogs is based on telephone interview with gups, the heads of the gewogs.[3] In instances where the gup had less information on the adoption and prevalence of rice variety, few Tshogpas[4] were also approached over telephone to properly identity villages that had adopted four specific traditional rice varieties.[5] They were asked whether they adopted the four specific local varieties; whether these adoption villages were in higher altitude; and whether they faced water scarcity. On the basis of the telephone interview, 5 out of 11 gewogs were retained for the analysis. These comprise 120 households from 8 chiwogs that adopted the traditional rice variety and 968 households from 25 chiwogs that did not adopt[6] it. The 968 households were retained because they are from the same gewogs, which makes the comparison more realistic. The households in the same gewogs

---

[1] These specific traditional variety includes Yangkum, Jama, Janaap and Jakaap in the local language. These varieties have two distinct characteristics. It grows in high altitude and requires short window for plantation and harvesting.

[2] A gewog is a sub-district. A district is a composition of several gewogs.

[3] Gups who heads a gewog in 11 gewogs in the district were approached over telephone. Gewogs are sub-district areas. Gups are referred to as the Chairman.

[4] Tshogpa heads a chiwog which is a sub-gewog.

[5] "A farmers' variety is defined as a variety which has been traditionally cultivated by farmers in their fields, or is a wild relative of a variety about which farmers possess the common knowledge" (Ragavan & O'shields, 2007).

[6] Non-adopter includes a village which may use either modern improved variety or other local varieties which are grown in lower altitude or valley wetlands.

Table 6.1 Showing number of adopters and non-adopters of traditional rice varieties within different gewogs of Punakha

Adoption of traditional rice	Gewog					
	Barp	Guma	Lingmukha	Shelnga-Bjemi	Toedwang	Total
No	245	257	138	118	210	968
Yes	8	10	12	46	44	120
Total	253	267	150	164	254	1,088

Note Figures represent count of households

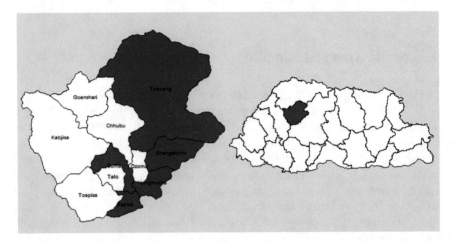

Fig. 6.1 Showing study sites where green shadings indicates selected gewogs for analysis

were found both in the lower valley and high altitude in the hilly areas, where they also faced water shortage resulting in irrigation problems (Table 6.1 and Fig. 6.1).[7]

The FGD[8] interviews are with the adopters, and some discussion from the field with these adopters makes the case study interesting:

> We grow Janaap which is originally brought from Shelngana, a village located in higher altitude in Punakha, and another variety, Jakaap brought from Lingmukha which is located in a higher altitude than our village. Due to customary laws regarding water usage, we get the

---

[7] Irrigation problems in these villages can exist in two forms. First, irrigation problems can exist because of a complete lack of irrigation water, which forces farmers to rely on monsoon rain. Second, irrigation water can be constrained because of customary law, which entitles households in upland to user rights first and households in lowland to later part of the plantation season. Due to decreasing volume in spring water and late arrival of monsoon, households in lowlands face only a few months (usually 4–5 months) for planting and harvesting, as compared to upland who get longer duration (7–8 months). This constraint forces farmers to rely on traditional varieties that require short duration for ripening.

[8] FGDs were mostly carried out on the phone due to covid-19 issues. Discussions were also held with few farmers in person during a visit to some of the villages.

water-use rights e late, in fourth and fifth month of the traditional calendar,[9] while irrigation begins in the second month in the lower valley. We cultivate a mix of Jakaap and Janaap, because we have only 4–5 months, till the ninth month which is the time for harvest. These rice varieties are suitable for short duration cropping system, and they ripen in the short duration. We even find that Jakaap is better than Janaap.

Village Tshogpa, Adopters

Yangkum and Jama are usually grown in the high altitude villages in Punakha. The plantation is carried out late and is completed by end of July. These villages start cultivation late when other chiwogs have already completed. Although the modern varieties have high yield and are early maturing, they are less tasty and have less demand in the market. Farmers prefer the local varieties because they mature late, are tasty and have high preference and demand in the market. They also rely on rain for irrigation which usually arrives late in the valley. They leave their land fallow if they miss this short window of rain water.

Extension Officer

We grow Jama and Yangkum which are the most preferred and suitable in our village. Jama is also cultivated in other village like Tamidamchu and Jojogoenpa. Jama is mostly grown because it is comparable in both taste and price to Ngabja which is found in the lower valley. These varieties are chosen because of the altitude and water shortage. They were passed down over generations. Yangkum takes longer duration to ripe and is planted during the fifth month and takes relatively more time as compared to Jama. Jama can be cultivated in the sixth month and ripens fast which is why they are adopted in our place which has irrigation constraints as we have to completely rely on rain. We cultivate these varieties as long as we have rain and leave our land fallow if there is shortage of rain. The production is half of what we usually cultivate because half of them do not ripe, when there is shortage of monsoon rain. We consume most of produce and sell some in the market.

2 Village women, Yuesikha, Adopters

6.3 Methods and Variables

The analysis of the study is based on the theory that links adoption of traditional rice varieties to changes in livelihood capitals, which enables households to build resilience to climate impacts using the sustainable livelihood approach (SLA). Asset or capacity building within the SLA model focuses on developing the resources and capacities that are required to secure livelihood on a sustainable basis. This framework provides a simple but well-developed way of thinking about complex issues (DFID, 2000).[10] A description of the constraints faced by the farmers, along with other important variables such as economic activity, demographic conditions, land holding, irrigation and plantation methods, crop activities, income and labour hiring, will precede the analysis to provide an understanding of the institutional setup.

---

[9] Bhutanese traditional month is two months lag of the modern calendar. May (5th month) is only the 3rd month in the traditional calendar. The modern calendar is usually two months ahead of traditional calendar.

[10] The DFID defines in the following sense: "a livelihood comprises the capabilities, assets (including both material and social resources) and activities required for a means of living. A livelihood is sustainable when it can cope with and recover from stresses and shocks and maintain or enhance its capabilities and assets both now and in the future".

For the first objective, we try to provide a justification of whether farmers in the district are homogenous. For this, we try to compare the socioeconomic conditions and agricultural practises between the adopters and non-adopters of traditional rice varieties.

For the second objective, this paper examines how this adoption has contributed to the households' resilience and livelihood achievements (Rajan et al., 2015). Resilience is measured in terms of the household's ability to meet food requirements, percentage of income earned from agricultural production, area irrigated under paddy, quantity of paddy production, wetland area left fallow, and number of man days employed. These metrics of resilience are also the objectives pursued by the institutions and policies in Bhutan for promoting climate change adaptation and mitigation activities (CIAT & World Bank, 2017). Although resilience should normally be studied over a long period of time requiring numerous observations, this study will use cross section data and attempt to understand household resilience at one time period, a robust and a comprehensive approach to understanding the impacts of the adoption, particularly in the absence of baseline information (Rajan et al., 2015).

6.4 Results

Table 6.2 shows the institutional settings under which the sampled farmers operate. These institutional settings show the constraints faced by the farmers from five gewogs. Among the constraints, irrigation is the dominant constraint (44%) followed by labour shortage (24%) and crop damage by wildlife (19%). The sampled households represents other farmers in Bhutan who face shortage of land (15%), crop damage by insects and disease (10%), high labour charges (9.7%) and lack of machinery (5%). Most farmers have access to market and a little over 10% of the farmers have availed of credit.

Table 6.3 shows the descriptive statistics of the variables used in the analysis. More than half of the sampled farmers (65.7%) engage in crop production as their main economic activity and 21.7% of the farmers are subsistence growers. During the time of interview, only 11% of the farmers had adopted four of the traditional rice varieties, with most of them using modern high-yielding rice introduced by the government through the extension centres, and other local varieties found in the lower valley. Most of the sample farmers have irrigated their field (92%) using surface water irrigation. Out of mean land holding of 3.14 acres, two-thirds of the land was cultivated with paddy (2.03 acres) and one-third was left fallow (0.92 acres).

Apart from paddy cultivation as the main economic activity, farmers also engaged in the production of other cereals (94%), legumes and oil (46.5%), vegetables (74.6%), root plants (13%) and permanent crops (53%). With 87% of them having employed labour, the mean average man days used was 55 days. These farmers also owned cattle (72%). A small fraction of the farmers (7.7%) also had protected land, and most of them (72%) collected non-timber forest products. A majority (93.8%)

Table 6.2 Constraints facing farmers in food production and asset creation

Variable	Obs	Mean	Standard deviation
Irrigation constraint	1088	0.44	0.49
Unproductive land	485	0.01	0.12
Labour shortage	615	0.24	0.42
High labour wages	623	0.09	0.29
Crop damage by wild animals	530	0.19	0.39
Crop damage by insect/diseases	662	0.10	0.30
Drought	498	0.00	0.06
Excessive rain	595	0.00	0.04
Hailstorm and wind	498	0.00	0.06
Landslides and soil erosion	596	0.00	0.07
Shortage of land	592	0.15	0.35
Limited access to market	590	0.01	0.10
Difficulty in getting machinery	583	0.05	0.22
Availed credit	1088	0.10	0.30

Note All the variables have minimum value of 0 and a maximum value of 1 since they are all dummy variables

of the farmers were able to meet all food requirements during the sampled year, and 42% of the sampled farmers earns 51–75% of their total income from crop production while 17, 19.7 and 21.7% of those farmers) earn between 0–25%, 26–50% and 76–100% of their income from crop production, respectively. Among the household heads, only 25% of them were male with an average age of 54 years and 78% of them were married.

6.4.1   Are Farmers Homogenous?

The main point of distinction between the adopters and non-adopters is that the former are mainly found in the high altitude areas lying on top of the valleys. The non-adopting farmers are found in the lower valley. Farmers in the lower valley areas also face water scarcity (Table 6.4). Farmers in the lower valleys are dependent on irrigation fed by surface running water. However, with climate change, paddy cultivation has become very difficult. The customary rules of sharing irrigation water have also made equitable sharing of water difficult. The government has responded

Table 6.3 Description of variables and their statistics

Variable	Description of variable	Obs	Mean	Std. dev
Crop_activity	= 1 if crop production is the main economic activity	1088	0.65	0.47
Adopts_2TRV	= 1 if the household adopts traditional rice variety	1088	0.11	0.31
Food_security	= 1 if household is able to meet food requirement	1088	0.93	0.24
Irrigated	= 1 if the household irrigated field	1088	0.92	0.26
Irrigation_surface	= 1 if surface irrigation practised	1002	0.98	0.10
Irrigation_source	= 1 if surface water is the source of irrigation	1002	0.97	0.14
Male	= 1 if household head is male	1088	0.25	0.43
Age	Age of the household head	1088	54.37	13.87
Married	= 1 if head of household is married	1088	0.78	0.41
Literate	= 1 if head of household is literate	1088	0.21	0.41
Hsize	Number of members in the household	1088	3.47	1.72
Irripaddy	Area irrigated in acres	966	2.03	1.43
Paddy_prod	Paddy produce in kilograms	966	4792.52	3698.06
Land_holding	Area of land holding in acres	1088	3.14	2.85
Fallow	Area of wetland left fallow in acres	233	0.92	1.17
Cereals	= 1 if cereal in grown	1088	0.94	0.23
Legumes	= 1 if legumes and oilseeds grown	1088	0.46	0.49
Vegetables	= 1 if vegetables grown	1088	0.74	0.43
Roots	= 1 if roots and tubers grown	1088	0.13	0.34
Permanent_Crops	= 1 if permanent crops grown	1088	0.53	0.49
Protected	= 1 if has protected land presence	1088	0.07	0.26
Livestock	= 1 if bovine animals owned	1088	0.72	0.44
LabourEmployed_	= 1 if labour employed	1088	0.87	0.33
	Number of man days employed	950	55.88	37.40
NWFP	= 1 non-timber forest products collected	1088	0.72	0.44
Income_RNR	0–25%	1088	0.20	
	26–50%	1088	0.17	
	51–75%	1088	0.42	0.40
	76–100%	1088	0.19	0.37
Subsistence	= 1 if production is only for self-consumption	1088	0.21	0.49

Source RNR Census (2019)

Table 6.4 Mean comparison of social, economic and agricultural practises between adopters and non-adopters of traditional rice varieties

Socioeconomic conditions	Non-adopters	Adaptors	P-value
= 1 if household faces irrigation constraint	0.45	0.35	0.04**
= 1 if crop production is main economic activity	0.63	0.87	0.00***
= 1 if head of household is married	0.78	0.76	0.64
= 1 if head of household is female	0.25	0.28	0.42
Age of household head	54.35	54.55	0.88
Household size	3.46	3.50	0.80
= 1 if household collects NWFP	0.71	0.82	0.01**
Land holding in acres	3.12	3.30	0.52
= 1 if household grows Legumes and oil	0.46	0.43	0.46
= 1 if household grows vegetables	0.75	0.69	0.14
= 1 if household grows roots	0.14	0.09	0.14
= 1 if household grows permanent crops	0.52	0.60	0.07*
= 1 if household has land under protection	0.07	0.07	0.92

All estimates were tested for a significance level of ***(10%), **(5%), *(1%)

to these problems by encouraging modern high-yielding varieties. Farmers in the lower valleys prefer to use varieties that are also sold easily in the market and are preferred by customers visiting local market.

Paddy rice cultivation comprises the majority of the occupational crop of most farmers in the Punakha valley. Farmers in the hilly areas have been exclusively subsistence crop growers and have refrained from engaging in cash crops. This could be due to distance from the market. In contrast, farmers in the lower valleys, because of proximity to market, have found growing oilseeds, legumes, vegetables and root crops such as potatoes more profitable. Field data also supports this argument (Table 6.4). Farmers in both the groups are also members of community forest and are surrounded by protected areas. These community forests are mostly in the hilly regions and have provided farmers there with non-timber forest products like mushrooms, wild fruits, firewood and fodder for their domesticated animals. Data shows that the adopting group collects relatively more of these resources from the protected forest (Fig. 6.2 and Table 6.4).

Field visits to the valley have shown that farmers in the study site can be considered as homogenous. There is plenty of evidence against social heterogeneity among farmers in the district. Data shows that there is no significant difference between farmers who adopt traditional rice varieties when compared to non-adopters. In terms of asset ownership, both the groups of farmers have similar land holding measured in acres (Fig. 6.2 and Table 6.4). These are all smallholder farmers with none holding land above 5 hectares. The household size, age, gender and marital status also play

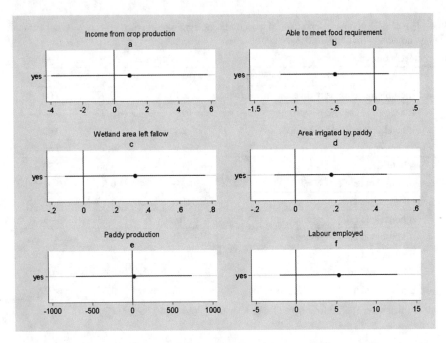

Fig. 6.2 Coefficient estimates of livelihood measures from adoption of traditional rice variety. *Note* **a** estimated using interval regression (Long & Freese, 2006), ($P = 0.71$), **b** estimated using logit regression (Wooldridge, 2016) (P value $= 0.15$), **c** (P value $= 0.15$), **d** (P value $= 0.21$), **e** (P value $= 0.96$) and **f** (P value $= 0.15$) estimated using linear regression (Wooldridge, 2016)

an important role in decision-making that can have an impact on livelihood achievements. Data shows that there is no significant difference between growers and nongrowers of traditional rice varieties for these socioeconomic conditions too (Fig. 6.2 and Table 6.4).

6.4.2 Are Livelihood Achievements Comparatively Similar?

It has been argued that farmers in both the groups are homogenous except that those in the higher hilly areas have resorted to traditional varieties that were passed on by ancestors while modern high-yielding and high-demand rice varieties are preferred by farmers in the lower valleys. If farmers adopting traditional varieties are a rational decision maker, then these decisions should also offer farmers comparable livelihood achievements. These livelihood achievements can be compared and tested for how farmers are able to meet food requirements, income earned and rice production since it is the primary employment of these farmers.

A cursory glance over their ability to meet food requirement shows that nonadopters are able to meet more food requirements compared to the adopters (Fig. 6.2).

This could be true because farmers in the lower valleys not only grow paddy but also engage in production of oilseeds, legumes, vegetables and root crops such as potatoes. Rice and potato constitute a major staple diet in Bhutan along with legumes such as beans and peas. But statistically the differences are not so huge or significant (Fig. 6.2). This implies that since rice is the primary production crop and the staple diet, and since farmers perceive food security in terms of their ability to meet three meals of rice, traditional rice growers are no less than the group which grows modern varieties.

This can be further substantiated by understanding income earned by farmers. Farmers in the lower valleys are closer to market and can sell their rice and vegetables in the market that opens on a weekly basis. But, mean comparison data shows that there is hardly any income difference between these groups (Fig. 6.2). This could be due to two reasons. First, the traditional rice, if brought to the market, fetches higher price due to superior taste over the rice grown in the lower valleys. Second, farmers in the hills sell their labour to farmers in the lower valleys because the formers' paddy cultivation starts later in late August and September, two months later than the time of the paddy cultivation in the valleys.

Farmers in the hills also cultivate more area of the field, which explains higher number of labour employed (Fig. 6.2) and yet they produced same quantity of paddy like the lower valley farmers. One explanation could be that the rice varieties in the lower valley that are provided by the government are high yielding. So, there is a quantity and quality trade-off here. However, more land were left fallow in the lower valley area too. The land ownership system offers an explanation to this. In the lower valleys, most of the farmers are sharecroppers although they also own some land. When farmers cultivate on land that belongs to others, they face more incentive to cultivate as much as they can, contrary to farmers in the hilly region where lands are privately owned and farmers being subsistence farmers, grow only what will be sufficient for the year. The phenomenon of out-migration could also offer another explanation. However, the differences are not huge or significant enough (Fig. 6.2); otherwise we would be tempted to conclude that traditional rice varieties are inferior to the modern varieties.

Finally, higher labour employed by adopters could be related to higher area of land irrigated compared to the non-adopting group (Fig. 6.2). This also has no significant statistical difference (Fig. 6.2). All these evidence shows that the two groups are comparable in livelihood measures.

6.5 Discussion and Conclusion

Much of research in rice adoption in Bhutan has focused on adoption alone, except one which has gone beyond and estimated the impact of adoption on livelihood measures such as poverty, to provide better policy recommendations (Bannor et al., 2020). This is important because simply adopting a variety does not guarantee resilience. This study followed a similar approach and compared livelihood impacts between adopters and non-adopters of traditional rice varieties. This paper tested the question whether farmers adopting traditional rice varieties in the hilly regions have the same adapting capacity and offer resilience from climate change, as compared to modern rice improved varieties which are often the most propagated. Estimated results show that they both have the same capacity. The result from this paper is in line with studies that have shown that traditional food crops have enabled households to meet food requirements and maintain food security (Shava et al., 2009).

Traditional agriculture has also helped to increase production and reduce food insecurity in communities in China, Kenya and Bolivia (Swidersk et al., 2011). In contrast, replacing traditional variety by modern high-yielding variety is found to have resulted in more rice production and has improved food security in Bangladesh (Shew et al., 2019). This is also in contrast to a study which shows yield production is more in modern variety as compared to traditional variety in the western region of Punakha and Wangdue Phodrang district, where 56% of the farmers adopted modern rice variety through promotion by the government (Chhogyel & Bajgai, 2015).

Farming communities who use community seed variety have also empowered poor farmers and women and have increased their income by 30% (Swidersk et al., 2011). In the Indian State of Uttarakhand, a study shows that adoption of traditional rice like basmati earned more net income compared to non-adopters (Jena & Grote, 2012). Although some studies report higher income earnings, we found insignificant difference between different rice adopters. In contrast to these results, Bannor et al. (2020) show that compared to traditional variety, adoption of modern improved variety leads to reduction in poverty gap and incidence. They also show that modern variety households have higher monthly household expenditure.

Although we find studies comparing traditional and modern improved rice varieties in terms of adoption and impact on food production, income and food security, there is a lack of studies showing the effect of adoption on the area of wetland left fallow and number of man days employed. This study uses a multitude of resilience measurements to find a robust comparison of adopters of traditional rice variety in comparison with modern rice variety along with other local varieties.

Although traditional agriculture along with their crops are facing threat from modern commercial food systems, this paper argues that traditional knowledge is capable of providing resilience towards climate change (Shava et al., 2009). There is a strong evidence how traditional knowledge makes a community reluctant to adopt modern varieties. Swidersk et al., (2011) argue that "Indigenous peoples and local communities often live in harsh natural environments, and have had to cope with extreme weather and adapt to environmental change for centuries in order to survive. They have done this using long standing traditions and practices relating to adaptive ecosystem management". They further argue that it is important to develop and promote context-dependent education and awareness on the coping opportunities provided by traditional crops (Shava et al., 2009; Swidersk et al., 2011).

This paper presented a case of paddy growers in Punakha valley, who have resorted and continue to adopt four rice varieties which have proved to provide a way forward in sustaining their livelihoods in the face of climate change. Specifically, this paper provides an account of the institutional context in which these farmers operate, and the farming technology that they use and compares it with communities that face similar situations but have access to other options.

Acknowledgements I would like to thank the villagers of Yuesikha who have first shared their resilience story and how irrigation has seriously constrained their way of life. I also want to thank them for hosting my students in 2018 and colleagues in 2017 who were together with me eager to learn about their livelihood and story. Special appreciation also goes to the Gups of all the gewogs and Tshogpas of few chiwogs, and extension officers in Punakha dzongkhag, who were willing to answer the questions I have asked them through telephone. Special appreciation goes to the editors for giving me this opportunity to write and share this incredible story from Bhutan.

References

Balasubramanian, R., & Saravanakumar, V. (2022). Climate sensitivity of groundwater systems in South India: Does it matter for agricultural income? In A. K. E. Haque, P. Mukhopadhyay, M. Nepal, & M. R. Shammin (Eds.), *Climate change and community resilience: Insights from South Asia* (pp. 143–156). Springer.

Bannor, R. K., Kumar, G. A. K., Oppong-Kyeremeh, H., & Wongnaa, C. A. (2020). Adoption and impact of modern rice varieties on poverty in Eastern India. *Rice Science, 27*(1), 56–66.

Bharti, N., Khandekar, N., Sengupta, P., Bhadwal, S., & Kochhar, I. (2020). Dynamics of urban water supply management of two Himalayan towns in India. *Water Policy, 22*(S1), 65–89.

Bojang, F., Traore, S., Togola, A., & Diallo, Y. (2020). Farmers perceptions about climate change, management practice and their on-farm adoption strategies at rice fields in Sapu and Kuntaur of the Gambia, West Africa. *American Journal of Climate Change, 9*(01), 1.

Chhogyel, N., & Bajgai, Y. (2015). Modern rice varieties adoption to raise productivity: A case study of two districts in Bhutan. *SAARC Journal of Agriculture, 13*(2), 34–49.

CIAT; World Bank. (2017). Climate-smart agriculture in Bhutan. CSA country profiles for Asia series. International Center for Tropical Agriculture (CIAT); The World Bank. Washington, D.C. 26 p.

Cline, W. R. (2007). *Global warming and agriculture: Impact estimates by country*. Peterson Institute.

Dar, M. H., Chakravorty, R., Waza, S. A., Sharma, M., Zaidi, N. W., Singh, A. N., & Ismail, A. M. (2017). Transforming rice cultivation in flood prone coastal Odisha to ensure food and economic security. *Food Security, 9*(4), 711–722.

DFID, G. S. (2000). Sustainable livelihoods guidance sheets, Section 2. *Framework*.

Galloway-McLean, K. (2017). Advance guard: Climate change impacts, adaptation, mitigation and indigenous peoples: A compendium of case studies.

Gurung, G. B., & Bhandari, D. (2009). Integrated approach to climate change adaptation. *Journal of Forest and Livelihood, 8*(1), 91–99.

IPCC. (2014). Summary for policymakers. In O. Edenhofer, R. Pichs-Madruga, Y. Sokona, E. Farahani, S. Kadner, K. Seyboth, …, & J. C. Minx (Eds.), *Climate Change 2014: Mitigation of Climate Change. Contribution of Working Group III to the Fifth Assessment Report of the Intergovernmental Panel on Climate Change*. Cambridge University Press.

Jena, P. R., & Grote, U. (2012). Impact evaluation of traditional Basmati rice cultivation in Uttarakhand State of Northern India: What implications does it hold for geographical indications? *World Development, 40*(9), 1895–1907.

Kabir, M. H., & Baten, M. A. (2019). Climate change adaptation practices to water sector in South-Western coastal area of Bangladesh.

Kattel, R. R., & Nepal, M. (2021). Rainwater harvesting and rural livelihoods in Nepal. In A. K. E. Haque, P. Mukhopadhyay, M. Nepal, & M. R. Shammin (Eds.), *Climate change and community resilience: Insights from South Asia* (pp. 159–173). Springer.

Long, J. S., & Freese, J. (2006). *Regression models for categorical dependent variables using Stata.* Stata press.

Ragavan, S., & O'shields, J. M. (2007). Has India addressed its farmers' woes-a story of plant protection issues. *Georgetown Environmental Law Review, 20,* 97.

Rai, R. K., & Nepal, M. (2022). A tale of three Himalayan towns: Would payment for ecosystem services make drinking water supply sustainable? In A. K. E. Haque, P. Mukhopadhyay, M. Nepal, & M. R. Shammin (Eds.), *Climate change and community resilience: Insights from South Asia* (pp. 357–367). Springer.

Ranjan, R., Pradhan, D., Reddy, V. R., & Syme, G. J. (2015). Evaluating the determinants of perceived drought resilience: An empirical analysis of farmers' survival capabilities in drought-prone regions of South India. In *Integrated assessment of scale impacts of watershed intervention* (pp. 253–285). Elsevier.

Rath, N. C., Das, L., Mishra, S. K., & Lenka, S. (2007). Adoption of upland rice technologies and its correlates. *ORYZA-an International Journal on Rice, 44*(4), 347–350.

Renewable Natural Resources Statistics Division Directorate Services. (2019). RNR Census of Bhutan, Thimphu. Ministry of Agriculture and Forests.

Shava, S., O'Donoghue, R., Krasny, M. E., & Zazu, C. (2009). Traditional food crops as a source of community resilience in Zimbabwe. *International Journal of African Renaissance Studies, 4*(1), 31–48.

Shew, A. M., Durand-Morat, A., Putman, B., Nalley, L. L., & Ghosh, A. (2019). Rice intensification in Bangladesh improves economic and environmental welfare. *Environmental Science & Policy, 95,* 46–57.

Swiderska, K., Reid, H., Song, Y., Li, J., Mutta, D., Ongogu, P., & Barriga, S. (2011). *The role of traditional knowledge and crop varieties in adaptation to climate change and food security in SW China, Bolivian Andes and coastal Kenya.* IIED.

Wooldridge, J. M. (2016). *Introductory econometrics: A modern approach.* Nelson Education.

Chapter 7
Autonomous Adaptation to Flooding by Farmers in Pakistan

Ajaz Ahmed

Key Messages

- Farmers in Pakistan are very vulnerable to the impacts of natural hazards such as floods and droughts.
- Farm households in northern Pakistan use autonomous adaptations to adjust to monsoon flooding.
- Farm households' vulnerability, their knowledge of adaptation options and expected benefits, and prior experience with adaptation motivate the uptake of adaptation.

7.1 Introduction

Pakistan has faced increasing frequency of natural disasters such as floods and droughts in the past few decades due to its vulnerability to climatic changes (Abid et al., 2016; Ullah et al., 2019). In the 2020 Global Climate Risk Index report, Pakistan ranks 5th among the 10 countries most affected by natural hazards from 1999 to 2018 (Eckstein et al., 2020). Climate projections indicate a further increase in temperature by 2°–3° and a significant variation in the distribution of rainfall in Pakistan by 2050 (Gorst et al.,). Pakistan is not only among the world's regions that are highly prone to climate change-induced natural disasters (Ullah et al., 2019), it has also sustained greater damage because of its low adaptive capacity (Stocker et al., 2013).

According to 2017 population census, roughly 64% of the population of Pakistan resides in rural areas and is predominantly involved in agriculture. The sector, which contributes 20% to the country's total GDP and employs 43% of its workforce (Ali & Erenstein, 2017; Jamshed et al., 2019), is also highly vulnerable to natural hazards

A. Ahmed (✉)
Institute of Business Administration, Karachi, Pakistan
e-mail: ajazahmeda23@gmail.com

© The Author(s) 2022
Haque et al. (eds.), *Climate Change and Community Resilience*,
https://doi.org/10.1007/978-981-16-0680-9_7

that are intensifying due to climate change (Abid et al., 2015; Ali & Erenstein, 2017; Bakhsh & Kamran, 2019; Ullah et al., 2019). The resulting crop failure causes loss of livelihood and damage to agricultural investment (Jamshed et al., 2019; Ullah et al., 2019) and the impact on staple food crops threatens food security (Bakhsh & Kamran, 2019).

Unfortunately, the situation is exacerbated by the fact that relatively poorer populations tend to reside in the most flood-prone rural areas (Qasim et al., 2015; Rana & Routray, 2016) with limited access to off-farm income opportunities, skills, and basic amenities (Deen, 2013).

While the frequency of natural hazards has gradually increased in Pakistan (Wester et al., 2019), monsoon floods occur almost annually (Ahmad and Ma, 2020; Jamshed et al., 2019; Rana & Routray, 2016) causing substantial damage to the Pakistani economy (Rana & Routray, 2016). However, the government response to monsoon flooding has been both inadequate and inefficient owing to a lack of resources, reactive planning, focus on post-disaster relief, poor coordination between the responsible agencies, the absence of local-level disaster preparedness, and a lack of community involvement (Ahmad et al., 2019; Deen, 2013; GoP, 2016; Rahman & Khan, 2011). This is also because government investments have prioritized urban infrastructure to support the increasing urban population and industrial activities (Bakhsh & Kamran, 2019).

Since roughly two-thirds of the country's population reside in rural areas and relies on agriculture, it is vital to develop rural farming communities' resilience to the climate change-driven monsoon floods using farm level adaptation measures (Ali & Erenstein, 2017; Bakhsh & Kamran, 2019) which can increase food security. Until now the use of farm-level adaptation measures has, however, been very limited— mainly due to a lack of knowledge and resources (Abid et al., 2015, 2016; Ahmad et al., 2019; Rauf et al., 2017).

Farm households in Pakistan have used various adaptation measures such as structural modifications to buildings and savings as precautionary measures in response to floods (Shah et al., 2017). Other measures include changes in crop varieties introducing more diversification (Adnan & Lohano, 2022, Chap. 28 of this volume) and drought-resistant varieties; changes in use of fertilizer and manure use; plowing methods; planting trees, changes in date of sowing; off-farm income participation and crop–livestock interaction; changes in water management; off-farm employment, consumption smoothing, credit, and migration in response to drought (Abid et al., 2015, 2016; Ali & Erenstein, 2017; Ashraf & Routray, 2013; Bakhsh & Kamran, 2019; Rahut & Ali, 2017; Ullah et al., 2019). Using traditional crop varieties (Tshotsho, 2022, Chap. 6 of this volume) and providing seasonal agro-met advisories (Manjula et al., 2021, Chap. 18 of this chapter) are additional methods of prompting climate change adaptation.

This research uses a primary survey and binary logistic regression to investigate the factors explaining farm household decisions in Nowshera district to invest in various adaptation measures in response to flooding. Located in the Khyber Pakhtunkhwa province in northern Pakistan, Nowshera is subject to monsoon flooding due to its proximity to the Kabul River (Ahmad et al., 2011; Khan et al., 2013). The results of this study will help in understanding the nature of autonomous adaptation measures uncovering the drivers and constraints.

7.2   Study Context and Data

Nowshera is an agricultural district with limited off-farm income opportunities and the majority of the population is involved in agriculture for food, fodder, and livelihood. The main crops are maize, wheat, barley, tobacco, and sugarcane; however, some farmers also grow vegetables on a commercial scale.

Floods are a big risk to farming, causing crop failure and loss of livelihood in addition to damages to housing and infrastructure. The fundamental assumption is that rational farmers will use flood adaptation strategies to mitigate the flood damages.

7.2.1   Identification of Adaptation Options

The selection of adaptation options for this study was done following three steps. The first step involved the survey of literature on flood adaptation measures being used in developing countries to prepare a broad list of adaptations. This was followed by focus group discussions (FGDs) with the District Agriculture Office and Field Extension Office to shortlist the most likely adaptation options. In the third step, flood-affected farm households were consulted, and they confirmed the use of two adaptation options. The first adaptation option is the elevated ground floor, i.e., elevated base column of the ground floor, which helps reduce the exposure to flood water. The second adaptation measure is maintaining food stock (by storing surplus wheat) to avoid shortage in case of crop failure. The use of both adaptation options has been reported in previous research (Shah et al., 2017).

7.2.2   Data Collection

A multistage sampling approach was adopted to select representative households. Initially, three flood-affected and two non-affected union councils[1] were selected from the total of 27 flood-affected and 20 non-affected union councils using secondary information provided by the local Field Extension Office. After this,

---

[1] A union council forms the third tier of local government, after district and tehsil, within a province in Pakistan. Size of the union council varies from district to district.

homogenous villages from both subpopulations were identified. Lastly, the households were sampled considering their distance to the river, farm size[2] and their location in five zones along the Kabul River, to account for spatial heterogeneity. The sample size of this study is 500 farm households, 300 of which belonged to flood-affected villages and 200 to non-flood-affected villages. A pilot questionnaire containing questions on household socioeconomic characteristics, flooding, agricultural practices, and other pertinent information was used to collect the data. The survey was administered by a team of trained enumerators who conducted supervised face-to-face interviews in Pashto, the local language.

7.2.3 Data Analysis

Descriptive analysis is used to present the information on flood impact, i.e., damages, and adaptation uptake, whereas empirical analysis is used to investigate the factors affecting farm households' flood adaptations. Empirical analysis is conducted using a binary logistic regression as it is an appropriate technique for a dichotomous choice-dependent variable (0/1). Two different regression models have been estimated to explore the drivers of uptake of each adaptation option. Each dependent variable in the regression is a dummy variable, where one is assigned a household, who adopted specific measure and zero otherwise. Adaptation uptake is a dichotomous choice that means adaptation or no adaptation. The underlying assumption is that farm households that adapt perceive adaptation as a risk reduction mechanism and thus beneficial in the face of floods. However, the farm households that do not adapt refrain from doing so maybe due to their personal conditions and various socioeconomic constraints.

7.3 Descriptive Results

7.3.1 Flood Impacts

Monsoon floods in Nowshera have a negative impact on crops, livestock, farmhouse infrastructure, and local businesses. As discussed in Aftab et al. (2021), data shows that in the last main flood about 60% households suffered crop damage, 28% reported damage to their farmhouses, 11% farm households lost their livestock, and 1% farm households reported damage to their businesses. Farmhouse infrastructure includes roofed spaces for livestock, storage, and farm machinery. The reported monetary value of average losses per household in Nowshera in the last main flood is 193,770 Pakistani rupees (PKRs) (1206.21 USD) for crops, PKRs 111,660 (695.138 USD)

---

[2] Small and large farms were categorized depending on whether they were below or above 1 ha, respectively.

Table 7.1 Flood damages and their monetary value

Damages	Affected households (%)	Average monetary value (PKRs) (USDs)
Crops	60	193,770 (1206.21)
Farm housing infrastructure	28	111,660 (695.138)
Livestock	11	91,650 (570.536)
Local businesses	1	101,750 (633.387)

Source Field data

Table 7.2 Basic services disruption and restoration time

Disrupted services	Affected households (%)	Restoration time (months)
Water supply	56	5
Electricity	95	6
Transport infrastructure	94	7
Health service	94	7

Source Field data

for damage to the housing infrastructure, PKRs 91,650 (570.536 USD) for livestock losses and PKRs 101,750 (633.387 USD) for business losses (see Table 7.1). These households are heavily reliant on farming for their livelihood and often do not have access to financial instruments such as loans or other income support. Since agriculture is the main source of livelihood, fewer households have suffered from business enterprise losses.

It is not just the farm households which have sustained flood damages, but flooding has also disrupted the supply of basic public services such as water, electricity, transport, and health in Nowshera. Survey responses show that roughly 56% of households lost access to domestic potable water and more than 90% reported that transportation and health and electricity services were disrupted during the last big flood. The minimum time to restore the basic services was anywhere between 5 and 7 months (see Table 7.2).

7.3.2 Flood Adaptations Uptake

As indicated in Aftab et al. (2021), about 45% of flood-affected farm households use an elevated ground floor to reduce the exposure to flood water. This is indisputably a flood adaptation as it is only reported in the flood-affected sample households. The maintenance of food stock, which has other benefits as well, is being used in both the flood-affected and non-affected samples. However, the use of food stock is 16% higher in the flood-affected areas which shows that it is clearly also an adaptation to monsoons floods. This is similar to the consumption smoothing adaptation reported in climate adaptation literature (Ashraf & Routray, 2013).

7.4 Empirical Results

Empirical results of factors affecting the uptake of flood adaptations strategies using binary logistic regression technique are presented in Table 7.3. The independent variables used in two estimated models include farm household sociodemographic characteristics, farming-related information, and institutional features. Marginal effects are presented for the ease of interpretation which shows the probability of implementing each adaptation option in percentage terms. The main results that regression analysis has uncovered in this research are divided into three contentions. First, the adaptation process is influenced by the known vulnerability of agents. Second, knowledge of adaptation options and expected benefits drive the adaptation uptake. Third, learning from the past such as prior experience with adaptation needs and their benefits motivates the uptake of adaptation. Each of the three arguments is contextualized below using the results of this research.

Table 7.3 Binary logistic estimates of factors affecting the adaptation decision

Variables	Margins	
	Elevated ground floor	Food stock
Off-farm work	− 0.036	− 0.075
	(2.02)*	(3.49)**
Agriculture extension	0.068	0.070
	(2.03)*	(1.58)
Farm to river distance	− 0.059	0.084
	(4.88)**	(5.80)**
Flood inundation	0.182	–
	(4.08)**	–
Flood duration	–	0.019
	–	(3.18)**
Farming experience	–	0.004
	–	(2.46)*
Tribal diversity	0.013	–
	(2.14)*	–
Flood-affected villages	–	0.648
	–	(5.89)**

Source Field data
*** p < 0.01, ** p < 0.05, * p < 0.1

While the aforementioned three arguments enrich the results' discussion, they are also supported by the findings of Aftab et al. (2021). Furthermore, unlike the analysis in Aftab et al. (2021), this chapter reports two regression models.

7.4.1   Difference in Known Vulnerability

Empirical results of this research have clearly indicated that adaptation is moderated by the difference in vulnerability of the flood-affected communities. For example, findings show that farm households with more family members working off-farm are less likely to elevate the building ground floor or stock the food (Table 7.3). The most plausible interpretation of this result is that perhaps farm households with additional sources of income are more resilient and less vulnerable to the impact of floods. Results show that each additional off-farm worker reduces the farm households' likelihood to elevate the ground floor by roughly 4%. Likewise, an additional off-farm household worker reduces the farm households' probability of keeping the stock of food by about 8%.

Findings reveal that the greater the area of land of a farm household inundated in the last flood the more likely it is that the household will elevate the ground floor of the house. The chances of the use of elevated ground floor as an adaptation to flooding increases by nearly 18% with every additional hectare of inundated land. On the other hand, the further a farm household is from the river, the lower the chances of elevating the ground floor of the house. Each additional kilometer of distance from the river reduces the farm household's probability of elevating the ground floor by about 6%. Findings are comparable with the results of Mulwa et al. (2017), Choloet al. (2018), Boansi et al. (2017) and Tessema et al. (2019) studies.

Another variable which shows the difference in vulnerability and thereby the adoption of adaptations among farm households in the study area is the flood-affected dummy variable which is used to differentiate the uptake of food stock between flood-affected and non-affected villages. Results show that flood-affected villages have almost 65% higher probability to use food stock than non-affected villages.

7.4.2   Knowledge and Communal Learning

Findings of this research support the contention that knowledge drives adaptation uptake. Findings reveal that farm households with access to agricultural extension services have a 7% higher likelihood of elevating the ground floor of their houses. Access to agricultural extension services has an almost equal contribution in the uptake of food stock as in adaptation to flooding. These findings are also consistent with previous studies such as Nhemachena and Hassan, (2007), Mulwa et al. (2017), Boansi et al. (2017), Tessema et al. (2019), and Bedeke et al. (2019).

Interestingly, tribal diversity has a positive correlation with the use of elevated ground floor, which suggests that farm households from villages with more tribal diversity have a higher uptake of an elevated ground floor as an adaptation option. It is reasonable to believe that villages with more tribal diversity have greater communal learning among the farm households in study areas. This also signifies the role of tribal diversity, and hence intertribal competition in the adoption of agricultural technologies that provides a safety net as well as comparative advantage in rural tribal societies such as in district Nowshera.

7.4.3   Learning from the Past

It is a well-established fact that past experience helps communities in dealing with disasters, for example, farmer experience is always a significant driver for adoption of different farming practices. However, this research reveals that farmer experience is even more crucial owing to low literacy and education and the limited role of agricultural extension. Results show that farming experience is positively correlated with food stock as an adaptation to monsoon flooding in Nowshera, signifying that farm households with greater farming experience are more likely to use food stock as an adaptation to monsoon flooding. It is plausible that farmers with greater experience have faced multiple shocks; including unfavorable weather conditions for farming and thereby their livelihood. Hence, they have a higher tendency to adapt than their counterparts with less farming experience.

A similar result shows that flood duration has a positive correlation with food stock, indicating that households which have experienced longer floods in the past are more likely to stock the food as an adaptation to floods. Each additional day of standing flood water stood in the last flood increases the likelihood that a farm household stocks food by about 2%. These results indicate that prior knowledge and experience play a vital role in the adaptation process.

7.5   Conclusion and Policy Implications

This study investigates farm households' autonomous adaptation to monsoon flooding using a primary survey of 500 flood-affected and non-affected farm households in one flood-affected district, i.e., Nowshera, in northern Pakistan. It uses binary logistic regression technique to analyze farm households' autonomous adaptation to monsoon flooding.

The findings reveal that monsoon flooding in Nowshera caused damage to crop, livestock, farm housing infrastructure, and local businesses in addition to disrupting the supply of basic public services such as water, electricity, transport, and health in the last major flood. Unfortunately, government departments took long to restore the supply of basic services which had a very negative impact on the resilience of the flood-affected communities. Timely restoration of basic gateway services could facilitate normalcy and reduce the bounce back time for flood-affected communities. Survey data shows that farm households in the study area use elevated ground floor and food stock as autonomous adaptation measures to adapt to the monsoon floods. While food stock is also used in non-affected areas, it is an adaptation in flood-affected villages. Evidently, farm households' adoption of autonomous adaptations indicates that they understand flood risks and are willing to make investment in resilience building measures. This indicates that farm households would be willing to adopt the policy interventions initiated to manage the flood risk. Findings of this research show that the adaptation process in study areas is moderated by three key phenomena.

First, it is the difference in known vulnerability that enables farm households to adapt to the monsoon flooding. For example, off-farm work has a negative correlation with both adaptation options, elevated ground floor and food stock, which suggests that farm households with multiple sources of income have better adaptive capacity, and hence better resilience to the floods. The policy implication of this result is that livelihood diversification can reduce the farm households' risk from flooding. The role of farm to river distance in the decision to implement flood adaptation implies that flood impact has a spatial aspect and the interventions including post-flood support should be designed considering the location of the communities, their chances of exposure, and the level of risk.

Second, knowledge and communal learning supports the adaptation process in study areas. For example, access to agricultural extension services has a positive role in farm households' decision to elevate the ground floor of their houses to mitigate potential damages from future flood. This is an important result which shows that effective delivery of agricultural extension service has the potential to increase farmers' resilience to flooding. Interestingly, tribal diversity has a positive role in farm households' uptake of adaptation in study areas which is possibly due to communal learning. This indicates that there is a social dimension to adoption of technologies that provides an economic safety net to vulnerable farm households in rural and tribal settings. Hence, policymakers must take this into account while designing disaster risk management policies and strategies.

Third, farm households' prior experience informs their adaptation decisions. Findings show that flood inundation, flood duration, and farming experience are positively correlated with the decision to implement flood adaptations. These factors allow policymakers to distinguish between households, target adaptation incentives, and implement activities while considering households' risk perceptions and their prior experience with flooding and farming practices.

This research has unveiled that there is a suboptimal use of autonomous flood adaptation measures in flood-affected villages as farm households have not exploited the full potential of available autonomous adaptation. While some adaptation options have limited uptake, others have not been explored at all. For instance, hardly any households in the study areas grow short duration crops that are suited to flooding. The findings highlight the need to facilitate and encourage flood adaptations through a program of agriculture extension services and other soft interventions.

Acknowledgements This research was funded by the South Asian Network for Development and Environmental Economics (SANDEE), Nepal and partial results of the survey reported in this study have been published in Aftab et al. (2021). I am thankful to SANDEE for their financial and technical support without which this research would not be possible.

References

Abid, M., Scheffran, J., Schneider, U. A., & Ashfaq, M. (2015). Farmers' perception of an adaptation strategies to climate change and their determinants: The case of Punjab Province, Pakistan. *Earth System Dynamics, 6*, 225–243.

Abid, M., Schneider, U. A., & Scheffran, J. (2016). Adaptation to climate change and its impacts on food productivity and crop income: Perspectives of farmers in rural Pakistan. *Journal of Rural Studies, 47*, 254–266.

Adnan, N., & Lohano, H. D. (2022). Resilience through Crop diversification in Pakistan. In A. K. E. Haque, P. Mukhopadhyay, M. Nepal, & M. R. Shammin (Eds.), *Climate change and community resilience: Insights from South Asia* (pp. 431–442). Springer.

Aftab, A., Ahmed, A., & Sacrpa, R. (2021). Farm households perception of weather change and flood adaptations in Northern Pakistan. *Ecological Economics, 182*, 106882.

Ahmad, D., Afzal, M., & Rauf, A. (2019). Analysis of wheat farmers' risk perceptions and attitudes: Evidence from Punjab, Pakistan. *Natural Hazards, 95*(3), 845–861.

Ahmad, F., Kazmi, S. F., & Pervez, T. (2011). Human response to hydro-meteorological disasters: A case study of the 2010 flash floods in Pakistan. *Journal of Geography Regional Planning, 4*(9), 518–524.

Ali, A., & Erenstein, O. (2017). Assessing farmer use of climate change adaptation practices and impacts on food security and poverty in Pakistan. *Climate Risk Management, 16*, 183–194.

Ashraf, M., & Routray, J. K. (2013). Perception and understanding of drought and coping strategies of farming households in north-west Balochistan. *International Journal of Disaster Risk Reduction, 5*, 49–60.

Bakhsh, K., & Kamran, M. A. (2019). Adaptation to climate change in rain-fed farming system in Punjab, Pakistan. *International Journal of the Commons, 13*(2), 833–847.

Bedeke, S., Vanhove, W., Gezahegn, M., Natarajan, K., & Damme, P. V. (2019). Adoption of climate change adaptation strategies by maize-dependent smallholders in Ethiopia. *NJAS—Wageningen Journal of Life Sciences, 88*, 96–104.

Boansi, D., Tambo, J. A., & Müller, M. (2017). Analysis of farmers' adaptation to weather extremes in West African Sudan Savanna. *Weather and Climate Extremes, 16*, 1–13.

Cholo, T., Fleskens, L., Sietz, D., & Peerlings, J. (2018). Is Land fragmentation facilitating or obstructing adoption of climate adaptation measures in Ethiopia? *Sustainability, 10*, 2120.

Deen, S. (2013). Pakistan 2010 floods. Policy gaps in disaster preparedness and response. *International Journal of Disaster Risk Reduction, 12*, 341–349.

Eckstein, D., Künzel, V., Schäfer, L., & Winges, M. (2020). *Global climate risk index 2020 who suffers most from extreme weather events? Weather-related loss events in 2018 and 1999 to 2018.* Briefing Paper, German Watch.

Gorst, A., Groom, B., & Dehlavi, A. (2015). *Crop productivity and adaptation to climate change in Pakistan.* Centre for climate change economics and policy. Working paper No. 214.

Government of Pakistan. (2016). *Annual flood report 2016.* Federal Flood Commission, Ministry of Water and Power, Government of Pakistan.

Jamshed, A., Rana, I. A., Mirza, U. M., & Birkmann, J. (2019). Assessing relationship between vulnerability and capacity: An empirical study on rural flooding in Pakistan. *International Journal of Disaster Risk Reduction, 36*, 101109.

Khan, A. N., Khan, B., Qasim, S., & Khan, S. N. (2013). Causes, effects and remedies: A case study of rural flooding in district Charsadda, Pakistan. *Journal of Management Science, VII*(1, 2).

Manjula, M., Rengalakshmi, R., & Devaraj, M. (2021). Using Climate information for building small holder resilience in India. In A. K. E. Haque, P. Mukhopadhyay, M. Nepal, & M. R. Shammin (Eds.), *Climate change and community resilience: Insights from South Asia* (pp. 275–289). Springer.

Mulwa, C., Marenya, P., Rahut, D. B., & Kassie, M. (2017). Response to climate risks among smallholder farmers in Malawi: A multivariate probit assessment of the role of information, household demographics, and farm characteristics. *Climate Risk Management* 16208–16221.

Nhemachena, C., & Hassan, R. (2007). *Micro-level analysis of farmers' adaptation to climate change in Southern Africa*. IFPRI Discussion Paper No. 00714. International Food Policy Research Institute.

Qasim, S., Khan, A. N., Shrestha, R. P., & Qasim, M. (2015). Risk perception of the people in the flood prone Khyber Pukhthunkhwa province of Pakistan. *International Journal of Disaster Risk Reduction, 14*, 373–378.

Rahman, A. U., Khan, A. N. (2011) Analysis of flood causes and associated socioeconomic damages in the Hindukush region. *Nature Hazards, 59*, 1239–1260

Rahut, D. B., & Ali, A. (2017). Coping with climate change and its impact on productivity, income, and poverty: Evidence from the Himalayan region of Pakistan. *International Journal of Disaster Risk Reduction, 24*, 515–525.

Rana, I. A., & Routray, J. K. (2016). Actual vis-à-vis perceived risk of flood prone urban communities in Pakistan. *International Journal of Disaster Risk Reduction, 19*, 366–378.

Rauf, S., Bakhsh, K., Abbas, A., Hassan, S., Ali, A., & Kächele, H. (2017). How hard they hit? Perception, adaptation and public health implications of heat waves in urban and peri-urban Pakistan. *Environmental Science and Pollution Research, 24*(11), 10630–10639.

Shah, A. A., Ye, J., Abid, M., & Ullah, R. (2017). Determinants of flood risk mitigation strategies at household level: A case of Khyber Pakhtunkhwa (KP) province, Pakistan. *Natural Hazards, 88*, 415–430.

Stocker, T. F., Dahe, Q., & Plattne, G. (2013). *Climate change 2013: The physical science basis. Working Group-I contribution to the fifth assessment report of the intergovernmental panel on climate change*. Summary for Policymakers.

Tessema, Y. A., Joerin, J., & Patt, A. (2019). Climate change as a motivating factor for farm-adjustments: Rethinking the link. *Climate Risk Management, 23*, 136–145.

Tshotsho. (2022). Indigenous practices of paddy growers in Bhutan: A safety net against climate change. In A. K. E. Haque, P. Mukhopadhyay, M. Nepal, & M. R. Shammin (Eds.), *Climate change and community resilience: Insights from South Asia* (pp. 87–100). Springer.

Ullah, W., Nafees, M., Khurshid, M., & Nihei, T. (2019). Assessing farmers' perspectives on climate change for effective farm-level adaptation measures in Khyber Pakhtunkhwa, Pakistan. *Environmental Monitoring and Assessment, 191*, 547.

Wester, P., Mishra, A., Mukherji, A., & Shrestha, A. B. (Eds.). (2019). *The HinduKush Himalaya assessment—Mountains, climate change, sustainability and people*. Springer Nature Switzerland AG.

Chapter 8
Resilience to Climate Stresses in South India: Conservation Responses and Exploitative Reactions

P. Indira Devi, Anu Susan Sam, and Archana Raghavan Sathyan

Key Messages

- State patronage and collective action ensure sustainable management of post-disaster agroecosystems.
- State participation is limited in the case of long-term climate change effects (like water scarcity) as opposed to the case of climate extremes (floods).
- Absence of state presence and collective action leads to resource-depleting practices which are socially, ecologically and financially undesirable.

8.1 Introduction

The responses of communities to biotic/abiotic stresses and their ability to adapt, adjust and configure to original or improved positions are largely influenced by their psychological, educational and financial capacities (personal or private attributes), the extent of state support and collective action with community groups (social and public support). However, some of these resilience mechanisms are exploitative and unsustainable in the long run while others are sustainable and improvements over existing ones. The level of social and public support plays a crucial role in adoption

P. Indira Devi (✉)
Kerala Agricultural University, Thrissur, Kerala, India
e-mail: indiradevi.p@kau.in

A. S. Sam
Regional Agricultural Research Station, Kerala Agricultural University, Kumarakom, Kerala, India
e-mail: anu.susan@kau.in

A. R. Sathyan
College of Agriculture, Kerala Agricultural University, Vellayani, Kerala, India
e-mail: archana.rs@kau.in

© The Author(s) 2022
Haque et al. (eds.), *Climate Change and Community Resilience*,
https://doi.org/10.1007/978-981-16-0680-9_8

of sustainable practices, as evidenced by the responses of communities to climate change risks.

The Intergovernmental Panel on Climate Change (IPCC) defines resilience as: "the ability of human communities to anticipate, absorb, accommodate and recover from the effects of disturbances" (IPCC, 2012). Though community is considered as an interconnected system, the integration across several disciplines has trailed in the case of community resilience thinking (Cutter, 2016; Hoque et al., 2019). Recently, the need has arisen for the development of a robust resilience assessment tool that can cover various dimensions of community resilience viz. environmental, social, economic, infrastructural and institutional (Cimellaro et al., 2016). Hence, resilience assessments (RAs) have emerged as a key method of understanding human responses to disasters and help them to prepare better strategies to reduce the subsequent negative effects, thus empowering a population that can withstand and adapt to various future disasters (Burton, 2015). Resilient communities are able to avoid or minimize the negative impacts or even gain from the situation. However, it is understood that the resilience mechanisms of individuals and communities differ across regions, nations and societies as impacts also differ in their characteristics across these scales. Understanding the resilience mechanisms and their pros and cons helps to improve and redesign the same to ensure societal acceptance and sustainable development.

We analyse the resilience mechanism to climate change events (floods and droughts) by communities in three different social settings that differ in the level of state intervention and community participation. The situations vary with respect to the nature of shock, level of state intervention and collective action. While two situations are based on the drought-induced water scarcity impacts (long term) where collective action and state intervention is limited, the third is on the management of the impact of severe floods through state participation and collective action. The former leads to exploitative practices and unsustainable resource management and lower social welfare. On the contrary, the latter, where there has been state intervention through farmer collectives, has led to improvement in the fertility status of soils through the ecosystem-based adaptation (EBA) approach, which has produced better yields. Ahmed (2022, Chap. 7 of this volume) documents similar adaptation efforts in Pakistan while Kattel and Nepal (2021, Chap. 11 this volume) in Nepal and Bari et al., (2022, Chap. 12, this volume) in Bangladesh show how technology adoption by communities is helping them build climate resilience.

EBA is defined as combining biodiversity and ecosystem services into an adaptation and development strategy that increases the resilience of ecosystems and communities to climate change through conservation, restoration and sustainable management of ecosystems (Colls et al., 2009). Key benefits of EBA have been identified as securing water resources, ensuring provisional services (food) and buffering people from natural hazards, erosion, and flooding (Munang et al., 2013). An approach that aims at conservation of natural resources through collective action is preferred, as it ensures continuous provision of ecosystem services and welfare. However, such an approach is to be facilitated through technological and financial support and awareness creation among the communities. When the communities are left to market forces, the adaptation strategies are prone to be resource exploitative and unsustainable. It may also lead to further reduction in household welfare through financial burden and widening social disparities despite the short-term financial gains.

We depict three case studies: one on the flood management situation in the *Kuttanad* rice ecosystem (Kerala, Southern India), through active state intervention for EBA and collective action by farmer collectives. The other two case studies are on addressing water scarcity in two agriculturally important districts of Kerala where state's presence is limited, and farmer collectives are not functioning.

8.2   State Interventions for Resilience to Weather Extremes

The southern Indian state of Kerala is a narrow strip of land extending from the Western Ghats into the Arabian Sea. Though 14.5% of the state's land area is prone to floods, the 2018 August floods were the worst in about a century, resulting in the death of 433 persons and destroying infrastructure and livelihood worth USD3.8 billion. Over 65,000 ha of land was inundated and 1259 out of 1664 villages across all the 14 districts of Kerala were affected by the flood (Government of Kerala, 2019).

Kuttanad, the wetland zone situated around the *Vembanad* Lake spreads across *Alappuzha, Kottayam* and *Pathanamthitta* districts of Kerala. Most of the ecosystem is spread over Alappuzha and Kottayam districts, and it is one of the major flood-prone areas of the state. The region is very ecologically sensitive, thickly populated and one of the main rice producing tracts in the state spread over 1100 km^2 of area in the fertile deltaic region of the five Western Ghats river basins. The paddy farming system in *Kuttanad*, which is situated 0–3 m below mean sea level, is acknowledged as a Globally Important Agricultural Heritage System by the Food and Agriculture Organisation (FAO) (Koohafkan & Altieri,). Farming in *Kuttanad* is made possible by constructing a series of artificial embankments which prevents saltwater intrusion and flood water entry into the fields. Rice production in this ecosystem is characterized by the presence of strong community institutions, i.e. a collective of farmers who cultivate in the continuous stretch of rice paddies demarcated by manmade earthen bunds (*padasekharams*) that prevent sea water intrusion. The cultivation is done by the cooperation and collective efforts of the farmer collective (*Padasekhara Samithi*).

In July 2018, rainfall exceeded the normal levels by 18% and by mid-July, *Kuttanad* was flooded. Before the complete withdrawal of flood water from *Kuttanad*, the second flood in August 2018 hit this region, making the lives miserable. Over 17,300 families lost their houses completely and more than 170,000 houses were partially destroyed. A number of other public and private buildings were destroyed including 1613 schools. *Kuttanad's* biodiversity, agriculture, animal husbandry, fisheries, infrastructure and water supply systems were also severely affected. As the floods occurred after the sowing of the additional/kharif crop,[1] the seedlings were

---

[1] Generally, in Kerala rice is grown during three different seasons. First season is *virippu* (I crop) when the crop is planted in April–May and harvested in August–September. Second season is

completely washed away and the protective bunds that prevented sea water intrusion were destroyed. The estimated loss in *Kuttanad* was equivalent to 6.6% of the state's income in 2018.

The state government (State Department of Agriculture Development and Farmer's Welfare) was prepared with a definitive action plan for post-flood recovery strategies and action was initiated immediately, once the floods receded. The action plan was prepared with scientific, technical, and social consultations. The recovery vision for the agriculture sector was to develop sustainable, responsible, integrated, inclusive, eco-friendly and resilient agriculture.

The state's post-flood plan of action was mainly under three strategies: short, medium and long term. Short-term activities addressed the urgent requirements for reinstating agricultural production. Medium- and long-term activities included resilience build up in each subsector through environmentally sustainable integrated farming systems, community-based management of water resources, improvements in the value chain, setting up of early warning systems, and effective communication with enhanced geographic information system and other technology-backed capabilities (Government of Kerala, 2018).

8.2.1 Short-Term Interventions

The immediate intervention by the state in this region was to facilitate the replanting of the crop. The strategies involved: confidence-building process among farmers, investing in reconstruction of damaged bunds, soil quality analysis and support mechanism for corrections and farm input supply (seeds, labour, machines, chemical fertilizers and soil ameliorants and organic manures) with technological support and easy access to information. The technological, financial and facilitating role was taken over by the State Department of Agriculture (SDA). The scientific prescription for crop management was developed by the State Agricultural University (SAU).

Local self-governments were also actively involved in the process of rebuilding. The District Disaster Management Authority, which coordinated the rebuilding process repaired and restored the damaged public assets like roads, buildings, flood protection structures like weirs, gates and dykes, coastal protection structures, irrigation and drainage canals and also removed the silt/debris deposits. The Department of Agriculture took the lead to dewater the inundated fields and construct the damaged bunds. This was also supported through the Mahatma Gandhi National Rural Employment Guarantee Programme, which also ensured income to poor households in the area.

---

mundakan (II crop) and the period is from September–October to December–January. *Puncha* (III crop) is the third season and it is from December–January to March–April.As Kuttanad is in special ecological zone, the rice cultivation is only in two seasons. First season is known as the *additional/kharif crop* and season is May–June to August–September. The main crop season is *puncha,* and it is from October–November to February–March.

Table 8.1 Soil parameters of *Kuttanad* during pre- and post-flood periods

Parameters	Pre-flood status	Post-flood status
Soil pH	3.0–4.0	4.0–5.0
Organic carbon	0.97%	1.68%
Soil electrical conductivity	0.2–0.8 dS/m	0.02–0.04 dS/m
Phosphorous	70–100 kg/ha	8–15.0 kg/ha
Potassium	350–500 kg/ha	50–100 mg/kg
Calcium	100–200 mg/kg	400–600 mg/kg
Magnesium	50–60 mg/kg	20–30 mg/kg

Source KAU (2019)

Simultaneously arrangements for soil testing were made, and scientific prescriptions were prepared. The silt and sand deposits had resulted in changes in mechanical, physical and chemical properties of the agricultural soils of paddies and created anaerobic conditions. However, silt deposits had also enriched the soil in these stretches with organic matter and certain nutrients that facilitated crop growth. Silt deposition in some areas of upper *Kuttanad* was up to 4 in. thickness while in lower *Kuttanad* it was up to 2 in. This silt and clay deposit had major impacts on soil aeration and crop growth.

Details of soil parameters during pre- and post-flood periods are given in Table 8.1. The scientific soil analysis revealed wide variation in the levels of soil nutrients across the regions. There was insufficient amount of some of the macronutrients (phosphorus, potassium, magnesium) and boron (micronutrient) while some of the micronutrients like calcium and zinc content increased in the soil after the flood. The soils of upper *Kuttanad* became low in organic carbon while it was high in the lower *Kuttanad*. Heavy metals and pesticide residues were absent, and the pH level improved and was near neutral.

The Department of Agriculture arranged for the corrections in soil quality (through application of soil ameliorants) and supply of quality seeds (fully subsidized) planting operations were also facilitated and closely monitored by experts (development workers and agricultural scientists). The educational programme focussed on the scientific crop management practices to be followed based on soil test results.

The *padasekhara samithis* (farmer collectives with members who own/operate within a *padasekharams*) were provided with 20% of the total cost of cultivation as advance credit, and the technical information was also provided to them. The Department of Agriculture and SAU were active in monitoring the crop situation as well as educating and supporting the *padasekhara samithis.*

These interventions have helped in increasing the production (75%) by bringing more area under farming (28%) and improving productivity (see Table 8.2 for details). The average productivity was 6–6.7 t/ha during the pre-flood period (2017–2018) which registered a quantum jump to 8.75–9.4 t/ha, post-flood (2018–2019). The scientific management, monitoring and support that helped to develop community

Table 8.2 District-wise area, production and productivity of rice in *Kuttanad*

Particulars	Alappuzha (*Puncha* rice)		Kottayam (*Puncha* rice)	
Period (year)	2017–2018	2018–2019	2017–2018	2018–2019
Area (ha)	24,000	28,800	6868	10,646
Production (tonnes)	115,000	191,000	34,000	71,000
Productivity (tonnes/ha)	4.79	6.63	5.0	6.7

Source KAU (2019)

resilience facilitated this. The management of deposited silt, correction of soil acidity and subsequent soil nutrient and pest management ensured better crop performance.

There are reports that the productivity levels during 2019–2020 are also on par with post-flood levels though official data is yet to be published. The community resilience development that focussed on an ecosystem-based approach (EBA) through social and knowledge capital built up by state participation has proved to be sustainable.

8.2.2 Medium- and Long-Term Strategies

The medium- to long-term measures to build up the resilience measures include flood forecasting based on flood-modelling studies. The main lead for this initiative is under the National Hydrology Project in close coordination with the ongoing interventions under river rejuvenation, lift irrigation stations, regulators and flood bunds in the *Kuttanad* region (Government of Kerala, 2019). Thus, multisector interventions along with systematic planning are expected to strengthen flood resilience in the region. The Kerala State Planning Board has also recommended a Rs. 2448-crore (USD340 million) special package and the state government has taken the decision to implement the same. The special *Kuttanad* package aims to improve the sanitation, water supply, flood control, management of water bodies, promotion of organic farming and to ensure responsible tourism and sustainable development.

Community resilience is becoming an effective strategy for enhancing community-level disaster preparedness, response and recovery. It is the ability of communities to withstand hazards and is also intertwined with individual resilience (Norris et al., 2008). Rural agricultural communities are food producers and source of environmental and social functionalities who build resilience through increased robustness and redundancy (Wilson, 2010). Hence, the most effective way to ensure a safe and healthy society is to train and equip them to mitigate and adapt to natural disasters. Capacitating the communities for EBA approaches makes the social system move towards welfare with equity and quality.

8.3   State Silence About Water Scarcity in *Chittur* and *Wayanad*

The responses of individuals and communities to stresses differ according to the nature of the climate event (weather extreme or climate change). The response to water scarcity is a slow process wherein the individual/community responses develop slowly, over a period of time depending on the adaptive capacity. Adaptive capacity differs among people, communities and countries across space and time. Community's adaptive capacity is a dynamic function of local processes and conditions, which are influenced by wider socio-economic and political scenarios, and access to resources (Smit & Wandel, 2006). Population pressure and resource depletion leading to limited access to resources may progressively lessen a system's coping ability, while economic growth and improvements in technology may lead to an increase in community's adaptive capacity (Folke et al., 2002; Smit & Pilifosova, 2003). Here, we discuss the resilience behaviour in response to water scarcity wherein collective action by farmers and the state's active presence are absent. The *padasekhara samitis* were historically evolved farmer collectives in *Kuttanad* rice farming due to the ecosystem peculiarities and political economy of the region. Unlike that situation, such strong and active farmer collectives are not present in *Chittur* and *Wayanad*.

The climate change impacts include both slow effects such as the reduced availability of natural resources like water as well as sudden extreme weather events like floods. However, the extreme events get more state attention and intervention compared to the long-term slow and steady adverse effects. This may be due to political reasons as climate extremes get more public attention and the state response is widely acknowledged and appreciated whereas measures against water scarcity issues are often less visible as they are more general in nature.

8.4   Water Scarcity and Community Responses

Kerala has been often perceived as a 'water-rich state' with an average annual rainfall of 3000 mm/year and a large number of freshwater bodies. Recently, the state has been experiencing more rainfall deficit years and acute water scarcity. With high runoff losses (40%) and progressively declining groundwater levels, the scarcity is getting intensified. While the agricultural practices and policies support irrigated agriculture, the resource scarcity poses challenges in agricultural production and food security. Agriculture is the major consumer of fresh water resources in the state, as is the case globally. *Chittur* Block in Palakkad district of Kerala represents the case of increasing water scarcity and farmer responses as it is a predominantly agriculture area situated in the North Eastern side of Kerala. According to the State Ground Water Department, *Chittur* is a groundwater over-exploited area which faces severe water scarcity, perhaps the worst in the state.

Wayanad is a hilly, ecologically sensitive agricultural district of Kerala. The region is reported to be experiencing lower rainfall and has had a falling water table, over the years. The main source of irrigation is from open wells and the average depth of open wells in the study area was 10.75 m whereas the decline in water table was to the tune of 4–5 m during the peak summer season. About 47% of the sample respondents in a study in *Wayanad* reported a 75% deviation in the water table (Rinu, 2012). It is evident that the situation is getting worse, as the deviation was 36.08% in 2005 and only 6.6% in 2000.

The major source of drinking and irrigation water in Kerala is open wells, with a very high well density of 200 wells per km^2 in the coastal region, 150 wells per km^2 in the midland and 70 wells per sq.km in the high land. As reported in the *Wayanad* study, the traditional practice was of depending on common wells, ponds or water sources from the neighbourhoods for domestic as well as irrigation purposes. There was a gradual shift from common sources to individually owned sources during a span of 10 years from 2000. Nearly 84% of the homesteads own either open wells, tube wells or ponds as compared to the years 2005 and 2000, when it was 79% and 75%, respectively. In 2000, only 63% had access to their own open wells and six % owned tube wells. However, only two-thirds of marginal farmers have their own sources compared to 95% for both small farmers and large farmers. This implies that the existing inequality in land holdings also leads to inequity in access to groundwater, which in turn further skews the divergence in assets and income distribution.

The farmers' immediate response to the declining water table in their own wells was to deepen them and subsequently, when the wells dried up, to dig new open wells. *Chittur* area experienced growing levels of water scarcity and severe and recurrent drought during the 2002–2005 period and a majority of the wells dried up. Since then the strategy shifted to digging borewells and later on deepening them. There were instances wherein the farmers used dried up open wells as storage structures for water pumped from borewells and pumping water from these open wells for irrigation (Fig. 8.1). Eventually, these borewells also dry up. The farmer's decision to opt for borewells and competitive well deepening is often influenced by the private borewell operators who extend the services as a package linking with credit support from institutional/non-institutional agencies. As the water scarcity worsens and borewells also dry up, farming activities are adversely affected, and farmers become defaulters. There are informal reports of social and domestic problems and farmer suicides on account of such pressures. Balasubramanian and Saravanakumar (2022, Chap. 10 this volume) document how state policy may itself have triggered such outcomes as adverse incentive emerging from subsiding electricity.

Table 8.3 provides details on various adaptation patterns and the economic costs of each adaptation in the *Chittur* area obtained/collected through a field survey conducted in 2015. During a span of five years from 2005, only 14% farmers who owned open well-irrigated farms decided to deepen the existing open wells. This could be done with an investment of USD 51 per farm per year (amortized cost). However, these wells also dried up after 2–3 years. Only 2% of the farmers dug new open wells at an annual amortized cost of USD 22 per farm. Deepening of existing

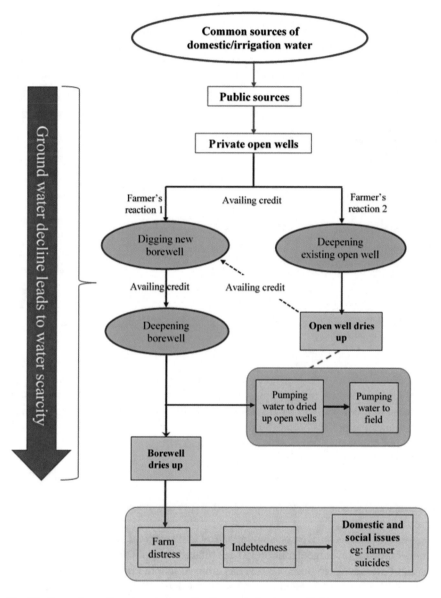

Fig. 8.1 Farmer's reactions to water scarcity. *Source* Author's compilation

wells was found costlier than digging new open wells, as the existing wells had hard rock layers at deeper levels (Table 8.3).

The majority of the famers (52%) opted for digging new borewells which irrigated nearly 59% of cultivated area. The technological advancement in this sector coupled with credit support has facilitated the spread of borewells. Moreover, a large number

Table 8.3 Adaptation to water scarcity and economic costs of adaptation in *Chittur* area

Particulars	Samples of open well owned farmers	Samples of borewell owned farmers
Existing wells deepening		
Farmers adoption rate (%)	14	6
Area benefitted (%)	19	5
Cost (USD/ha/year)	32	45
Cost (USD/farm/year)	51	48
Digging of new borewells		
Farmers adopted (%)	52	58
Area benefitted (%)	59	61
Cost (USD/ha/year)	69	116
Cost (USD/farm/year)	86	169
New open wells digging		
Farmers adoption rate (%)	2	0
Area benefitted (%)	3	0
Cost (USD/ha/year)	30	0
Cost (USD/farm/year)	22	0

Source Seenath (2017)

of private operators who constantly updated the technology prompted the farmers to opt for borewells. Often the borewell operators arranged the institutional or non-institutional credit support to the farmers. Due to the weak regulatory mechanism by the state, the private borewell operators created an environment for widespread digging of borewells. They facilitated credit, technological and institutional support for digging new borewells.

Though there are strict guidelines and licensing systems for borewell digging, the implementation is not strict due to poor monitoring and governance. The absence of farmer collectives leads to actions based on private motives, often at competitive levels.

A farmer invests an average amount of 169 USD per year (amortized) for digging a new borewell (average depth of 177 m), mainly with credit support. Though the Central Ground Water Board (2013) has advised that the maximum feasible depth of borewells in *Chittur* region as 150 m below ground level (mbgl), the depth of borewells was increasing at an annual rate of 2.9 m/year. In 2000, the average depth of borewells in *Chittur* region was 140 mbgl which had increased to 180 mbgl during 2014. Furthermore, the water reserve for open wells is unconfined aquifers, while borewells depend on confined aquifers. The recharge of confined aquifers takes many years (while unconfined aquifers get recharged at a faster rates) and hence resource replenishment is not in tune with extraction rates. Thus, intensive usage of water from borewells may lead to drying up of borewells because of the quicker extraction and slow recharge. This may further intensify water scarcity and thus proves to be unsustainable. Digging new wells and competitive deepening of existing open wells to adapt to water scarcity are common among the farmers across India (Shaheen & Shiyani, 2005).

Apart from deepening, the water extraction methods have also been modernized, viz. centrifugal, submersible and compressor pumps fitted with electric motors. Except a few (6%), all the farms have either replaced the conventional centrifugal pumps with new ones or upgraded to submersible pumps. Submersible pumps are energy efficient and even suitable for summer months, if the wells have adequate water to submerge the motor and pump set. In the case of borewells with low water yield, compressor pump is appropriate. In *Chittur,* the wells are fitted mostly with compressor pumps. Most farms (72%) in the area used compressor pumps with valve systems with switching over facility between open well and borewell. Usually, the farmers resorted to continuous pumping for 7–18 h during early summer months and stored the water in existing open wells. Subsequently, water is pumped from the open wells. The average pumping time in farms with both open and borewells was 1286.8 h/ha/year. This was three times greater than that of the pumping time in farms with only open wells. This system though exploitative and energy intensive is not reflected in the private cost accounting, as fuel charges are fully subsidized. These case studies highlight the unscientific and uneconomical practices followed by individual farmers as well as the weak government regulatory mechanism to control the exploitative responses of the individuals towards common property rights (CPRs) of water. Poor state interventions, absence of collective action and faulty policies have led to financially weak, unsustainable and resource-depleting practices. There are clear evidences of rigorous groundwater extraction from other Indian states like Punjab, Haryana and Uttar Pradesh, which is facilitated by the public policies (energy and chemical input subsidies) and green revolution technologies (Chand and Parappurath 2011; Jeevandas et al., 2008).

While the consumption of water is on the increase, the efforts to conserve and improve the supply are rather limited. Water conservation structures are one of the solutions to improve groundwater availability by enhancing recharge. However, low-cost conservation methods such as conservation pits and water harvesting tanks were adopted by only less than 10% farmers. Those who followed the practice failed to follow the scientific approach as well. These water conservation techniques are slow to show the results, though they are very cost effective (USD61 and USD89 per ha per year). Burying of coconut husk and mulching with coconut leaves were also not very popular, despite its proven results. The time lag in realizing the returns and sensitivity to discount rates limit the level of adoption of such measures (Pande et al., 2011). The main reasons for low adoption of private conservation investments are incomplete property rights and the externalities (high cost of exclusion and non-rivalry) associated with public goods. Leach et al. (1999) have argued that stewardship over natural resources is the state's responsibility and conservation measures could be initiated by the state. This view was shared by the respondents as well. These measures are not promoted by private agencies because of low profitability which has also contributed to limited adoption.

At the same time, roof water harvesting, adopted by 14%, was installed with technical and financial support from grass-roots-level government organizations and the National Bank for Agricultural and Rural Development (NABARD) at a cost of USD34/unit/year with 50% subsidy. The presence of the public sector thus facilitates adoption of natural resource-based practices that ensures sustainability. Thus, it is very clear that collective action with support from government and other public sector organizations contributes to resilience strategies which are sustainable and efficient. Behavioural responses of communities to resource scarcity and resilience are largely governed by the technical and financial backing from the public sector. Thus, in the absence of collective action nurtured by the state, the individual actions are resource depleting and exploitative in nature. Poor governance and monitoring lead to the operation of private market players who prompt resource-depleting unsustainable consumption practices. It is also noticed that state's interventions in addressing such issues (resource conservation) where results are visible only slowly and steadily are rather limited. This may be due to political reasons as well. While sudden and extreme weather events attract public attention, the climate change impacts of long-term nature do not attract massive public attention. The situation is further aggravated by absence of bargaining power, as collectives are not present.

8.5   Conclusions

Community resilience denotes the sustained capability of a community to employ available resources to respond to, withstand, and recover from the adverse effects of any disquiets. This chapter gives a clear overview on the importance of state interventions and regulatory mechanisms to ensure sustainability and judicious utilization of common property resources in individual responses to different stress situations (flood and drought). The first case study in *Kuttanad* flood region demonstrated how public investments, technological interventions and collective action built up the resilience to disasters among farmers that could lead to bumper harvest and sustainable farming culture.

The cases of *Chittur* and *Wayanad* areas depict adoption of supply-driven strategies to improve resource availability which were mainly exploitative in nature, viz. borewell digging, deepening and intensive extraction. This is perhaps due to the absence of public sector intervention and the exploitative approach of private market operators. At the same time, the conservation approach was found to be rather poor. The respondent's expectation of the state's role in resource conservation also restricts their conservation behaviour. A similar situation in *Wayanad* district shows an increased level of privatization of common property resources (water) with more farmers ensuring their own sources for water rather than traditional common sources. This has also led to widening social inequalities as the existing inequality in land holdings leads to an inequity in access to groundwater, which in turn widens the skewness in assets and income distribution where the state has chosen not to intervene.

The role of the state in capacitating and enhancing the preparedness and post-disaster management through policy shifts, capacity building and facilitating collective action is underlined. At the same time, the responses of the political system to climate risks are found to vary according to the nature of the risk. The catastrophes (floods) and pandemics trigger immediate and effective action while climate change-induced slow impacts (water scarcity) do not attract the same approach. Capacitating the communities on EBA through active state participation facilitates collective action that ensures sustainable outcomes.

References

Ahmed, A. (2022). Autonomous adaptation to flooding by farmers in Pakistan. In A. K. E. Haque, P. Mukhopadhyay, M. Nepal, & M. R. Shammin (Eds.), *Climate change and community resilience: Insights from South Asia* (pp. 101–112). Springer.

Balasubramanian, R., & Saravanakumar, V. (2022). Climate sensitivity of groundwater systems in South India: Does it matter for agricultural income? In A. K. E. Haque, P. Mukhopadhyay, M. Nepal, & M. R. Shammin (Eds.), *Climate change and community resilience: Insights from South Asia* (pp. 143–156). Springer.

Bari, E., Haque, A. K. E., & Khan, Z. K. (2022). Local strategies to build climate resilient communities in Bangladesh. In A. K. E. Haque, P. Mukhopadhyay, M. Nepal, & M. R. Shammin (Eds.), *Climate change and community resilience: Insights from South Asia* (pp. 175–189). Springer.

Burton, C. G. (2015). A validation of metrics for community resilience to natural hazards and disasters using the recovery from Hurricane Katrina as a case study. *Annals of the Association of American Geographers, 105*(1), 67–86. https://doi.org/10.1080/00045608.2014.960039

Central Ground Water Board. (2013). Annual report 2012–2013. Ministry of Water Resources, New Delhi.

Chand, R., & Parappurathu, S. (2011). Historical and spatial trends in agriculture: Growth analysis at national and state level in India. In *IGIDR Proceeding/Projects Series no. PP-069–3b*. November, 10–11 2011, India International Centre (on-line). Available at http://www.igidr.ac.in/newspdf/srijit/PP-069-03b.pdf. Accessed November 30, 2020.

Cimellaro, G. P., Renschler, C., Reinhorn, A. M., & Arendt, L. (2016). PEOPLES: A framework for evaluating resilience. *Journal of Structural Engineering, 142*(10), 4016063. https://doi.org/10.1061/(ASCE)ST.1943-541X.0001514

Colls, A., Ash, N., & Ikkala, N. (2009). Ecosystem-based adaptation: A natural response to climate change. In *International Union for conservation of nature and natural resources Switzerland*. Available at https://www.iucn.org/content/ecosystem-based-adaptation-a-natural-response-climate-change. Accessed November 30, 2020.

Cutter, S. (2016). *Social vulnerability and community resilience measurement and tools*. University of South Carolina.

Folke, C., Carpenter, S., Elmqvist, T., Gunderson, L., Holling, C.S., & Walker, B. (2002). Resilience and sustainable development: Building adaptive capacity in a world of transformation. *AMBIO: A Journal of the Human Environment, 31*(5), 437–440. https://doi.org/10.1579/0044-7447-31.5.437

Government of Kerala. (2018). United Nations, Asian Development Bank, The World Bank, European Union Civil Protection and Humanitarian Aid, Kerala Post Disaster Needs Assessment Floods and Landslides—August 2018, Government of Kerala. Available at https://rebuild.kerala.gov.in/reports/PDNA_Kerala_India.pdf

Government of Kerala. (2019). *Rebuild Kerala, development programme*. Government of Kerala.

Hoque, M. Z., Cui, S., Lilai, X., Islam, I., Ali, G., & Tang, J. (2019). Resilience of coastal communities to climate change in Bangladesh: Research gaps and future directions. *Watershed Ecology and the Environment, 1*, 42–56. https://doi.org/10.1016/j.wsee.2019.10.001

Intergovernmental Panel on Climate Change. (2012). Special Report on Managing the Risks of Extreme Events and Disasters to Advance Climate Change Adaptation. In C. B. Field, V. Barros, T. F. Stocker, D. Qin, D. J. Dokken, K. L. Ebi, M. D. Mastrandrea, K. J. Mach, G. K. Plattner, S. K. Allen, M. Tignor, & P. M. Midgley (Eds.), *A special report of working groups I and II of the intergovernmental panel on climate change* (pp. 582). Cambridge University Press.

Jeevandas, A., Singh, R. P., & Kumar, R. (2008). Concerns of groundwater depletion and irrigation efficiency in Punjab agriculture: A micro-level study. *Agricultural Economics Research Review, 21*, 191–199.

Kattel, R. R., & Nepal, M. (2021). Rainwater harvesting and rural livelihoods in Nepal. In A. K. E. Haque, P. Mukhopadhyay, M. Nepal, & M. R. Shammin (Eds.), *Climate change and community resilience: Insights from South Asia* (pp. 159–173). Springer.

Kerala Agricultural University. (2019). *Report of soil analysis in Alappuzha and Kottayam districts.* Prepared by RARS, Kumarakom. 2019. Thrissur, Kerala 1:3.

Koohafkan, P., & Altieri, M. A. (2010). *Globally important agricultural heritage systems: A legacy for the future.* UN-FAO.

Leach, M., Mearns, R., & Scoones, I. (1999). Environmental entitlements: Dynamics and institutions in community-based natural resource management. *World Development, 27*(2), 225–247. https://doi.org/10.1016/S0305-750X(98)00141-7

Munang, R., Thiaw, I., Alverson, K., Mumba, M., Liu, J., & Rivington, M. (2013). Climate change and Ecosystem-based Adaptation: A new pragmatic approach to buffering climate change impacts. *Current Opinion in Environmental Sustainability, 5*(1), 67–71. https://doi.org/10.1016/j.cosust.2012.12.001

Norris, F. H., Sherrieb, K., Galea, S., & Pfefferbaum, B. (2008). Capacities that promote community resilience: can we assess them? Paper presented at the 2nd annual Department of Homeland Security University Network Summit, Washington, DC. www.orau.gov/dhsresummit08/presentations/Mar20/Norris.pdf

Pande, V. C., Kurothe, R. S., Singh, H. B., & Tiwari, S. P. (2011). Incentives for soil and water conservation on farm in Ravines of Gujarat: Policy implications for future adoption. *Agricultural Economics Research Review, 24*(1), 109–118.

Rinu, T. V. (2012). Socio economic vulnerability and adaptive strategies to environmental risk: A case of water scarcity in agriculture. MSc (Ag) thesis, Kerala Agricultural University.

Seenath, P. (2017). *Groundwater irrigation: Management, adaptation and economic costs under declining resource conditions.* Ph.D. Thesis, Kerala Agricultural University.

Shaheen, F. A., & Shiyani, R. L. (2005). Water use efficiency and externality in groundwater exploited and energy subsidised regime. *Indian Journal of Agricultural Economics, 60*(3), 445–457.

Smit, B., & Wandel, J. (2006). Adaptation, adaptive capacity and vulnerability. *Global Environmental Change, 16*(3), 282–292. https://doi.org/10.1016/j.gloenvcha.2006.03.008

Smit, B., & Pilifosova, O. (2003). From adaptation to adaptive capacity and vulnerability reduction. In J. B. Smith, R. J. T. Klein, & S. Huq (Eds.), *Climate change* (p. 158p). Imperial College Press.

Wilson, G. (2010). Multifunctional "quality" and rural community resilience. *Transactions of the Institute of British Geographers, 8, 35*(3), 364–381. Available at http://www.jstor.org/stable/40890993. Accessed November 30, 2020.

Chapter 9
Climate Adaptation by Farmers in Three Communities in the Maldives

Fathimath Shafeeqa and Rathnayake M. Abeyrathne

Key Messages

- Farmers need a valid source of information at the island level on climate change and its impacts in order to adapt to the changing conditions and to be more resilient in the future.
- Women farmers who tend to the home gardens and the homestead plots with assistance from their families and those who have participated in agricultural workshops conducted in the island are more likely to contribute to the social capital of farming.
- Farmer cooperatives are an important institution for farmers to build trust and reciprocity which has been identified as one of weakest links among the surveyed communities.

With climate change, the growth of subsistence root crops and vegetables is likely to be affected. Moreover, sea-level rise and its consequent saline intrusion will have major impacts on crop production, especially in low islands and atolls in the Pacific, where all the crop agriculture is found on or near the cost (UNFCCC, 2005, p. 19).

F. Shafeeqa (✉)
Institute of Research and Development Pvt. Ltd., Malé, Maldives
e-mail: fathimathshafeeqa@yahoo.com

R. M. Abeyrathne
Department of Sociology, University of Peradeniya, Kandy, Sri Lanka
e-mail: abeyrathnayake1965@arts.pdn.ac.lk

© The Author(s) 2022   129
Haque et al. (eds.), *Climate Change and Community Resilience*,
https://doi.org/10.1007/978-981-16-0680-9_9

9.1   Introduction

The impacts of climate change are more evident in Small Island Developing States (SIDS) like the Maldives Islands (Baldacchino & Kelman, 2014). SIDS are vulnerable to coastal inundation and erratic rainfall. These events are further aggravated by El Ninõ, tropical cyclones, and hurricanes (Metz et al., 2007; Khushal, 2016). Sea-level rise has been identified as a major threat to the low-lying island nations in the SIDS and especially to island nation states such as the Maldives where critical infrastructure, housing, and settlement areas are near the coastline. Ideally, the farms near these areas could be moved inland to combat such coastal erosions. However, in the case of the Maldives due to the small size of the islands, there is no inland area away from the coast for such relocation. This threat is exacerbated by storms, tidal waves, and beach erosion (MEE, 2016); increase in the minimum and maximum temperatures; and decreasing quantity and increasing intensity of rainfall.

Since 80% of the land is one meter above sea level (MEE, 2016), the risks of an increase in sea level are immense and could change the soil and fresh groundwater resources making it unfit for human use even before total inundation of these areas occur. Sea-level records for the past 20 years show a rise of 3.753 and 2.933 mm per year in Malé and Gan, respectively (MEE, 2016).

Schleussner and Hare (2015) state that the effects of climate change on the agriculture sector in the SIDS will impact negatively on the national income. In the past, most Maldivians have practiced subsistence farming and used very few tools. Khushal (2016) states that: "Small farmers in SIDS are among the most powerless victims of climate change and the most immediately affected, but their voices and concerns are not adequately understood and supported in international negotiations."

There have been a series of policies published by the Ministry of Environment and Energy, the Disaster Management Authority, and other stakeholders that address the impacts of climate change on the Maldives (MEE, 2016, 2017). However, impact assessments have not been undertaken at the national and island level in an organized manner and need to be studied further in order to formulate effective mitigation and adaptation strategies at the local and national level. A crucial part of this process is to develop a better understanding of the impact of climate change on communities at the local level. In this study, we ask two questions: (1) what are the current adaptation practices of Maldivian farmers? and (2) what is the role of social capital in influencing these practices?

9.2   Adaptation and Climate Change

Dodman and Mitlin (2015) stress the importance of considering the local knowledge and community experience in the development of climate change policies and adaptation strategies. Local perceptions are critical in determining why communities are not taking appropriate adaptive action when faced with disasters (Tshotsho,

2022, Chap. 6 this volume). Lopez-Marrero and Yarnal (2010) have studied two communities in Puerto Rico and found that the communities were more interested in attending to their health, family well-being, and other economic factors even while these communities were facing floods and hurricanes. This shows that it is very critical to address adaptation within the context and situation of the communities and their well-being (Smit, 2000; Smit & Wandel, 2006). The authors have discussed that there are differences between autonomous (automatic, spontaneous, or passive adaptations) and planned (strategic or active) adaptation. Smit et al. (2000) have also added that adaptations may occur as an inadvertent outcome of other activities. They have stated the importance of contextualizing the adaptive capacity and actions of the respective communities and individuals.

Kelman (2007) has identified some advantages in SIDS to tackle these issues which include tight kinship networks, unique heritage, a strong sense of identity and community, sustainable livelihoods, remittances from islander diasporas supporting life on SIDS and local knowledge and experience of dealing with environmental and social changes throughout history (Kelman, 2007). Social capital is one of the major assets of small islands and has been discussed by many authors.

9.3 Social Capital and Adaptive Capacity

First defined in the early twentieth century, social capital is defined by Moser et al., (2010, p. 7): "Social capital is an intangible asset, defined as the rules, norms, obligations, reciprocity and trust embedded in social relations, social structures, and societies' institutional arrangements. It is embedded at the micro-institutional level (communities and households) as well as in the rules and regulations governing formalized institutions in the marketplace, political system and civil society." Any community or group that can self-organize to work together for a common challenge can be considered as having social capital (Moser et al., 2010). Although there are many explanations and interpretations of social capital, the main components highlighted in the above literature are trust, reciprocity, and interpersonal relationships (Pelling & High, 2005).

In the Maldives, small-holder farmers who rely on their own families to work in the homestead plots share information largely by word of mouth informally while sitting and relaxing in the community seating areas (*joalis* or *holhuashi* near the beach area) or while going to the homestead plots or working in the homestead plots. Although all the farmers use smart phones and text messaging, these are not used to communicate agricultural practices in a more practical and systematic way.

9.4   Study Area and Methodology

For data collection, three island communities were selected based on their agricultural production. The three sites selected for this study include the islands of Gan and Fonadhoo in Laamu atoll and Fuamulaku in Gnaviyani atoll. These sites lie in the southern end of the archipelago and is the one of biggest three islands of the Maldives in terms of land area. In Gan and Fonadhoo islands, 80% of the island is marshy and in the rainy season the entire area is flooded. In Fuamulaku, about 70% is marshy land.[1] The largest natural island of Maldives is Gan. The census of 2014 enumerates the total population of Gan as 2809, Fonadhoo as 1400 people while Fuamulaku's total population is 8095. Figure 9.1 shows the populated areas and the areas utilized for agriculture in each of the sites.

The study uses a qualitative approach using field data collection. Primary data was collected using in-depth interviews with a random sample of farmers in each of the three sites. The list of farmers registered in the council office was sought from the councillors in each community. First the sampling started by selecting an element from the list at random and then every kth element in the frame is selected, where k is the sampling interval and n is the sample size, and N is the farmer population size. After selecting the list, an exclusion criterion of active farmers who work on their homestead plots, who managed homestead plots and home gardens plus the availability of the selected famers was listed with the assistance of the councillors. The secondary data sources used were published and unpublished documents from the relevant agencies.

9.5   Case Studies

Selected narratives extracted from the in-depth interviews are used here to present the adaptation methods used by the farmers from Fonadhoo, Gan, and Fuamulaku. The social adaptive capacity and its role in complementing the identified agricultural adaptation practices in the three studied communities are examined and presented in these narratives. The kinds of adaptation practice the farmers employ are discussed below.

9.5.1   Adaptation Methods Used During the Dry Season

During the dry season, there is less rain and the temperature is hot. Daily temperature ranges from around 31 to 33 °C (Maldives Meteorological Service, 2020). All participants in the study use shade nets to cover the crops, especially the scotch bonnet

---

[1] Information provided by the respective island councils.

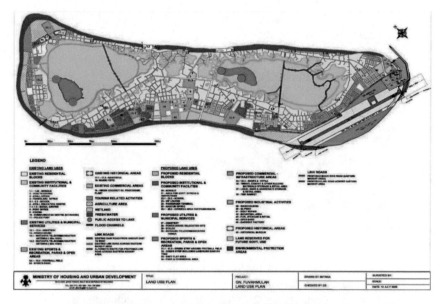

(a) Source: The map has been provided by Fuamulaku city council.

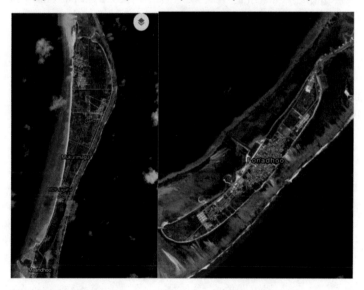

(b) Source: Google maps   **(c)** Source: Google maps

Fig. 9.1   a Map of Fuamulaku island; **b** Map of Gan island; and **c** Map of Fonadhoo island

chillies. In the dry season, cucumbers and pumpkins are planted on the surface as well. This season is utilized by the farmers to grow watermelon.

Narrative 1 (Respondent F-007)

I use a shade net to cover my plants during the dry season because the heat is too much for the plants. I cover the whole plot with a shade net to provide protection for my scotch bonnet chillies. I have invested a lot of money on the shade nets and the structure to hold the shade nets in place because it has to be sturdy and strong during the rainy season as well. The winds can be strong and it can damage the shade nets. Even during the rainy season there are days that are very sunny and the plants wilt, so it is better to have the shade net all the year round. The idea of shade nets has been shared by my friends who have participated in the trainings.

The adaptation method used by the farmers in the dry season as narrated by most farmers and as specified in narration 1 is using shade nets to minimize the harsh sun from the plants (Fig. 9.2). Some farmers have been using the shade nets since it was introduced about 5 years back. This practice has been introduced by the older farmers who participated in the trainings organized by the agriculture ministry in these study sites. These trainings were conducted by the agriculture extension officers based in the region. This information has been shared among the farmers, and those who participated mainly in the trainings share this knowledge with their family and friends. The shade nets are used to cover mainly the scotch bonnet chillies, a chilli popular to the Maldivians as an ingredient in local dishes, which fetches a good price in the market.

9.5.2 Adaptation Methods Used During the Wet Season

During the wet season, there is a lot of rain and the homestead plots get flooded. All the participants used containers to plant short crops like chillies and use improvised fences to grow creepers during this season (Fig. 9.3). Some farmers raised the bed to plant creepers on the surface as well.

Narrative 2 (Respondent G-004)

My homestead plots have suffered from the rains and flooding. 800 plants fell from my homestead plot early this year (2018) due to rain. All the chillies are planted in the homestead plots are in containers, because it is easily available, and in the home gardens chilli and local cabbage are planted in containers. I raise the beds to plant other crops because if there is a heavy shower the

Fig. 9.2 Using shade nets during the dry season. *Source* Waheedha @ Fuamulaku

Fig. 9.3 Container gardening. *Source* Waheedha @ Fuamulaku

surface gets flooded. The lady's finger and tomatoes are planted in wooden boxes. Container gardening is used in home gardens and homestead plots for local cabbage and other varieties because when planted on the surface ground the pests attack the plants. Container gardening is also good because I have to use less water. We mainly use five-liter water bottles, because we have been informed that yellow containers attract pests and diseases and some people paint the containers black. We have to use less fertilizers in containers. During the Hulhangu monsoon, the creepers such as cucumbers are not planted on the surface. Rather, I built improvised fences and support the creepers, to prevent the plants from getting destroyed by flooding. This information has been shared by my family and friends after they have tried it out to avoid damaging the crops due to flooding.

Container gardening used by farmers since the 90s in some home gardens because they have limited space and this is used as a flood adaptation strategy in the three sites. The container method has proven to be useful because it reduces the need for fertilizers and water and saves space in a country like the Maldives which has very little arable land. This method is practised now in homestead plots as well as home gardens for multiple purposes such as avoiding the wet season and this information has been shared among the farmers in this site. Figure 9.3 shows local cabbages grown in containers.

9.5.3 Changing of Crops to Rain-Fed Agricultural Crops as an Adaptation Technique

One of the participants stated that due to pests and diseases it was very difficult to grow the short crops he used to grow in the homestead plots and he changed to rain-fed crops, planting only coconuts in his homestead plots. In this community no land rent has to be paid to the council. It is easy to manage and not labor-intensive.

Narrative 3 (FN-004)

I own two homestead plots each of 100,000 ft^2. In Fonadhoo island, we do not have to pay any land rent on the homestead plots. In one homestead plot, I have planted 600 coconut palm trees which is mainly rain fed and does not require much input. I hired helpers during the planting of the coconut trees to dig the pits required for planting of the coconut trees when they were small. The coconuts are sold to the local markets and the boats and fetch a very good price. I sell the young coconuts and also the coconuts used in everyday cooking in the Maldivian homes to both the local market as well as to the boats that carry the goods to the Male' markets. I do not need to water the plantation since the coconut plantations are rain fed and I check the plantation on and off for any pests and diseases and cleaning of the coconut palms for a good harvest. I understand the impacts of climate change very much and I used to grow a variety of crops like pumpkin and scotch bonnet chillies I have now given up to concentrate only on the coconut plantation which is more beneficial.

Changing of the crops to rain-fed agriculture with more hardy and resilient crops have been practiced by some of the farmers as an adaptation strategy, because the changing climate and prolonged periods without rain have introduced many pests and diseases, and the chemical pesticides necessary are costly and difficult to manage. Organic farming in home gardens is an adaptation technique. All the home gardens which are inside the boundary walls of the homes and some homestead plots have changed to organic farming as an adaptation technique.

Organic farming is used as an adaptation strategy by farmers for multiple reasons. Due to the changing climatic conditions farmers in the three sites, many face problems such as unpredictable rains, soil degradation, and new or different pests and diseases. The farmers in the three sites have said that organic farming helps to maintain high soil organic matter content and soil cover which in turn helps to prevent nutrient and water loss. This makes soils more resilient to floods, droughts, and land degradation processes. Since organic farming does not use the imported fertilizers, it reduces energy consumption by eliminating the energy required to manufacture the imported fertilizers, and by using internal farm inputs, thus reducing fuel used for transportation. Similar studies have shown that organic farming is a low-risk farming method which reduces the cost of the required inputs, lowers the risk of crop failure in extreme weather events, and changes climatic conditions (Eyhorn, 2007).

9.6   Discussion

9.6.1   What Do the Local Farmers Know About Climate Change?

The finding from this study shows that farmers are aware of climate change in the study area and are trying to adapt. Most of the respondents understood that the impacts of climate change as causing increase in temperatures, lengthening of the dry periods, rainfall variations, storm surges and flooding due to rain, unpredictability of the weather and pests and diseases. The communities can no longer rely on observing the weather and climate by the traditional *Nakaiy* system—which is part of the formal calendar in the Maldives that depicts the weather.

9.6.2 What Do the Farmers not Know About Climate Change?

The findings of this study suggest that although there is some understanding of climate and its impacts among the local farmers there is still a dearth of knowledge on the causes of climate change consistent with a similar study undertaken in the Maldives. The Rapid Assessment of Perceptions (RAP) , undertaken in the communities by Live & Learn Environmental Education with regard to community-level adaptation of fishery and agricultural sectors, found there to be limited knowledge, awareness (and action) on climate variability, future climatic changes and their likely impact on agriculture, fisheries, and food security systems at large. All communities lack strong plans and communication systems for climate change and food security—highlighting the need for stronger links between the National Adaptation Program of Action (NAPA) and community-level planning (Shafeeqa, 2011).

This has been observed when some participants have changed the crops they were growing and experimented with new crops which were not based on any valid source of information. Thus, farmers need a valid source of information at the island level on climate change and its impacts in order to adapt to the changing conditions and to be more resilient in the future. A similar study undertaken by Foran (2014) suggests that there is a need to introduce new approaches to agriculture research which are more climate smart and the building up of decentralized knowledge networks that can address these complex issues at the community level to reach out to the small-holder farmers are very important.

9.6.3 How Do Farmers Adapt to Climate Change with the Knowledge They Have?

Participants had limited information on climate change and its impacts on the crops. Some of the participants learnt from his/her own experience and the other respondents received information from different sources including workshops, radio, visiting personnel from the agriculture ministry officials, field officers, and from other farmers. The respondents have expressed the reasoning and understanding based on their everyday experience with growing their crops which is very much tied to their livelihoods. In Gan and Fonadhoo, a lot of awareness programs on climate change and its impacts have been a part of the training given under programs for the United Nations Development Programme (UNDP) and other non-governmental organizations based in this atoll. The findings of this study show that the respondents are making efforts to adapt to the changing conditions and adapt based on their experiences. However, lack of technical know-how and information is a hindrance to make the farmer community more resilient.

9.6.4   What Are the Strengths and Weaknesses of Their Current Adaptation Practices?

The current adaptation practices that have been documented give the readers an overview of how farmers in the three sites are trying to cope with the changing climate by using local knowledge and experiences and other information gained from various sources. All current adaptation practices that have been captured in this study may not be considered as the best adaptation options. Some of the reasons are lack of technical know-how and information. However, these strategies will not suffice in the future in changing climatic and environmental conditions. The findings of this study show autonomous and experimental adaptation strategies being implemented by farmers.

9.6.5   What is the Role of Social Capital in Adaptation?

This study has shown that women farmers who tend to the home gardens and the homestead plots with the assistance from their families and those who have participated in the agricultural workshops undertaken in the island are more likely to contribute to the farming social capital. With regard to demographic features, the women farmers who are younger show a more positive relationship in contributing to social capital. The three studied communities are adapting to the climate change impacts at an individual and household level and not at the societal level, because agricultural plots and home gardens are owned by individual farmers and there are no informal or formal institutions formulated to deal with the challenges regarding climatic variations. Information is shared among some of the farmers especially the younger women farmers and critical resources accessed by sharing this valuable information with each other. Farmer organizations or non-governmental organizations working for the benefit of farmers were not visible in the three study sites. However, this study has shown that among individual farmers a lot of interactions and critical information are shared among the individual farmers and within families who work in the home garden and homestead plots and thus can be called social capital at the family and individual level. The trust factor was very low among and between the farmers working in the same communities.

9.7   Conclusion

Farmer cooperatives are an important institution for farmers to build trust and reciprocity which has been identified as one of weakest links among the surveyed communities. The adaptation policies need to take into account the building of social capital and such policies are critical for the farmers so that in the future the social capital can be harnessed to its full potential in the island communities. The farmers lack the technical knowledge and information on best adaptation practices and on this basis this study suggests designing of effective agriculture-related adaptation policies considering all the characteristics of the farmers in the island communities. These findings are important when designing agricultural policies in contributing to sustainable agriculture in the islands of the Maldives.

References

Baldacchino, G., & Kelman, I. (2014). Critiquing the pursuit of island sustainability. *Shima: The International Journal of Research into Island Cultures, 8*(2), 1–21.

Bourdieu, P. (1998). *Practical reason on the theory of action.* Stanford University Press.

Dodman, D., & Mitlin, D. (2015). The national and local politics of climate change adaptation in Zimbabwe. *Climate and Development, 7*(3), 223–234.

Eyhorn, F. (2007). *Assessing the potential of organic farming for sustainable livelihoods in developing countries—The case of cotton in India.* Accessed on October 24th, 2020. https://www.researchgate.net/publication

Foran, T., et al. (2014). Taking complexity in food systems seriously: An interdisciplinary analysis. *World Development, 61*, 85–101.

Kelman, I. (2007). The island advantage. *Id21 Insights, 70*, 6.

Kushal, S. (2016). Climate change, smallfarmers and sids—The Paris agreement and SIDS. In S. Sharma-Khushal (Ed.), *Climate Change and SIDS: A Voice at COP21 for Small Farmers (Annex 1)* (pp. 1–11).

López-Marrero, T., & Yarnal, B. (2010). Putting adaptive capacity into the context of people's lives: A case study of two flood-prone communities in Puerto Rico. *Natural Hazards, 52*(2), 277–297. https://doi.org/10.1007/s11069-009-9370-7

Maldives meteorological service Republic of Maldives. (n.d.). Retrieved October 19th, 2020 from https://www.meteorology.gov.mv

MEE. (2016) Second National Communication of Maldives to the United Nations Framework Convention on Climate Change: Ministry of Environment and Energy. In *Ministry of environment and energy* (p. 56)

MEE. (2017). *State of the environment 2016.* Male, Maldives.

Metz, B., Meyer, L., & Bosch, P. (2007). *Climate change 2007 mitigation of climate change.* IPCC.

Moser, C., et al. (2010). *Pro-poor adaptation to climate change in urban centres: Case studies of vulnerability and resilience in Kenya and Nicaragua.* Report No.54947-GLB (pp. 1–84).

Pelling, M., & High, C. (2005). Social learning and adaptation to climate change. *Disaster Studies* 1–19.

Schleussner, F. C., & Hare, B. (2015). *Briefing note on the report on the structured expert dialogue on the 2013–2015 review* (pp. 1–10). www.climateanalytics.org

Shafeeqa, F. (2011). *Madi Kilambu: A rapid assessment of perceptions: Part 2.* Live & Learn.

Smit, B., & Wandel, J. (2006). Adaptation, adaptive capacity and vulnerability. *Global Environmental Change, 16*(3), 282–292.

Smit, B., et al. (2000). An anatomy of adaptation to climate change and variability. *Climatic Change, 45*(1), 223–251.

Tshotsho. (2022). Indigenous practices of paddy growers in Bhutan: A safety net against climate change. In A. K. E. Haque, P. Mukhopadhyay, M. Nepal, & M. R. Shammin (Eds.), *Climate change and community resilience: Insights from South Asia* (pp. 87–100). Springer.

UNFCCC. (2005). Climate Change, Small Island Developing States. Climate Change Secretariat, Bonn, Germany. https://unfccc.int/resource/docs/publications/cc_sids.pdf

Chapter 10
Climate Sensitivity of Groundwater Systems in South India: Does It Matter for Agricultural Income?

R. Balasubramanian and V. Saravanakumar

Key Messages

- Flat or fully subsidized electricity pricing policy in Tamil Nadu negatively impacted groundwater table, implying that the subsidized pricing has lowered the water table.
- Increase in temperature had a negative impact on farm income, while increase in rainfall had a positive impact on net revenue. Increase in depth to water reduces farm income.
- An important policy implication of the analysis is the negative impact of well density beyond a threshold level on farm income.

10.1 Introduction

Groundwater is the source of about one-third of global water withdrawals and provides drinking water for a large portion of the global population. In many regions, it is subject to stress with respect to both quantity and quality (Kundzewicz & Döll, 2009). Climate change will lead to significant changes in groundwater recharge and thus renewable groundwater resources (Döll, 2009). Climate change impacts may add to existing pressure on groundwater resources directly through changes in groundwater recharge (supply) and indirectly through changes in demand for groundwater (Taylor et al., 2013). Even though groundwater is the major source of water in most parts of the world, particularly in rural areas in arid and semi-arid regions, there has been limited research on the potential effects of climate change on groundwater

R. Balasubramanian (✉) · V. Saravanakumar
Department of Agricultural Economics, Tamil Nadu Agricultural University, Coimbatore, India
e-mail: rubalu@gmail.com

V. Saravanakumar
e-mail: sharanu2k@gmail.com

© The Author(s) 2022   143
Haque et al. (eds.), *Climate Change and Community Resilience*,
https://doi.org/10.1007/978-981-16-0680-9_10

recharge and its quality (Alley, 2001; Bates et al., 2008). Therefore, understanding the impact of climate variability and change on the availability and sustainability of surface water and groundwater resources is of significant ecological and economic significance (Dragoni & Sukhija, 2008). There is an evolving literature that looks at factors that affect farmer adaptation to climate change (Bahinipati & Patnaik, 2021, Chap. 4 of this volume).

India is the largest user of groundwater globally with an estimated annual groundwater withdrawal of 230 km^3. More than 60% of India's irrigated acreage and as much as 85% of rural India's drinking water requirement is met from groundwater. Western and peninsular India is groundwater hotspots from a climate change point of view (Shah, 2009). In spite of the intricate interrelationship between climate variables and groundwater and the growing significance of groundwater in sustaining agricultural productivity, systematic studies on impact of climate and non-climate variables on groundwater balance and agricultural production are very limited in India. This study is an attempt to fill this void through a systematic analysis of climate-groundwater-agriculture nexus. The chapter is organized as follows: in the following section, we review key issues and challenges posed by climate change for groundwater in general and its specific implications for groundwater-dependent agriculture. The next section describes the study area and the sources of data. The section on methods deals with the theoretical model and empirical strategy for econometric estimation, while the subsequent section presents and discusses the key results of the study. The final section presents the conclusions and policy recommendations emerging from the study.

10.2   Climate Change and Groundwater Irrigation[1]

10.2.1   Impact of Climate Variables on Groundwater Dynamics

Global warming is expected to cause lower water tables and reduce groundwater availability, while the extraction of groundwater is likely to increase to meet the growing demand (Okkonen et al.,). Climate variability directly affects rain-fed agricultural production through changes in supply of soil moisture and indirectly affects irrigated agriculture through its impact on surface water runoff and groundwater recharge. Two broad approaches have been used to study the impact of climate change on groundwater resources. The first is a physical approach wherein the changes in groundwater reserves are quantified by physical measurements using hydrological modelling such as water balance method or GIS and simulation modelling. The second is a statistical modelling approach where the changes in groundwater

---

[1] A significant portion of the materials is drawn with permission from author's working paper: Balasubramanian (2015). Climate Sensitivity of Groundwater Systems Critical for Agricultural Incomes in South India, SANDEE Working Paper No 96–15.

levels are estimated through building statistical relationship between groundwater level and rainfall, temperature and other variables. Regression analysis incorporating climate- and/or non-climate variables has been used by many researchers to study the impact of these variables on water table levels (Balasubramanian, 1998; Ferguson and George, 2003; Ngongondo, 2006; Palanisami & Balasubramanian, 1993). Using regression analysis of groundwater levels with monthly rainfall data, Bloomfield et al. (2003)predicted groundwater levels under different future climates and found that even with a small increase in total annual rainfall, annual groundwater level could fall in the future due to changes in seasonality and increased frequency of drought events. Chen et al. (2001) used a log–log regression model to study the impact of temperature and rainfall on groundwater table.

10.2.2   Climate Change and Irrigated Agriculture

There are two distinct but interrelated issues concerning the treatment of irrigation in economic studies on climate change impact on agriculture: (a) the issue of climate sensitivity of irrigated vs. rain-fed farms, which is about carrying out separate or pooled analysis of climate change impact for irrigated and rain-fed farms; and (b) the inclusion of irrigation as one of the explanatory variables in the regression model. The early Ricardian models did not account for irrigation in analysing climate change impact and ignored both of these issues. As the supply and demand for irrigation water are affected by climate change, inclusion of irrigation as an explanatory variable is important. However, studies that have explicitly incorporated groundwater availability and/or withdrawal in climate change impact models in Indian agriculture are very limited. Though a few studies have incorporated groundwater irrigation in economic models of climate change impact on agriculture, these have not studied dual effects of climate change directly on production agriculture and its impact via groundwater on agricultural productivity and/or farm income using an econometric approach. For example, Schlenker et al. (2007) examined the impact of climate variables, surface water availability and depth to groundwater on irrigated agriculture in California using Ricardian analysis using farmland value for 2555 farms as the dependent variable. Their analysis shows that while surface water availability had significant positive impact on farmland values, depth of groundwater table was not statistically significant. Though a study by Chen et al. (2001) estimated both climate change impact on groundwater and its subsequent impact on regional economy including agriculture, it has used mathematical programming and simulation modelling to quantify economic impacts.

10.3   Study Area and Data

10.3.1   Description of Study Site

This study was carried out in the southern Indian state of Tamil Nadu where ground-water aquifers are under severe stress due to poor management caused by perverse incentives such as fully subsidized electricity for groundwater pumping. Tamil Nadu is located in the southernmost part of India and is divided into seven agro-climatic regions with a wide diversity in climate and crops cultivated. The average annual rainfall varies from 650 to 1350 mm in the plains, and the average annual maximum temperature varies from 31 to 34.50 °C. Groundwater irrigation in the state has expanded rapidly in the last five decades due to the decline and/or instability in surface irrigation sources, massive expansion in rural electrification, the advent of modern well-drilling technologies and subsidized supply of electricity for ground-water pumping. Groundwater overexploitation is reported in more than one-third of blocks[2] in Tamil Nadu (CGWB, 2012).

While the total area irrigated from surface irrigation sources such as tanks and canals decreased considerably from more than 1.80 million ha in 1960–61 to about 1.30 million ha in 2010, the total number of groundwater wells has increased from about 0.87 million to 1.90 million and the area irrigated by wells has increased from 0.6 million ha in 1965 to 1.6 million ha in 2016. Electricity pricing for agricultural pumping underwent significant changes over time from a pro-rata tariff until the early 1980s, to a system of flat tariff in late 1980s and finally to a fully subsidized (100% subsidy) electricity supply for agriculture from 1990 to 91 onward. The introduction of "zero-marginal-cost" pricing of electricity (after the flat-rate tariff was introduced), along with the advent of low-cost well-drilling technologies have provided added impetus to the drilling of deep bore wells in the last two decades. As a result, the State of Tamil Nadu has become one of the groundwater hotspot areas in India which makes it an ideal location to study the climate change impact on groundwater resources.

10.3.2   Data Sources

Time-series data on water-level data from 1740 observation wells over a period of 40 years from 1971 to 2010 and the corresponding data on rainfall, temperature, number of groundwater wells and surface water sources and area of various crops cultivated with groundwater and surface water irrigation, and other socioeconomic variables such as population and urbanization were collected from Government publications. Spatial distribution of observation wells across the entire state with different

---

[2] Blocks are the bottom-most unit in the administrative hierarchy in the State. The data on groundwater recharge and pumping volumes are estimated at block level.

endowments of surface water resources, aquifer formations and related hydrogeo-logical and socio-economic factors were used to divide the entire state into several cross-sectional units. The water-level data for 1740 individual wells was compressed into district-level averages which will help conduct panel data analysis of ground-water table fluctuations vis-à-vis climatic and non-climatic variables, with districts serving as cross-sectional units (panels). During the period of 40 years from 1971 to 2010, for which the water-level data is available, electricity pricing for groundwater pumping, institutional arrangements for groundwater management, availability and performance of other sources of irrigation like canals and tanks, and the technology for well-drilling have undergone significant changes. We incorporate these variables in our model.

The second part of the analysis, which is concerned with the impact of climatic factors along with groundwater level changes on agricultural crop production, relies on agricultural crop production data at district-level to estimate net revenue per hectare from crop production. Data on crop-wise, district-level average produc-tivity (yield in kg/ha) were sourced from Season and Crops Reports for Tamil Nadu, published by the Government of Tamil Nadu over the entire study period from 1971 to 2010. Data on input quantities and costs were assembled from the Government of India Scheme on Cost of Cultivation of Principal Crops being implemented in the Department of Agricultural Economics, Tamil Nadu Agricultural University. Using these two sets of data on input quantities, yield and input and output prices, estimates on average net income per hectare for 11 districts of Tamil Nadu were constructed for 40 years (1971–2010).

10.4 Methods

10.4.1 Empirical Model

Though several studies used Ricardian model to estimate climate change impact, Deschênes and Greenstone (2007) proposed an alternative to cross-sectional Ricar-dian approach by using random, year-to-year variability in weather parameters such as precipitation and temperature on farm profits. Consequently, this approach allows the use of panel data model, to estimate the effect of weather on farm profits, condi-tional on locations by year fixed effects. Kumar (2011) used this approach in his analysis of 271 Indian districts over a period of 20 years to estimate the impact of climate change on farm net revenue. Following these studies, the empirical, panel data econometric model of our study consists of a set of two equations—the first one concerning the impact of climate and non-climate factors on groundwater table and the second one concerning the impact of climate, water and other economic factors affecting farm income.

$$\text{Watlev}_{it} = \alpha_0 + \alpha_1\,\text{Lwatlev} + \alpha_2\,\text{Rain} + \alpha_3\,\text{Lrain}$$

$$+\,\alpha_4\,\text{Tmax} + \alpha_5\,\text{Watint} + \alpha_6\,\text{Lwatint}$$
$$+\,\alpha_7\,\text{Elecdum} + \alpha_8\,\text{Tankgia} + \alpha_9\,\text{Canalgia} + \alpha_{10}\,\text{Time} \qquad (10.1)$$

Equation (10.1) was estimated using dynamic panel data approach in view of the presence of lagged dependent variable as one of the regressors.

The equation for net returns is specified as shown below, and it was estimated using aggregate district-level data rather than farm-level data. In the net returns equation, we use the estimated depth to water table from Eq. 10.1. In addition to the depth to water table, climate variables, dummy for districts (coastal and non-coastal), dummy for electricity price, well density, indices of input and output prices were also used as explanatory variables.

$$\text{Return} = \gamma_0 + \gamma_1\,\text{Rain}_{it} + \gamma_2\,\text{Rain}^2 + \gamma_3\,\text{Tmax}$$
$$+\,\gamma_4\,\text{Tmax}^2 + \gamma_5\,\text{Distdum} + \gamma_6\,\text{Wellden}$$
$$+\,\gamma_7\,\text{Wellden}^2 + \gamma_8\,\text{Elecdum} + \gamma_9\,\text{Ewatlev}$$
$$+\,\gamma_{10}\,\text{Inprice} + \gamma_{11}\,\text{Outprice} + \gamma_{12}\,\text{Time} + \gamma_{13}\,\text{Surfgia} \qquad (10.2)$$

where

Return = Net farm income from crop production

Rain = Rainfall (mm)

Tmax = Max. temperature (°C)

Distdum = District dummy (0= Coastal district; 1= Non-coastal district)

Wellden = Well density (total number of wells per ha of geographical area of the ith district)

Elecdum = Dummy for electricity price (= 0 for pro-rata tariff; = 1 for flat rate or full subsidy)

Ewatlev = Estimated water level from Eq. 10.1

Inprice = Weighted average of input prices

Outprice = Weighted average of output prices

Time = Time (Trend variable)

Surfgia = Proportion of surface irrigated area to gross irrigated area by all sources

Tankgia = Gross irrigated area by tanks (ha)

Canalgia = Gross irrigated area by canals (ha)

Watlev = Groundwater level (in metres below surface)

Lwatlev = One-period lag of groundwater level

Watint = Share of water-intensive crops to gross cropped area

Lwatint = One-period lag of share of water-intensive crops to gross cropped area.

10.4.2   Estimation Strategy

The first equation was estimated using spatial dynamic panel method due to the presence of lagged dependent variable as one of the explanatory variables in the model. The model was estimated with two-period lag structure, using Arellano–Bond estimators for spatial dynamic panel model (Arellano & Bond, 1991). The net revenue equation was estimated using panel-corrected standard errors regression model using estimated water level from the first equation as one of the explanatory variables. Descriptive statistics of the variables are presented in Table 10.1.

Table 10.1 Definition of variables and their descriptive statistics

Variable	Definition	N	Mean	Std. dev	Min	Max
Watlev	Depth to water table (m)	440	7.46	2.69	3.25	16.44
Rain	Rainfall (mm)	440	969.24	344.87	305.1	2106.95
$Rain^2$	Square of rainfall	440	1,058,096	749,672	93,086	4,439,238
Lrain	Lagged rainfall	429	963.43	342.42	305	2107
Tmax	Max. temperature (C)	440	33.03	1.07	30.53	35.26
$Tmax^2$	Square of max temperature	440	1091.97	70.49	932.08	1243.27
Time	Trend variable (year)	440	20.50	11.56	1.00	40.00
Canalgia	Gross irrigated area by canals ('000 ha)	440	82.01	131.28	0.00	650.76
Tankgia	Gross irrigated area by tanks ('000 ha)	440	65.31	64.74	0.168	696.46
Watint	Share of area under water-intensive crops	440	0.44	0.21	0.047	0.98
Returns	Net returns	440	17,715	6961	4201	43,986
Elecdum	Dummy for electricity price (0 for pro-rata price; 1 = for flat or zero tariff)	440	0.65	0.48	0.00	1.00
Distdum	Dummy for coastal districts	440	0.73	0.45	0.00	1.00
Wellden	Well density (No. per km^2 of geographical area)	440	0.11	0.06	0.0018	0.28
$Wellden^2$	Square of well density	440	0.017	0.017	0.00001	0.08
Inprice	Weighted average input price	440	43.47	8.94	31.80	119.81
Outprice	Weighted average of output prices	440	21.00	16.49	1.32	89.71

Source Field data

10.5 Results and Discussion

10.5.1 Impact of Climate Change on Groundwater Dynamics

Climatic variables, viz., both current period and one-period lagged rainfall, as well as maximum temperature were found to be statistically significant in impacting the groundwater table. As expected, maximum temperature had a positive impact on depth to groundwater table which is the direct result of increased evapotranspiration which in turn would result in lower recharge as well as higher withdrawal of groundwater to compensate for evapotranspiration. Similarly, rainfall had negative impacts on depth to groundwater table as higher rainfall results in higher recharge thus resulting in reduced depth to (or a rise in) groundwater table. The estimated elasticity values indicate that temperature has a much higher impact on water levels than rainfall. A 1% increase in temperature is found to increase the depth of the water level by more than 1%, whereas a 1% increase in current period rainfall and lagged rainfall would reduce the depth of the water table only by a meagre 0.08 and 0.15%, respectively. Using the regression coefficients for current period rainfall and temperature, it could be seen that to offset an increase in temperature by 1 °C, the current rainfall increases by about 193 mm. A 1% increase in other variables such as share of water-intensive crops in the previous year, share of tank irrigated area to gross irrigated area, and the trend variable had 0.03–0.04% reduction in the depth of the water table. The dummy variable for electricity pricing results in an increased depth to the water table indicating that zero-marginal cost for water pumping induced farmers to pump more water thereby causing a drop in the water table. The overall explanatory power of the model as indicated by Wald Chi-square is found to be statistically highly significant at less than 1% level indicating that the model is a good fit for the data. The results of dynamic panel data regression model are presented in Table 10.2.

10.5.2 Climate Change, Groundwater Dynamics and Farm Income

The second part of the econometric exercise is concerned with the estimation of net returns equation in which the estimated values of change in depth to water table and tank irrigated area from the previous section of econometric analysis were used as explanatory variables along with other exogenous variables. The results of the panel-corrected standard error regression analysis are presented in Table 10.3. All the independent variables except the linear term of rainfall were found to be statistically significant. The coefficients of temperature and its quadratic terms indicate that net returns per hectare increase with increase in temperature up to a point beyond which it starts declining. The threshold level of temperature which results in maximum net returns is found to be 39.20 °C. In a study of rural income and climate change in the

Table 10.2 Dynamic panel estimation of depth to water table

Variables	Coef	Robust Std. err	z	Elasticity
Lwatlev	0.70^{***}	0.04	16.65	0.70
Rain	-0.001^{**}	0.0003	-1.98	-0.08
Lrain	-0.001^{***}	0.0002	-4.96	-0.15
Tmax	0.23^{***}	0.08	2.9	1.01
Watintsh	-0.53	0.77	-0.69	-0.03
Lwatintsh	0.71^{*}	0.43	1.67	0.04
Elecdum	0.50^{***}	0.14	3.61	0.04
Tankgia	-0.004^{**}	0.002	-1.96	-0.04
Canalgia	-0.001	0.001	-1.28	-0.01
Time	-0.025^{***}	0.008	-3.28	-0.07
_Cons	-3.115^{***}	2.95	-1.06	
Wald Chi2	42,915.35		Prob > chi2 = 0	

Arellano–Bond test for AR(2) in first differences: $z = 1.33$ Pr > $z = 0.182$
Sargan test of overid. restrictions: chi2(338) = 421.39 Prob > chi2 = 0.001

Source Field data

Table 10.3 Panel-corrected standard errors regression estimation of net returns

Variables	Coef	Std. Err	z	$P > z$	Elasticity
Rain	1.61	1.07	1.5	0.134	-0.0043
Rain^2	-0.00^{*}	0.00	-1.86	0.063	
Tmax	$44,537.90^{***}$	12,675.89	3.51	0	0.74
Tmax^2	-667.73^{***}	191.36	-3.49	0	
Distdum	4278.86^{***}	779.78	5.49	0	0.16
Ewatlev	-639.74^{***}	216.84	-2.95	0.003	-0.25
Inprice	-131.77^{***}	35.17	-3.75	0	-0.30
Outprice	195.25^{***}	34.04	5.74	0	0.21
Time	89.49^{*}	51.28	1.75	0.081	0.10
Elecdum	2341.08^{**}	1096.52	2.14	0.033	0.08
Surfgia	$-10,803.55^{***}$	1760.08	-6.14	0	-0.27
Wellden	$93,494.50^{***}$	19,150.87	4.88	0	0.061
Wellden^2	$-369,441.90^{***}$	49,564.53	-7.45	0	
Constant	$-722,676.20^{***}$	209,380.50	-3.45	0.001	
R-squared = 0.6270		Wald chi2 =.851		Prob > chi2 = 0.00	

Source Field data

US and Brazil, Mendelsohn & Seo (2007) found that both agricultural net income and total rural income are affected by climate, and regions with poorer climates have more rural poverty.

The estimated water level from the first equation, which was used as one of the regressors in this model, turned out to be statistically significant. The elasticity value indicates that an increase in depth to water table by 1% reduces the net returns by about 0.25%. The regression coefficient of trend (time) variable which could be expected to capture technological progress over time has turned out to be positive and significant indicating that net returns at constant prices have increased over time probably due to technological progress in agriculture. A dummy variable was used to differentiate between coastal and non-coastal districts in view of their significant differences in crop pattern, intensity and distribution of rainfall, and soil type. This variable has turned out to be statistically significant indicating that non-coastal districts have higher net returns as compared to their coastal counterparts. This is primarily because the coastal districts are predominantly paddy-based agricultural production systems where the net returns are lower. Further, distribution of rainfall is often skewed in coastal districts with heavy rains during the north-east monsoon. Increasing salinity and drainage problems in these districts could also have contributed to lower net farm incomes in coastal districts. We included both a linear and a quadratic term for well density as Tamil Nadu is witnessing a steep increase in the number of wells resulting in significant spatial externalities and poorer water yield per well. The regression estimates reveal that the coefficient for linear term was positively significant and the quadratic term was negatively significant. This implies that, *ceteris paribus,* an increase in number of wells per hectare of land might initially contribute for an increased net return, while the increase in well density beyond a threshold might result in reduced net returns per hectare of land due to intense rivalry in sharing scarce groundwater reserves. Using the results of the regression analysis, it was estimated that the optimal number of wells was 0.13 per ha of geographical area. Both the input and output price indices turned out to be statistically significant with expected signs.

The previous pro-rata pricing of electricity for groundwater pumping in the state has been replaced by the flat-rate system of electricity pricing in the late 1980s and fully subsidized supply of electricity for agricultural pumping from the year 1990–91 and has significantly altered the balance of economic access to groundwater. Large and medium farmers with access to capital started sinking deep bore wells thus depriving the small farmers of their reasonable share of groundwater resources. The race for groundwater has become further exacerbated by the advent of modern well-drilling technologies at a lower cost. This has, however, resulted in many of the small, but moderately well-off farmers joining the race for increasingly scarce groundwater resources. Steep increase in the number of deep bore wells with poor water yields, fitted with compressor pumps to enable continuous water extraction, with zero-marginal costs (thanks to flat-rate pricing or free electricity) has become a harsh reality in many parts of the state spelling doom for groundwater conservation efforts. It is therefore appropriate to include a dummy variable to capture the impact of introduction of zero-marginal cost pricing of electricity for irrigation. This variable has turned out to be statistically significant indicating that the provision of fully subsidized electricity to agriculture has in fact increased the net returns in agriculture probably because of the increased acreage under high-value crops even though some of these crops are water-intensive.

10.6   Conclusions and Policy Recommendations

This study has quantified the impact of climatic and non-climatic factors on groundwater level and its consequences for net farm income. It used panel data econometric analysis of data on groundwater levels from more than 1700 observation wells spread over the entire state, and the data on costs and returns from crop production for 11 districts (panels), over a period of 40 years from 1971 to 2010. The econometric analysis of depth to water table reveals that climate variables viz., current and one-period lagged rainfall had significant effect in reducing the depth to water table, while the share of area under water-intensive crops and maximum temperature had significant role in increasing the depth to (pushing down) water table. Flat or fully subsidized electricity pricing policy that resulted in zero-marginal cost of pumping has had a negative impact on groundwater table, implying that the recent subsidized pricing has lowered the water table. The analysis of climatic and non-climatic factors affecting net farm revenue revealed that both temperature and its quadratic term had expected signs and turned out to be statistically significant, while rainfall had a positive impact on net revenue. The depth to water table had expected sign and turned out to be statistically significant indicating that an increase in depth to water reduces farm income.

An important policy implication of the analysis is the negative impact of well density beyond a threshold level on farm income. The import of this result is the need for regulating the sinking of new wells, especially deep bore wells. Though free electricity has increased the average farm net revenues as indicated by the positive regression coefficient for electricity price dummy, there is a huge social cost due to full subsidy for electricity for pumping—both in terms of cost of electricity generation as well as the significant negative impact of electricity subsidy in increasing the depth to water table. This points to the need for considering the removal of free electricity as one of the mechanisms to regulate the unfettered growth of deep bore wells. The question of removal of electricity subsidy and regulation of well density puts both the policy makers and the farmers in a tight spot with regard to conserving groundwater in the current political climate. This is because the short-term interests of both farmers (especially the large-farmer lobby), and politicians will be better-served if the subsidies continue, while the continuance of subsidies could thwart groundwater conservation efforts. Therefore, convincing farmers to opt for pro-rata electricity pricing in exchange for increased public investments and/or subsidies for recharge programs and farm-level water conservation investments should receive top priority in the future. This has the potential to foster the development of sustainable groundwater management solutions and more equitable distribution of access and opportunities.

The shift to water-intensive crops such as coconut and sugarcane in response to increasing labour scarcity is a major contributing factor for increased groundwater extraction. The distributional implications of cultivation of water-intensive crops by mostly economically well-off farmers has to be further studied since the present study is based on average net revenues at district level, and hence cannot throw light on who gains and who loses from groundwater overexploitation. However, a few studies in the past found that it is the poor farmers who will lose in the long-run since reaching down to groundwater at deeper aquifers is a capital-intensive venture which is affordable only for large and affluent farmers. Groundwater overexploitation and the consequent spatial (inter-well), and inter-temporal (intra- and intergenerational) externalities need to be carefully analysed, since both increasing the number of wells as well as deepening of existing wells could further exacerbate the situation. In view of the intensifying race for groundwater and the attendant externalities, appropriate institutional arrangements to regulate digging new wells and deepening existing wells are needed in order to manage scarce groundwater resources in a sustainable way. Though farmers respond to market signals in deciding the crop pattern and regulatory mechanisms could play a very little role, appropriate incentive structure such as subsidies for water-saving crops and/or technologies could be considered as alternative mechanisms to discourage the cultivation of water-intensive crops in groundwater hotspot areas. Public and private investments in groundwater recharge such as watershed development, percolation ponds, recharge wells and farm ponds should be stepped up in future. Efforts to revive traditional systems like the Small Tank Cascade Systems (STCS) of Sri Lanka (Vidanage et al., 2022, Chap. 15 of this volume) and innovative rainwater harvesting technology in mountain villages in Nepal (Kattel & Nepal, 2021, Chap. 21 of this volume) are potential strategies that are available for replication.

Acknowledgements The research was supported financially and technically by the South Asian Network for Development and Environmental Economics (SANDEE) at the International Centre for Integrated Mountain Development (ICIMOD). We are grateful to Céline Nauges, Priya Shyamsundar, other SANDEE resource persons and peers for their critical comments and encouragement.

References

Alley, W. M. (2001). Groundwater and climate. *Groundwater, 39*, 161.

Arellano, M., & Bond, S. R. (1991). Some tests of specification for panel data: Monte Carlo evidence and an application to employment equations. *Review of Economic Studies, 58*, 277–297.

Bahinipati, C. S., & Patnaik, U. (2021). What motivates farm level adaptation in India? A systematic review. In A. K. E. Haque, P. Mukhopadhyay, M. Nepal, & M. R. Shammin (Eds.), *Climate change and community resilience: Insights from South Asia* (pp. 49–68). Springer.

Balasubramanian, R. (1998). *An enquiry into the nature, causes and consequences of growth of well irrigation in a Vanguard agrarian economy*, Ph.D. thesis submitted to the Department of Agricultural Economics, Tamil Nadu Agricultural University, Coimbatore, India

Balasubramanian, R. (2015). *Climate sensitivity of groundwater systems critical for agricultural incomes in South India*. SANDEE Working Paper No 96-15.

Bates, B. C., Kundzewicz, Z. W., Wu, S., & Palutikof, J. P. (Eds.). (2008). Climate change and water. Technical Paper of the Intergovernmental Panel on Climate Change (pp. 210). IPCC Secretariat.

Bloomfield, J. P., Gaus, I., & Wade, S. D. (2003). A method for investigating the potential impacts of climate-change scenarios on annual minimum groundwater levels. *Water and Environment Journal, 17*(2), 86–91.

Central Groundwater Board. (2012). *Aquifer systems of Tamil Nadu and Puducherry*. Southern Regional Office, Ministry of Water Resources, Government of India, Chennai.

Chen, C., Gillig, D., & McCarl, B. A. (2001). Effects of climate change on a water dependent regional economy: A study of the Texas Edwards aquifer. *Climate Change, 49*, 397–409.

Deschênes, O., & Greenstone, M. (2007). The economic impacts of climate change: Evidence from agricultural output and random fluctuations in weather. *The American Economic Review, 97*(1), 354–385.

Döll, P. (2009). Vulnerability to the impact of climate change on renewable groundwater resources: A global-scale assessment. *Environment Research Letters, 4*, 1–12.

Dragoni, W., & Sukhija, B. S. (2008). Climate change and groundwater—A short review. In W. Dragoni & B. S. Sikhija (Eds.), *Climate change and groundwater* (vol. 288, pp. 1–12). Geological Society, Special Publications London.

Ferguson, G., & Scott, S. G. (2003). Historical and estimated ground water levels near Winnipeg, Canada, and their sensitivity to climatic variability. *Journal of the American Water Resources Association, 39*(5), 1249–1259.

Kattel, R. R., & Nepal, M. (2021). Rainwater harvesting and rural livelihoods in Nepal. In A. K. E. Haque, P. Mukhopadhyay, M. Nepal, & M. R. Shammin (Eds.), *Climate change and community resilience: Insights from South Asia* (pp. 159–173). Springer.

Kumar, K. S. K. (2011). Climate sensitivity of Indian agriculture: Do spatial effects matter? *Cambridge Journal of Regions, Economy and Society, 4*, 221–235.

Kundzewicz, Z. W., & Döll, P. (2009). Will groundwater ease freshwater stress under climate change? *Hydrological Sciences Journal, 54*(4), 665–675.

Mendelsohn, R., & Seo, N. (2007). Changing farm types and irrigation as an adaptation to climate change in Latin American agriculture, *World Bank Policy Research Working Paper, 4161*. The World Bank, Washington D.C.

Ngongondo, C. S. (2006). An analysis of long-term rainfall variability, trends and groundwater availability in the Mulunguzi river catchment area, Zomba mountain, Southern Malawi. *Quaternary International, 148*, 45–50.

Okkonen, J., Jyrkama, M., & Kløve, B. (2010). A conceptual approach for assessing the impact of climate change on groundwater and related surface waters in cold regions (Finland). *Hydrogeology Journal, 18*, 429–439.

Palanisami, K., & Balasubramanian, R. (1993) Overexploitation of groundwater resource—Experiences from Tamil Nadu. In M. Moench (Ed.), *Water management: India's groundwater challenge*. VIKSAT/Pacific Institute.

Schlenker, W., Hanemann, W. M., & Fisher, A. C. (2007). Water availability, degree days, and the potential impact of climate change on irrigated agriculture in California. *Climatic Change, 81*(1), 19–38.

Shah, T. (2009). Climate change and groundwater: India's opportunities for mitigation and adaptation. *Environmental Research Letters, 4*, 1–13.

Taylor, R., Scanlon, B., Döll, P., et al. (2013). Ground water and climate change. *Nature Climate Change, 3*, 322–329.

Vidanage, S. P., Kotagama, H. B., & Dunusinghe, P. M. (2022). Sri Lanka's small tank cascade systems: Building agricultural resilience in the dry zone. In A. K. E. Haque, P. Mukhopadhyay, M. Nepal, & M. R. Shammin (Eds.), *Climate change and community resilience: Insights from South Asia* (pp. 225–235). Springer.

Part III
Technology Adoption

Chapter 11
Rainwater Harvesting and Rural Livelihoods in Nepal

Rishi Ram Kattel and Mani Nepal

Key Messages

- Rainwater harvesting helps mountain farmers to overcome water scarcity during dry season and in diversification from traditional cereals crops to high-value vegetable farming.
- The investment on the technology could be recovered in two years time but adoption of the technology is low due to the high start-up cost in the absence of subsidy.
- Providing trainings and subsidy to the farmers would help increasing the adoption rates of the technology as a part of climate change adaptation.

11.1   Introduction[1]

Rainfed agriculture is one of the sectors most sensitive to climate change (Cline, 2007), and in many countries in South Asia, a decline in crop yield is observed due to rising temperature, rainfall variability and extreme weather events (Balasubramanian, & Saravanakumar, 2022, Chap. 10 of this volume; Cruz et al., 2007; IPCC, 2007a; IFAD, 2008). Water scarcity is expected to increase while heat stress is expected to

R. R. Kattel
Department of Agricultural Economics and Agribusiness Management, Agriculture and Forestry University, Rampur, Chitwan, Nepal
e-mail: rrkattel@afu.edu.np

M. Nepal (✉)
South Asian Network for Development and Environmental Economics (SANDEE), International Centre for Integrated Mountain Development (ICIMOD), Kathmandu, Nepal
e-mail: mani.nepal@icimod.org

[1] A significant portion of the materials is drawn with permission from author's working paper: Kattel (2015). Rainwater Harvesting and Rural Livelihoods in Nepal, SANDEE Working Paper No 102-15.

© The Author(s) 2022   159
Haque et al. (eds.), *Climate Change and Community Resilience*,
https://doi.org/10.1007/978-981-16-0680-9_11

contribute to reduction of area available for high-yielding wheat production in the Indo-Gangetic Plains (IPCC, 2014). If no adaptation strategies for climate change are implemented, agricultural productivity could decline by as much as 10–25% in South Asia by 2080. For some countries, the decline in crop yield in rainfed agriculture could be as much as 50% (IPCC, 2018). The adverse impact of climate change on agriculture will be especially detrimental to Nepal, where over 60% of the population is dependent on subsistence and mostly rainfed agriculture for its livelihood. With natural springs drying up in the hills and the mountains (Bharti et al., 2020; Rai & Nepal, 2022, Chap. 23 of this volume), it is important to explore how monsoonal rain water can be conserved better and used effectively for hill agriculture as ground water is difficult to obtain and expensive. In this paper, we look at rainwater harvesting, which is being increasingly used in mountain agriculture in Nepal.

Rainfed agriculture accounts for 65% of the total cultivable land area in Nepal. Since only 24% of the arable land is irrigated (mainly in the lowland Terai), crop productivity is significantly low in comparison to the rest of South Asia and the country relies heavily on food imports (Bartlett et al., 2010). Agriculture consumes around 96% of all water withdrawn in the country (CIA, 2010) and contributes slightly over 25% to the GDP (World Bank, 2019).

In Nepal, more than 80% of precipitation occurs during a short monsoon season (June to September) resulting in flooding, landslides and loss of topsoil (Malla, 2008) and leading to crop failure and increased food and livelihood insecurity (Gentle & Maraseni, 2012; Gurung & Bhandari, 2009; Kohler et al., 2010).[2] Mountain people in Nepal are subject to even more livelihood risks as water sources are drying up and rivers are located a considerable distance away, usually below the farmland making irrigation impossible without access to appropriate technology.

This makes rainwater harvesting increasingly important as an adaptation strategy, which is a traditional technology used in highland pastures for generations in Nepal for collecting rainwater for animals, which has been re-designed and re-introduced in the farming system in recent years. Cascade tanks are popular for rainwater harvesting in other South Asian countries (Vidanage et al., 2022, Chap. 15 of this volume). There are two types of rainwater harvesting (RWH) practices: surface rainwater harvesting and rooftop rainwater harvesting. The rooftop system is mainly used for collecting rainwater for either household use or recharging ground water, while the surface rainwater collecting system is used for supporting agriculture. The focus of this study is the individually managed plastic or cemented RWH ponds, which have been promoted in the hills of Nepal for collecting surface rainwater.[3]

Since 2003, the government has been promoting plastic and cemented RWH ponds (MoI, 2014). However, till date, only about 5% farmers were found to have adopted RWH for crop production in the area surveyed for this study. If RWH technologies

---

[2] Though livestock of buffalo, dairy cattle, goat and sheep are feasible enterprises in the hilly areas of Nepal, the lack of a reliable water supply can restrict the extensive use of grazing lands and create pressure on scarce water resources (Zomer et al., 2014).

[3] Harvesting rainwater for animals is an old-age tradition in the hills of Nepal but rainwater harvesting for agriculture is relatively a new phenomenon.

are to be scaled up as a climate adaptation strategy, it is important to understand the impetus behind farmer's adoption decisions and their profitability. In this study, we ask two interrelated questions: (1) who adopts the RWH technology? and (2) what is the impact of RWH technology on farm income?

To answer these questions, we use farm, household and community-level information from four districts of Nepal. We find that the training received by farmers regarding agriculture and livestock production as a part of extension services is a strong determinant of RWH technology adoption, which significantly increases annual household income from agriculture and livestock. Benefit–cost analysis suggests that the RWH technology is viable in rainfed agricultural systems because adopters can diversify from cereal crops into high-value off-season vegetable crops for enhancing household income and farm profits. However, farm households face a large start-up cost (almost 30% of their annual income) for adopting the technology and also lack knowledge, which can be overcome by providing related trainings and subsidizing the RWH technology as a part of climate change adaptation strategy.

11.2   Technology Adoption in Agriculture

Technology adoption models are generally based on the theory that farmers make decisions in order to maximize their expected profits or utility (Feder et al., 1985). Subsistence farmers may maximize utility but not necessarily maximize profits at the same time (Sadoulet & de Janvry, 1995).[4] For this study, we use a utility maximization framework since the farmers in our study area are mostly subsistence farmers who have small parcels of land for producing agricultural crops for their own consumption, with little or no surplus for selling.

Risk is generally viewed as a major factor that influences the rate of adoption of any kind of innovation (Jensen, 1982; Just & Zilberman, 1983).[5] There are two types of associated uncertainties: the perceived risk associated with farm yield after adoption and production; and uncertainty related to the costs of farm inputs and outputs. Koundouri et al. (2006) propose that farmers adopt new technology in order to hedge against production risk and that human capital plays a significant role in the decision to adopt more efficient irrigation technology. In this context, Adesina and Zinnah (1993) and Getnet and MacAlister (2012) emphasize the importance of farmer's perception of the innovation-related characteristics of the technology in making an adoption decision. Our study provides evidence on the impact of RWH technology on farm income and how the adoption rate could be improved for mountain agriculture in Nepal.

---

[4] Despite its failure to identify the psychological processes that determine preferences, the framework is considered to be less restrictive than profit maximization approach (Lynne et al., 1988).

[5] Optimizing utility may also include considerations such as health benefits, environmental concerns, food security and risk (Napier et al., 2000; Ribaudo, 1998).

11.3   Study Area and Sampling

We conducted the study in 15 villages from four mid-hill districts of Nepal, namely Makwanpur, Palpa, Gulmi and Syangja. We chose these four districts deliberately for two main reasons: firstly, they have recorded the highest rates[6] of individually managed RWH technology adoption, and secondly, they allow us to capture variation in rainfall and elevation across the hilly districts of Nepal. Rainfed agriculture is predominantly practised in these areas and is associated with the cultivation of major staple crops such as maize, wheat, rice, millet and vegetables. For this study, we selected six Village Development Committees (VDCs)[7] from Makwanpur, four VDCs from Palpa, three VDCs from Gulmi and two VDCs from the Syangja district.

We used a multistage sampling technique to select four districts from two regions. We then selected VDCs from each district based on the RWH technology adoption rates. Secondly, we stratified farmers in each VDC into two groups, namely adopters and non-adopters. We identified adopters in each sampled village with the help of the District Agriculture Development Office (DADO) that keeps records of the adopters. In each sampled village, there were fewer numbers of RWH technology adopters than non-adopters.[8] We oversampled the RWH technology adopters such that the proportion of adopters and non-adopters is the same in our sample, and we applied the probability proportion to size technique to ensure that farmers in the large village clusters had the same probability of getting into the sample as those in the smaller village clusters. We sampled at least 10–15 households[9] from each VDC among the population of individually managed RWH adopters and non-adopters. The farm household survey was conducted between August and November 2012 through a structured survey interview of 282 farm households comprising 141 RWH adopters and 141 non-adopters (Fig. 11.1).[10]

We also conducted three community-level Focus Group Discussions (FGDs) of around 5–10 RWH adopters/non-adopters from different castes, genders and economic backgrounds in each district for obtaining qualitative information for understanding farmers' perspective on adopting/not adapting the RWH technology.

---

[6] We used the National Population and Housing Census Report (2011), District Profile (2011) and the District Agriculture Development Office Report (2011) to gather secondary information, including information on the RWH adopters' list, household population size and the occupational diversity of the households living in the selected villages in order to develop the sampling frame. We obtained the list of villages and adopters from the 2011 Census of Nepal.

[7] A Village Development Committee (VDC) was the lowest-level administrative unit in Nepal (till 2015) comprising small villages (wards).

[8] We found 20–120 RWH adopters and 100–410 non-adopters in each sample village.

[9] We selected 10–20 households randomly from each VDC where 5–10 were RWH adopting HHs (Treatment HHs) and 5–7 were RWH non-adopters/participants (Control HHs).

[10] We define a farm household as one where a group of individuals related by blood or marriage live on the same premises, share a kitchen and practice agriculture farming system.

Fig. 11.1 Location of the
four districts. *Source* Authors

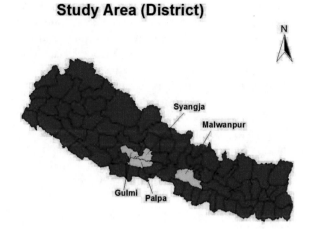

11.4   Methods and Variables

Household income and adoption of RWH technology may affect each other, wherein households with higher income may adopt RWH technology and adoption of RWH technology may also increase farm income due to the availability of irrigation water for off-season agriculture. The problem can be resolved technically using either a treatment-effects model as in Maddala (1983) or an instrumental variable (IV) approach as in Angrist (2000). We estimate farmers' adoption decisions and farm income simultaneously using a treatment-effects model to control for self-selection (Heckman, 1978, 1979; Heckman & Navarro-Lozano, 2003). We also use an IV approach for checking robustness of the estimates obtained from treatment-effect models.

11.5   Results and Discussion

11.5.1   Rainwater Harvesting Technology

In this section, we discuss the key findings from analysing the survey data. Among the 141 RWH adopters, approximately 82% had constructed a plastic pond (cheaper option) and the rest (18%) had cement ponds (more expensive). The size of each pond ranged from 1000 to 75,000 L. Nearly half of the RWH adopters in the study area had received a subsidy from the government (i.e. District Agricultural Development Office) and/or non-governmental organizations (NGOs) (mainly materials like plastic, pipe or cement) amounting to between 30 and 50% of the total construction costs, while other farmers adopted the technology of their own cost.

Approximately, 94% of the RWH ponds are located near the homestead of the farmers. While 45% of the adopters used only rainwater, about 31% of them used both rainwater and stream water in their RWH ponds. Almost all (97%) adopters stated that they installed a RWH pond for producing vegetable and high-value crops. We found most of the RWH ponds to be actively used (93%) and a majority (85%) of RWH adopters reported that their cropping pattern had changed after RWH pond construction. A quarter of the respondents reported problems with RWH pond management and water-holding capacity of the pond due to seepage and weeds and 7% ponds were inactive because of this. About three-fourth of the adopters used RWH pond water in the field for irrigation purposes through a pipeline connection. The age of the RWH ponds varied from 1 to 14 years. Kattel (2015) provides further information about the technology used in the study area.

11.5.2 Socio-demographic and Economic Characteristics

In the sample, the average age of the household head was approximately 50 years. Though RWH adopters are on average younger than non-adopters, the difference (between the two means) is not statistically significant. Approximately three-fourths of the household heads are male with very little education (two years of schooling on average). The level of education is higher for adopters than for the non-adopters. Among other variables, the total number of spades in the house (an indicator of agricultural tools), the training received with regard to agriculture and livestock production and the social network (membership of the household head in any group, organization or cooperative) were found to be greater among RWH adopters than among non-adopters. There was a greater proportion of higher caste households and those with knowledge of climate change among the RWH adopters than among the non-adopters. The percentage of people below the poverty line is lower for adopters (34%) sub-sample than for non-adopters (45%). This indicates that there is a negative association between RWH technology adoption and household poverty. The difference between the sample means in the adopter and non-adopter sub-groups is not statistically significant for the following variables: the livestock standard unit, the availability of extension services at the farm, access to credit, size of landholding, total cultivated land and per cent of upland in total cultivated land. The latter indicates that the two groups of farmers are mostly comparable. Table 11.1 presents the socio-demographic and economic characteristics of the sampled farmers.

The proportion of farmers who had diversified their cropping pattern was higher among RWH adopters with more adopters growing cauliflower and cabbage (39% compared to 22% in non-adopters 22%), tomatoes (55% compared to 27% in non-adopters) as well as beans, pea, broadleaf mustard, and gourds. Cereal crop production was more common among the non-adopters than among the adopters. Most RWH adopters were growing a high-value crop (i.e. vegetable) due to the availability of water during the dry season. Figure 11.2 shows the distribution of different types of vegetables and cereals produced by the RWH adopters and non-adopters.

Table 11.1 Socio-demographic characteristics of RWH adopters and non-adopters

Particular	Full sample ($N = 282$)	RWH adopters ($n = 141$)	Non-adopters ($n = 141$)	Mean difference (t-test)
Age of household head (in years)	50.21	48.91	51.50	− 2.59
Gender of the household head (if male = 1)	0.74	0.81	0.67	0.13***
Years of schooling	4.51	5.31	3.69	1.62***
Caste (if higher caste = 1)[11]	0.48	0.53	0.43	0.099*
Family size	6.48	6.58	6.39	0.19
Economically active household members (15–60 years old)	4.08	4.22	3.93	0.29
Upland cultivated (in *ropani*)	6.47	7.03	5.91	1.12
Lowland cultivated (in *ropani*)	1.94	1.89	1.99	0.10
Total cultivated land (in *ropani*)	8.41	8.93	7.91	1.29
Per cent shared by upland in total land	79.5	80.2	78.7	0.41
Livestock standard unit (LSU)[12]	3.13	3.34	2.93	0.41
Number of spades (type of physical asset)	4.7	5.2	4.3	2.88***
Extension service (if yes = 1)	0.43	0.46	0.40	0.06
Agriculture and livestock production-related training received (if yes = 1)	0.46	0.67	0.24	0.43***
Access to credit (if yes = 1)	0.77	0.81	0.74	0.08
Social network (membership in any group, cooperative and/or organization [if yes = 1)]	0.63	0.72	0.55	0.16***

(continued)

---

[11] Brahmin, Chettri and Takuri are the higher castes in Nepal.

[12] LSU is livestock standard unit (based on cattle equivalent: 1 cow/cattle = 10 goats/lambs = 4 pigs and = 143 chicken/ducks).

Table 11.1 (continued)

Particular	Full sample (N = 282)	RWH adopters (n = 141)	Non-adopters (n = 141)	Mean difference (t-test)
Knowledge of Climate Change (if yes = 1)	0.49	0.54	0.43	0.11*
Poor (if yes = 1)	0.39	0.34	0.45	0.11*

Note Figures in parentheses are standard deviation. ***Significant at 1% level, *Significant at 10% level
Source Field survey

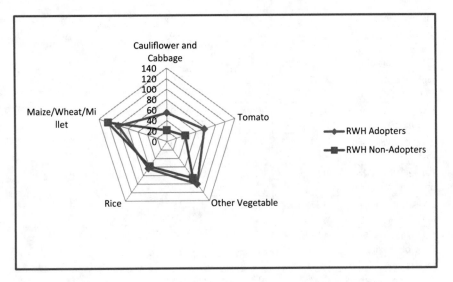

Fig. 11.2 Number of vegetable and cereal crop producers among RWH adopters and non-adopters. *Source* Field survey data

Table 11.2 presents the major crops produced and the revenue from the marketed crops in both the lowland and the upland during different cropping seasons. The production of cereal crops (mainly, rice, maize, and wheat) in upland and lowland areas is not statistically significant. With regard to farm revenue from different crops, revenue from tomato and other vegetables is significantly higher among the RWH adopters.

We found income from vegetables and fruits and from agriculture and livestock sectors as well as total annual household income to be significantly higher among RWH adopters than non-adopters. However, income from off-farm activities was significantly higher for non-adopters. The annual income from the agriculture and livestock sectors for RWH adopters (NRs. 104,969 (US$1049))[13] was almost double

---

[13] US$1 = NRs. 100.

Table 11.2 Crops production and revenue of RWH adopters and non-adopters

Particular	Full sample ($N = 282$)	RWH adopters ($n = 141$)	Non-adopters ($n = 141$)	Mean difference (t-test)
Crop production (quintal = 40 kg)				
Cauliflower/cabbage	20.08 (5.08)	22.05 (6.41)	14.69 (7.43)	7.35
Tomato	50.39 (26.82)	69.82 (39.95)	11.04 (2.21)	58.77
Other vegetable	8.33 (0.89)	9.76 (1.32)	6.66 (1.17)	3.09*
Rice	19.93 (6.75)	9.78 (1.94)	22.61 (13.95)	− 12.82
Wheat/maize	17.36 (9.04)	28.11 (19.93)	8.47 (1.22)	19.63
Crops-based HH revenue (in NRs.) from				
Cauliflower/cabbage	36,284 (7682)	42,559 (9932)	19,252 (8562)	23,306
Tomato	49,179 (5629)	59,175 (7557)	28,945 (6423)	30,220***
Other vegetable	19,711 (1877)	22,647 (2833)	16,297 (2337)	6350*
Rice	23,712 (2669)	26,379 (4546)	20,723 (2457)	5656
Maize/wheat	18,373 (2109)	19,151 (2533)	17,729 (3242)	1421

Note Figures in parentheses are standard deviation. ***Significant at 1% level, *Significant at 10% level
Source Field survey

as compared to the non-adopters (NRs. 53,876 (US$538)). The RWH technology adopters thus appeared to benefit from an increased supply of irrigation water during the dry season which allowed them to diversify their cropping system from cereal crops to high-value vegetable crops (see Table 11.3).

11.5.3   Results and Discussion

Results from econometric analysis indicate that RWH technology adoption significantly increases household income from agriculture and livestock. RWH technology adopters have earned around 270% more annual household income from agriculture and livestock sectors due to availability of irrigation water than the non-adopters farmers in the study area.

Table 11.3 Annual household income from different sectors among RWH adopters and non-adopters (NRs)

Particular	Full sample (*N* = 282)	RWH adopters (*n* = 141)	Non-adopters (*n* = 141)	Mean difference (*t*-test)
Cereal crops	23,671 (2050)	23,796 (2697)	23,546 (3111)	250
Vegetable and fruit	44,148 (4663)	67,446 (8414)	20,850 (2960)	46,595***
Livestock	11,603 (1322)	13,726 (2247)	9480 (1379)	4246*
Employment/services	40,604 (6532)	38,082 (9313)	43,127 (9191)	− 5045
Off-farm	34,881 (4311)	26,134 (5201)	43,627 (6815)	− 17,492**
Foreign employment	40,730 (6899)	47,368 (12,313)	34,092 (6232)	13,276
Agriculture and livestock (cereal + vegetable + livestock)	79,423 (5739)	104,969 (9988)	53,876 (4812)	51,092***
Total annual household income	195,640 (11,201)	216,555 (17,713)	174,724 (13,551)	41,831*

Note Figures in parentheses are standard deviation. ***Significant at 1% level, **Significant at 5% level, *Significant at 10% level
Source Field survey

Our results suggest that the most important factor affecting farmers' adoption decision is extension service such as trainings related to farming and livestock rearing. The age of the household head, annual household income from off-farm activities and poverty status have significant but negative impacts on RWH technology adoption, whereas the training received by the farmer and the gender of the household head (i.e. being a male) have significantly positive impacts on the RWH adoption decision. However, other variables like the economically active members in the household, the share of upland cultivated land, and education of the household head had no impact on the adoption decision.

If a farmer receives agriculture and livestock production-related training, the probability of RWH technology adoption decision increases by 28.5%. The training provides knowledge and skill to farmers to adopt innovative technology at the farm. Figure 11.3 illustrates household farm income by training received and RWH technology adoption. The probability of adopting RWH technology is 18.7% lower if the farmer is poor compared to a relatively better off farmer.

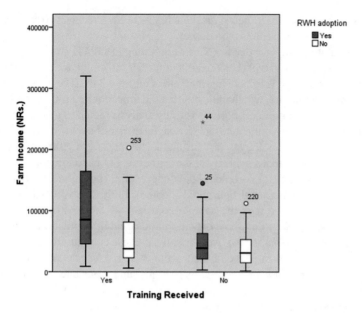

Fig. 11.3 Box plot of farm income by training received and RWH technology adoption. *Source* Authors

11.5.4   Cost–Benefit Analysis of RWH Pond Adoption

For cost–benefit analysis, we have chosen plastic ponds as they are cheaper and more popular than cement ponds. We calculate the annual benefit and cash flow for a ten-year period as a plastic pond functions well up to ten years. We estimate the benefit–cost (*B/C*) ratio, net present value (NPV), internal rate of return (IRR) and payback period (PBP) using the standard discount rate of 12% that is used in financial analysis in Nepal.[14] We perform sensitivity analysis to check the robustness of the results with alternative discount rates.

Cost calculations are based on the total investment cost including labour, plastic, other equipment and the opportunity cost of the land on which the pond is constructed. The initial average investment cost to construct a RWH plastic pond on 3 *ropani* of upland area (with a water-holding capacity of approximately 45,000 L) is NRs. 55,372 (US$543). In addition, there is a maintenance cost of NRs. 913 per year per household based on survey information.

We estimate the incremental income from RWH pond construction and crop diversification from the second year onwards using the benefits estimates. An average farmer using RWH technology obtains an annual incremental benefit of NRs. 69,456

[14] ADB (2013) proposes a higher discount rate (about 12%) for cost–benefit analysis in developing countries considering higher production and market risks and uncertainty to introduce new technologies at farm.

(US$700). This number reflects the benefits a farmer receives from using this technology relative to farmers who do not.

Our calculations suggest that the NPV of investing in a RWH pond with a capacity of 45,000 L is NRs. 276,649 (US$2766) to a farmer, assuming a 12% discount rate. The benefit to cost ratio is 6.1 and internal rate of return is 34%. The payback period is approximately two years, which indicates that the time required for the repayment of the initial investment is rather short. Sensitivity analysis indicates that the investment is viable even if the investment costs increase by 20% or benefits decrease by 20%.

Although RWH technology is very profitable, findings from stakeholder meetings and field survey show that a majority of the farmers are not adopting this technology. This is because of lack of technical knowledge, large start-up cost (NRs. 55,000, which is 28% of total annual household income), limited subsidies (only 45% of the households received it in the sample) and lack of labour in the communities to construct ponds due to massive out-migration of adult population for better job opportunity (Karki Nepal, 2016). Additionally, in these communities, farming is not commercialized and most of the farmers seem to be risk averse and do not want to shift from cereal-based farming systems to high-value crops due to production and market risks.

Our analysis indicates that the training received by farmers on farm management, agriculture and livestock production helps to increase RWH adoption. The cost of a three-day community-wide training package is about NRs. 3000 (US$30) per farmer (including 30% organizational overhead costs). With training, the probability of RWH technology adoption increases by approximately 30%. Thus, in any district in our study area, if approximately 10% of households from the communities (i.e. some 7000 households), are trained, then we can expect 2030 trainees to adopt RWH (see Kattel, 2015 for more detail on Table 7). The net annual benefits from training on adoption of RWH technology are NRs. 66,457 per farmer. Thus, per district annual benefits from providing training to 10% of farm households from the communities is expected to be approximately NRs. 131 million (1.3 million US$). These benefits, however, would require substantial initial investment in building the RWH ponds and providing extension services to the farmers.

11.6   Conclusion

Mountain springs are drying up and disappearing rapidly, putting mountain agriculture at stake. Adapting rainwater harvesting technology helps mountain communities to address irrigation water scarcity to some extent allowing them to diversify their cropping system from subsistence cereal crops to high-value commercial vegetables that helps increase their resilience in the face of climate change.

Although RWH technology is highly profitable in rainfed mountain agriculture systems, a majority of farmers seem reluctant to adopt this technology. Our study suggests that at least some of the constraints for adoption of RWH technology can be reduced by providing appropriate training to the farmers. Thus, policy makers and extension service providers need to play a more proactive role in promoting RWH technology in the rainfed hilly region of Nepal by providing credits or subsidies and appropriate training to the potential adopters that helps building community resilience in the face of climate extremes such as droughts.

Acknowledgements The research was supported financially and technically by the South Asian Network for Development and Environmental Economics (SANDEE) at the International Centre for Integrated Mountain Development (ICIMOD). We are grateful to Céline Nauges, Priya Shyamsundar, other SANDEE resource persons and peers for their critical comments and encouragement. We are also deeply indebted to the farmers in the study area, who are too numerous to mention individually, but without whose cooperation this study would not have been possible. The International Centre for Integrated Mountain Development (ICIMOD), where the corresponding author is affiliated, acknowledges with gratitude the support of the Governments of Afghanistan, Australia, Austria, Bangladesh, Bhutan, China, India, Myanmar, Nepal, Norway, Pakistan, Sweden and Switzerland. However, the views as well as interpretations of the results presented in this research are those of the authors and should not be attributed to their affiliated organizations, their supporters or the funding agency.

References

Adesina, A., & Zinnah, M. M. (1993). Technology characteristics, farmers' perceptions and adoption decisions: A Tobit model application in Sierra Leone. *Agricultural Economics, 9*, 297–311.

Angrist, J. (2000). *Estimating of limited dependent variable models with dummy endogenous regression: Simple strategies for empirical practices.* National Bureau of Economic Research, Massachusetts Avenue Cambridge, NBER Working Paper No. 248.

Asian Development Bank (ADB). (2013). *Cost benefit analysis for development: a practical guideline.* Mandaluyong City.

Balasubramanian, R., & Saravanakumar, V. (2022). Climate sensitivity of groundwater systems in South India: Does it matter for agricultural income? In A. K. E. Haque, P. Mukhopadhyay, M. Nepal, & M. R. Shammin (Eds.), *Climate change and community resilience: Insights from South Asia.* Springer.

Bartlett, R., Bharati, L., Pant, D., Hosterman, H., & McCormick, P. (2010). Climate change impacts and adaptation in Nepal. Working Paper 139, International Water Management Institute (IWMI), Colombo, Sri Lanka. Accessed December 2012. Available at http://www.iwmi.cgiar.org/Publications/Working_Papers/working/WOR139.pdf

Bharti, N., Khandekar, N., Sengupta, P., Bhadwal, S., & Kochhar, I. (2020). Dynamics of urban water supply management of two Himalayan towns in India. *Water Policy, 22*(S1), 65–89.

Central Intelligence Agency (CIA). (2010). *The world factbook.* Accessed February 2013. Available at http://www.cia.gov/library/publications/the-world-factbook/goes/np.html

Cline, W. R. (2007). *Global warming and agriculture: Impact estimates by country.* Centre for Global Development and the Peterson Institute for International Economics.

Cruz, R. V., Harasawa, H., Lal, M., Wu, S., Anokhin, Y., Punsalmaa, B., Honda, Y., Jafari, M., Li, C., & HuuNinh, N. (2007). Asia, climate change 2007: Impacts, adaptation and vulnerability. In M. L. Parry, O. F. Canziani, J. P. Palutikof, P.J. van der Linden, & C. E. Hanson (Eds.), *Contribution of working group II to the fourth assessment report of the intergovernmental panel on climate change* (pp. 469–506). Cambridge University Press.

Feder, G., Just, R. E., & Zilberman, D. (1985). Adoption of agricultural innovations in developing countries: A survey. *Economic Development and Cultural Change, 33*, 255–298.

Gentle, P., & Maraseni, T. N. (2012). Climate change, poverty and livelihoods: Adaptation practices by rural mountain communities in Nepal. *Environmental Science and Policy, 21*, 24–34.

Gurung, G. B., & Bhandari, D. (2009). Integrated approach to climate change adoption. *Journal of Forest and Livelihood, 8*(1), 91–99.

Heckman, J., & Navarro-Lozano, S. (2003). Using matching, instrumental variables and control functions to estimate economic choice models. Working Paper, IFAU—Institute for Labour Market Policy Evaluation, No. 2003:4.

Heckman, J. J. (1978). Dummy endogenous variables in a simultaneous equation system. *Econometrica, 46*(6), 931–959.

Heckman, J. (1979). Sample selection bias as a specification error. *Econometrica, 47*, 153–161.

International Fund for Agricultural Development (IFAD). (2008). Climate change and the future of smallholder agriculture; how can rural poor people be a part of the solution to climate change? Discussion paper prepared for the round table on climate change at the 31st session of IFAD's Governing Council, February 14, 2008. IFAD Policy Reference Group on Climate Change.

IPCC. (2014). Climate change 2014: Impacts, adaptation, and vulnerability, Part B: Regional aspects. In V. R. Barros, C. B. Field, D. J. Dokken, M. D. Mastrandrea, K. J. Mach, T. E. Bilir, M. Chatterjee, K. L. Ebi, Y. O. Estrada, R. C. Genova, B. Girma, E. S. Kissel, A. N. Levy, S. MacCracken, P. R. Mastrandrea, & L. L. White (Eds.), *Contribution of working group II to the fifth assessment report of the intergovernmental panel on climate change* (pp. 5–10). Cambridge University Press.

IPCC. (2018). Global warming of 1.5 °C. An IPCC special report on the impacts of global warming of 1.5 °C above pre-industrial levels and related global greenhouse gas emission pathways, in the context of strengthening the global response to the threat of climate change, sustainable development, and efforts to eradicate poverty. In V. Masson-Delmotte, P. Zhai, H.-O. Pörtner, D. Roberts, J. Skea, P. R. Shukla, A. Pirani, W. Moufouma-Okia, C. Péan, R. Pidcock, S. Connors, J. B. R. Matthews, Y. Chen, X. Zhou, M. I. Gomis, E. Lonnoy, T. Maycock, M. Tignor, & T. Waterfield (Eds.), In Press.

Jensen, R. (1982). Adoption and diffusion of an innovation of uncertain profitability. *Journal of Economic Theory, 27*, 182–192.

Just, R. E., & Zilberman, D. (1983). Stochastic structure, farm size and technology adoption in developing agriculture. *Oxford Economic Papers, 35*, 307–328.

Karki Nepal, A. (2016). The impact of international remittances on child outcomes and household expenditures in Nepal. *The Journal of Development Studies, 52*(6), 838–853.

Kattel, R. R. (2015). Rainwater harvesting and rural livelihoods in Nepal. South Asian Network for development and environmental economics (SANDEE) Working Paper No. 102-15. Accessed September 2020. Available at http://www.sandeeonline.org/uploads/documents/public ation/1078_PUB_Working_Paper_102_Rishi.pdf

Kohler, T., Giger, M., Hurni, H., Ott, C., Wiesmann, U., von Dach, S. W., & Maselli, D. (2010). Mountains and climate change: A global concern. *Mountain Research and Development, 30*(1), 53–55.

Koundouri, P., Nauges, C., & Tzouvelekas, V. (2006). Technology adoption under production uncertainty: Theory and application to irrigation technology. *American Journal of Agricultural Econonomics, 88*(3), 657–670.

Lynne, G. D., Shonkwiller, J. S., & Rola, L. R. (1988). Attitudes and farmer conservation behavior. *American Journal of Agricultural Economics, 70*, 12–19.

Maddala, G. (1983). *Limited dependent and qualitative variables in econometrics.* Cambridge University Press.

Malla, G. (2008). Climate change and its impact on Nepalese agriculture. *The Journal of Agriculture and Environment, 9*(7), 62–71.

MoI. (2014). *Irrigation policy, 2013.* Ministry of Irrigation (MoI), Government of Nepal, Official Site. Accessed October 2014. Available at http://www.moir.gov.np/pdf_files/Irrigation-Policy-2070-final.pdf

Napier, T. L., Tucker, M., & McCater, S. (2000). Adoption of conservation tillage production systems in three Midwest watersheds. *Journal of Soil and Water Conservation, 53*, 123–134.

Intergovernmental Panel on Climate Change (IPCC). (2007a). Summary for policymakers in climate change 2007: impacts, adaptation and vulnerability. In M. L. Parry, O. F. Canziani, J. P. Palutikof, P. J. van der Linden, & C. E. Hanson (Eds.), Contribution of working group II to the fourth assessment report of the intergovernmental panel on climate change (pp. 7–22). Cambridge University Press.

Rai, R. K., & Nepal, M. (2022). A tale of three himalayan towns: Would payment for ecosystem services make drinking water supply sustainable? In A. K. E. Haque, P. Mukhopadhyay, M. Nepal, & M. R. Shammin (Eds.), *Climate change and community resilience: Insights from South Asia.* Springer.

Ribaudo, M. O. (1998). Lessons learned about the performance of USDA Agricultural non-point source pollution programs. *Journal of Soil and Water Conservation, 53*, 4–10.

Sadoulet, E., & deJanvry, A. (1995). *A Quantitative development policy analysis.* The Johns Hopkins University Press.

Vidanage, S. P., Kotagama, H. B., & Dunusinghe, P. M. (2022). Sri Lanka's small tank cascade systems: Building agricultural resilience in the dry zone. In A. K. E. Haque, P. Mukhopadhyay, M. Nepal, & M. R. Shammin (Eds.), *Climate change and community resilience: Insights from South Asia.* Springer.

World Bank. (2019). *World development update.* Accessed December 2020. Available at https://www.worldbank.org/en/country/nepal/publication/nepaldevelopmentupdate

Zomer, R. J., Trabucco, A., Metzger, M. J., Wang, M., Oli, K. P., & Xu, J. (2014). Projected climate change impacts on spatial distribution of bioclimatic zones and ecoregions within the Kailash Sacred Landscape of China, India, Nepal. *Climatic Change, 125*, 445–460.

Chapter 12
Local Strategies to Build Climate Resilient Communities in Bangladesh

Estiaque Bari, A. K. Enamul Haque, and Zakir Hossain Khan

Key Messages

- Creating alternative market chains for fuel (e.g. LPG) reduces forest dependency for fuelwood and is supportive towards forest conservation.
- *Bandalling*—an indigenous bamboo structure still has the merit to reduce river erosion in smaller river basins and to reclaim agricultural lands.
- Floating agriculture (e.g. *Baira)* enables communities to become resilient against waterlogging and also offers income generating opportunities.

12.1 Introduction

Erratic changes in climate parameters such as precipitation rate, temperature, wind pressure and solar radiation threaten communities in Bangladesh by exposing them to various natural disasters like floods, riverbank erosion, drought, waterlogging, cyclones and earthquakes (Banholzer et al., 2014; Kharin et al., 2007). Despite having achieved some degree of resilience in flood management through community adaptation alongside supportive flood management policies (Sultana et al., 2008; Younus & Harvey, 2013), the country is likely to face risks of greater frequency and intensity in the future. One estimate suggests that a 1 m increase of sea-level rise (SLR) may inundate one-sixth of total land area displacing more than 13 million people from

E. Bari (✉) · A. K. E. Haque
Department of Economics, East West University, Dhaka, Bangladesh
e-mail: estiaque@ewubd.edu

A. K. E. Haque
e-mail: akehaque@gmail.com

Z. H. Khan
Fellow, Asian Center for Development, Dhaka, Bangladesh
e-mail: zhkhaneco@gmail.com

© The Author(s) 2022   175
Haque et al. (eds.), *Climate Change and Community Resilience*,
https://doi.org/10.1007/978-981-16-0680-9_12

Southern Bangladesh (Huq et al., 1995). The intrusion of salinity in agricultural lands will lead to a shift in cropping patterns (Dasgupta et al., 2014). Agricultural lands may turn into a waterbody in one place while in another there may be barren lands emerging in the absence of water. Communities (farmers in particular) must adapt with the evolving climate situations (Rounsevell et al., 1999). While afforestation is a means of combating climate risks (Noss, 2001), the forest cover in Bangladesh is only 10.7%, well below the 17.5% area designated forest land in the country (Forest Department of Bangladesh, 2017). The high density of population causes severe pressure on forests due to the demand of fuelwood and timber for other uses (Iftekhar, 2006; Iftekhar & Hoque, 2005; Salam et al., 1999). This chapter will highlight three cases of local adaptation strategies to combat climate change.

The first case is an example of community engagement using external assistance to reduce pressure on a protected forest area. With the influx of over 700,000 Rohingya refugees threatening the forest due to demand for fuelwood, support from the United Nations High Commissioner for Refugees (UNHCR) was used to develop a market chain for gas supply (United Nations High Commissioner for Refugees, 2020) that resulted in significant reduction on collection of fuelwood from the forests. The second case illustrates how communities use indigenous technology to effectively protect their village from river erosion. The third case presents a unique scenario where farmers used indigenous technology—the floating agricultural bed—to adapt to permanent waterlogging. These strategies are supported by the government and the non-government organizations (NGOs) working in Bangladesh.

12.2   Case I: Market Development and Forest Conservation

This case is based on research that the authors carried out with the International Union for Conservation of Nature and Natural Resources (IUCN), Bangladesh (International Union for Conservation of Nature and Natural Resources, 2019) and estimates the impact of supply of liquefied petroleum gas (LPG) to the refugee families on replacing fuelwood demand. The project began in 2018 and by 2019, nearly all the refugee families residing in the camps of Ukhiya and Teknaf received LPG for cooking from the UNHCR (Inter Sector Coordination Group et al., 2018).

Nearly 711,000 Rohingya refugees have entered Bangladesh since 25 August 2017, the majority arriving in the first three months of this crisis (United Nations High Commissioner for Refugees, 2020). They were settled in the refugee camps of Ukhiya and Teknaf of Cox's Bazar district of Bangladesh—occupying nearly 6% of the land (Inter Sector Coordination Group, 2019) of the 650 km^2 forest (Bangladesh Bureau of Statistics, 2015). While the UNHCR and the International Organization for Migration (IOM) distributed food, clothing and shelters, the supply of fuel initially remained outside their lens. The demand of fuelwood was so high that a market for fuelwood began to appear (Fig. 12.1). With both locals and the refugees extracting fuelwood, there was degradation of the forests and reduction of water supply in the streams which affected the livelihood of the local people.

Fig. 12.1 Fuelwood shop inside the Rohingya refugee camp. *Photograph credit* A. K. Enamul Haque

Estimates from our study suggest that immediately after the arrival of Rohingya refugees the demand for fuelwood was nearly 4.76 kg per household per day and the annual biomass extraction rose from a regular 95,000–462,000 tons in a year (International Union for Conservation of Nature and Natural Resources, 2019), well beyond the sustainable yield of the forest. With depletion of forest land, human–wildlife conflicts began in the area. Attacks by elephants and snakes were on the rise and IUCN, Bangladesh, was invited by the UNHCR to step in to devise strategies to reduce such incidents.

The situation also demanded a change in the strategies of managing refugees inside a national forest and the UNHCR agreed to supply LPG in addition to other supplies. The programme designed by UNHCR in collaboration with other partners was primarily aimed at protecting the forest area and reducing carbon emission (Inter Sector Coordination Group et al., 2018).

12.2.1   Case I: Intervention

As per the arrangements, the UNHCR and the IOM provided LPG to every refugee household every 26–45 days (determined by family size). All refugee households

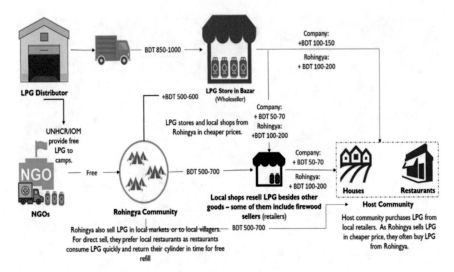

Fig. 12.2 LPG supply chain in Ukhiya Upazila, Cox's Bazar. *Source* Developed by Authors. *Note* 1 USD = 80 Taka

were also given a stove to switch to LPG-based cooking instead of using fuelwood in their regular stoves. The LPG supply programme initially also included local non-refugee families (called host families) living nearby for at least six months in order to maintain the social cohesion between two communities.

In addition to more than 0.7 million refugees inside the 29 camps, there are hundreds of restaurants and markets, local communities, NGOs working with the refugees, government offices and security personnel who are the potential users of LPG. As such, a network of supply chains began to emerge once LPG distribution started in the camps. The distribution centres for LPG were developed in the camp areas as well as in the local markets and a supply chain developed to ensure stable supply of LPG and repair and maintenance of LPG stoves. The supply chain of LPG is shown in Fig. 12.2.

12.2.2   Case I: Impact

We conducted a randomized survey[1] of 1200 refugees and 200 local households (International Union for Conservation of Nature and Natural Resources, 2019) and found

---

[1] Randomization was done based on two stages: a) randomized the camps by numbers; b) randomized the refugee households by their unique ID. Host households were selected using proximity to the selected camps.

(a)   Among the refugee households there has been an 80% drop in the demand for fuelwood and among the local households it fell by 53%. It is important to note that local households were not given free LPG refills after the first six months of the programme.
(b)   Total demand for fuelwood fell from 462,000 tons to 38,000 tons, much below the pre-refugee demand for fuelwood in the area (which was 95,000 tons/year). The chain of events that led to such fall in demand for fuelwood is more interesting as a lesson learned.

First, development of the LPG supply chain created local demand for LPG—restaurants, local communities (living outside the camp area) began to take part in the market and their demand for fuelwood fell sharply. Second, a private market for unused LPG gas began where refugee families also supplied their unused LPG at a discounted price to receive free refills from the camp. Shops which used to sell fuelwood became resellers of unused LPG supply in the market.

Third, wide-scale adoption of LPG for cooking in and outside the camp area resulted in a significant reduction in CO_2 equivalent emissions. Estimates show that there was a net reduction of 0.82 million tons of CO_2 (equivalent) due to the introduction of the LPG use among the refugees and in local communities (Fig. 12.3).

Fig. 12.3 LPG distribution centres inside the camps. *Photograph credit* A. K. Enamul Haque

Fourth, because of the drop of demand for fuelwood, the need for locals and refugees to enter the reserve forest area has been reduced. It is now expected that the forest department will be able to restore the degraded forest lands and the natural barriers against the cyclones and storms may be rehabilitated. Also, the wildlife will once again find an undisturbed habitat.

Building resilience has been one of the primary goals for adaptation in coastal areas across the world. Most often threats to this come from the pressure of economic development and increased economic activities. In this particular case, the threat was external—through an influx of refugees which outnumbered local communities by 4:1. As such, the scale of the threat was unprecedented and abrupt and the ecosystem was incapable of dealing with such massive extraction with no respite. The huge pressure of population created conflicts and caused degradation of forest land which dried up many natural streams on which local people were dependent.

Introduction of LPG distribution in the camps created a market for LPG in the local area. This resulted in a reduction in demand for fuelwood from the refugee households as well as significant changes in the demand for fuelwood from local communities. Our study also revealed that the average monthly cost of LPG and that of fuelwood are very similar in the area and it created incentives for local people to adopt LPG as a new fuel for cooking (International Union for Conservation of Nature and Natural Resources, 2019). It shows the power of markets—development of market chains and how it led to reduction of fuelwood collection from the forest area. In addition, it has created incentives to reduce carbon emissions without large-scale public subsidies. Thomas et al., (2022, Chap. 13 of this volume) have also shown how it can also be part of government strategy to increase rural employment.

12.3   Case II: Bandalling—A Traditional Approach to Reduce Riverbank Erosion

The river in its pristine state is supposed to flow through a main channel and gradually bring changes in its depth, width and riverbanks by adjusting to the changes in sediment layers. Both flooding and river erosion are natural processes in a vibrant river life cycle, and it carries sediments that changes the landscape and the climate regime (Smith & Roy, 1998). However, several man-made interventions such as dams, barrages, embankments, as well as economic activities at upper-stream and lower-stream areas disturb the natural flow of rivers forcing the river to change course abruptly, causing erosion of its banks (Intergovernmental Panel on Climate Change, 2007). Climatic changes are likely to further aggravate the situation as it causes changes in the rainfall pattern in the watershed region of the river and/or because of glacier-melts in the Himalayas. While governments spend a significant amount to protect economic hubs, communities living far-off from the major hubs of economic activities are left on their own. River-dependent communities must, therefore, adjust to the changes in river flows to reduce their vulnerability against bank erosion with or without external resources.

The latest survey on riverbank erosion by the Bangladesh Bureau of Statistics (BBS) shows that from 2009 to 2014 about 215,880 households were affected by river/coastal erosion (Bangladesh Bureau of Statistics, 2016). Loss of crops, livestock, poultry and fishery and damage of land, houses, homestead and forestry alone amounted to 36,410-million-taka in damage and losses. On aggregate, it accounts for nearly 20% of all disaster-related damage and losses caused in this period (Bangladesh Bureau of Statistics, 2016). River erosion not only forces millions of people to migrate but also adversely affects rural agriculture and its economy (Bhuiyan et al., 2017). Though riverine communities have continuously adjusted to the hazards of riverbank erosion though different adaptation strategies, there is not much mention of these mitigation strategies in literature (Haque, 1988). One measure being used by communities for centuries is the traditional bamboo-made structures to protect the banks. The practice is more common in communities living in far-off from cities and towns. These structures are locally known as 'bunds' and as such the word 'bund' also refers to river-front areas in a community. Bunds are horizontal to the river and is filled with soil.

12.3.1   Case II: Intervention

Bandalling is a temporary traditional structure made of local materials, e.g. bamboo and wooden log, mainly to maintain or improve navigation channels in lean periods and protect riverbanks from erosion (Nakagawa et al., 2011). Similar to a fence with cross-pieces, the structures reduce the speed of the monsoon water flow creating a natural flood barrier and channelling the water midstream. There are two parts to this structure; while the top part is blocked diverting the high velocity water flow, the lower part is opened to secure the flow of water (Rahman & Osman, 2015; Nakagawa et al., 2011). This structure allows the lower velocity flow to pass through the bottom and deposit sediments at the bottom of the *Bandals* (Rahman & Osman 2015; Nakagawa et al., 2011). The angle of bamboo struts and the space between bamboos mats placed at the top for blocking the water flow are critically important to gain structural stability (Rahman, 2019) and vary according to the intensity of the river flow.

Despite attaining certain structural stability, the bandals need to be rebuilt almost every year. However, the cost of construction is significantly low when compared to alternative means of protecting riverbanks. Alternatives which are also used in Bangladesh to protect townships, ports and other valuable installations are much costlier. They include: spurs, groynes (permeable and impermeable), sand-filled mattresses, revetments or a combination of these to build embankments to mitigate riverbank erosions and safeguard thousands of houses, agricultural lands and other administrative and non-administrative structures.

Fig. 12.4 Construction of bandalling. *Photograph credit* Bokhtiar Hossain Shishir

Villagers in Rowmari upazila have recently used 'bandalling' to protect the bank of Jinjiram—a distributary of the Brahmaputra River, from erosion. Namapara is a village on the bank of the river with nearly 200 families who depend on agriculture for their livelihood. The river has been a constant threat because of bank erosion in rainy seasons, and to protect against this, they have got together and built several 'bandals' at different places. During 2017, the villages of Namapara raised a community fund to develop nearly twenty-four single-layer bandal structures (Fig. 12.4). The width of each bandal structure was between 10 to 30 feet. The combined length of these structures was about 400 feet long and primarily helped to protect villagers from riverbank erosion (Shishir, 2020; Siddique, 2020).

These types of fences are built every few hundred metres in the river, and it eventually shifts the main course of the river away from the banks, stops erosion and recovers land. By bringing in more water into the main channel of the river, it also increases navigability.

The Bangladesh Water Development Board (BWDB) which is responsible for building and maintaining large structures also encourages local communities to build 'bandalling' as a solution to bank erosion on smaller rivers. It is argued that *bandalling* may be a feasible solution only to protect riverbanks in relatively smaller river basins (Rahman, 2019). The Bangladesh River Research Institute (RRI) has initiated three pilot projects including one project in Brahmaputra basin in an attempt to find a feasible bandalling structure that will sustain against the physical characteristics of the river (River Research Institute, 2018).

12.3.2   Case II: Impact

The discussion with the local NGO officials in July 2020 reveals that due to lack of proper structural know-how the Namapara community did not achieve the optimal level of stability from bandalling in the first year. However, the experience helped them to learn and improve the design of the structure. They also dredged the riverbed and piled the sand near the bandal. Less bank erosion was observed in the Namapara village but as the seasonal river Jinjiram carries relatively lower sediment the expected reclamation of agriculture land from bandalling was not observed in this case.

The success of the methodology, however, has some caveats. First, the communities do not get the rights on the reclaimed land—which is a bone of contention in the community. The right, as per the law, belongs to the owner of the land who had lost it previously or it automatically becomes a government land. Second, communities need technical know-how to build a stable structure and for this support may be needed from government agencies such as the BWDB which are mandated to provide this. This case shows how communities used local knowledge and used community participation to halt bank erosion. Udayakumara (2021, Chap. 19 of this volume) has shown using training to promote adoption of erosion control technologies in rural areas.

12.4   Case III: Baira—The Floating Agriculture Technique

The Government of Bangladesh constructed many polders and embankments in the late 1960s to block the ingress and egress of tidal waters and facilitate irrigation for agricultural (Dev, 2013). Failure to manage locks or gates of these structures results in the gates of these enclosures often getting silted and the agricultural land inside becoming waterlogged (Rahman, 1995). As a result, thousands of acres of land in coastal areas have become permanently waterlogged. In Bangladesh, about 5% of potential agricultural land remains waterlogged during monsoon and the latest survey report shows that between 2009 and 2014, about 605,300 households were affected by waterlogging (Bangladesh Bureau of Statistics, 2016). Nearly 16,060-million-taka damage and losses were caused by waterlogging alone; of which, 50% was due to loss of crops (Bangladesh Bureau of Statistics, 2016).

The problem has been aggravated by increased precipitation due to climate change. As such, communities need to solve it. One solution is to improve maintenance of the gates but this often runs into conflict between shrimp and rice farmers. Climate change and potential sea-level rise may inundate more crop lands permanently. Prolonged waterlogging would create additional adaptation challenges such as salinisation of agricultural land in these areas and migration, and mitigating this requires urgent solutions. One such solution is known as *Baira*—a technique to build floating agricultural beds using water hyacinth and soil in order to cultivate agricultural crops on a wetland.

Describing the role of Baira, the Ministry of Agriculture of the Government of Bangladesh states '... the system generating goods and services sustainably for the locals and practitioners date back a few thousand years in southern Bangladesh. ... Without the system, cultivating only Aman rice in deep water would be still prevailing in this region' (Ministry of Agriculture, 2017). The Ministry states that the centuries-old system has undergone several environmental and socio-economic changes and includes the latest technologies to enable farmers to cultivate diverse crops. Adoption of new crop varieties and cultivation methods enables local farmers to ensure food and livelihood security by improving cropping intensity on these platforms. Initiatives from local and international NGOs have helped to scale this up as they disseminate this indigenous knowledge and technology among farmers through several training sessions. '... From 2011, the government also took part in dissemination collaborating with NGOs' (Ministry of Agriculture, 2017).

The technique requires farmers to use country boats to move along the floating platforms and to prepare the bed, do the weeding and also the harvesting (Fig. 12.5). It is an older form of modern 'hydroponics' and requires less inputs. The technique has several names—*Baira, Dhap, Gaota, Geto,* Floating Cultivation (*Vashoman Chash*) and others (Anik & Khan, 2012; Irfanullah et al., 2008). *Baira* is the most popular name among all.

Studies have shown that floating agriculture practices if effectively managed could reduce adverse impacts of waterlogging and turn a temporarily inundated waterbody into a potential soil-less agricultural land. Besides Bangladesh, floating gardens of Xochimilco in Mexico and Dal Lake in India also have historical backgrounds. Originally, Baira was used for seedbeds for seedlings of rainwater varieties of rice in Bangladesh. However, with interventions from the government and from the NGOs,

Fig. 12.5 Floating beds. *Photograph credit* Food and Agriculture Organization

the practice has been expanded for cultivation of vegetables such as cucumber, okra, ginger, bitter gourd, Arum, potato, turmeric, brinjal, Lal sak, Palang sak, Danta, cauliflower, pumpkin and chilli (Anik & Khan, 2012; Chowdhury & Moore, 2017; Islam & Atkins, 2007).

The process of preparing a floating bed requires several steps. First, a stack of mature water hyacinths (a weed which has a slow rate of decomposition) is used to prepare the base of the floating bed and to create stability and buoyancy (Irfanullah et al., 2008). In the second layer, farmers use other forms of manure to speed up the rate of decomposition. Usually within 8 to 10 days, beds are ready for farmers to transplant seedlings (Dev, 2013). Typically, I-shaped (narrow) floating agricultural beds are common, and however, the size and shape of beds vary upon the volume of the ditch or waterbodies (Chowdhury & Moore, 2017; Hasan et al., 2017; Islam & Atkins, 2007). In general, the length of the beds is between 60 and 10 m and the breadth is between 1.25 and 4.0 m. The fully prepared beds are anchored using bamboo poles to safeguard against the currents and wind. The residuals of floating beds prepared during monsoon are usually utilized to prepare winter gardens to grow vegetables (Irfanullah et al., 2008). The preparation requires about two to three weeks' time and involves only labour as water hyacinths are readily available in all wetlands in Bangladesh except in highly saline areas (Dev, 2013). Productivity is about five times higher than in traditional land-based agriculture (Dev, 2013). The production system not only builds resilience against waterlogging but also ensures greater food security, local employment and better management of water drainage systems (Dev, 2013). To improve the sustainability of the production system, farmers need training on creating the beds, selecting the crop and on maintenance of the bed during the season. These are done with support from the government and NGOs.

12.4.1 Case III: Intervention

In this section, we will primarily highlight a project-based case assessment documented in a journal article by Irfanullah et al. (2011). The project of promoting floating agriculture was implemented by the NGO Practical Action under the funding of UKaid. The assessment of the first project was carried out among the *haor* (a wetland ecosystem situated in the north-eastern part of Bangladesh) communities from 53 villages of Kishoreganj and Sunamganj districts of Bangladesh in 2007—a year which saw major flooding (Irfanullah et al., 2011). The selected households were based near relatively stagnant water and had access to water hyacinths to make floating agricultural beds. Despite being affected by monsoon floods these communities managed to cultivate the following six vegetable seeds amaranth (*data shak*), red-amaranth (*lal shak*), bottle gourd, hyacinth bean, kang kong (*gima kolmi*) and pumpkin in 177 floating platforms while 23 platforms were damaged in the rainy season (Irfanullah et. al., 2011). Similarly, the same communities managed to cultivate diverse vegetables in the winter season in early January 2008. About 83–90% of vegetables cultivated were consumed by the participants while the rest were sold in the marketplace or distributed among neighbours or relatives. This intervention highlights that despite being affected by floods—floating agriculture in relatively stagnant *haor* water was cost effective, generated employment, ensured food and nutrition and in the best-case scenario created profitability. Similar outcomes were found later in between 2010 and 2012 when Practice Action trained and promoted floating agriculture in 700 relatively poor families in the four northern districts of Bangladesh including Gaibandha. A total of 131,600 kg of vegetables, grown on about 1,500 floating beds, helped these communities to meet local food demand at the community level and generate additional income at the household level (Irfanullah, 2013).

12.4.2   Case III: Impact

Success in floating agriculture in the coastal and waterlogged areas in Bangladesh led Bangladesh Agriculture Research Institute (BARI) to research and develop improved strategies to produce more. It has also induced the Department of Agricultural Extension to develop strategies to promote such practices as a resilience building exercise in areas suffering from waterlogging in Bangladesh. In addition, several local, national and international NGOs are supporting projects to promote floating agriculture in Bangladesh as part of climate adaptation programmes (Chowdhury & Moore, 2017).

Though the floating agricultural production system is not particularly suitable for open water or extreme flood-prone areas, the scope of scaling up of this system in wetlands in different parts of Bangladesh is immense. As this system coincides with features of climate change adaptation, international donors, especially FAO, would be interested in providing funds for promotion of *'Baira'* across waterlogged lands in Bangladesh.

At present, floating agricultural practice in Bangladesh has been recognized as a successful strategy for building resilience in waterlogged areas and so it is seen as a possible adaptation strategy against climate threats. In addition, it is also linked to several sustainable development goals as it fulfils the objective of reducing hunger and poverty, increasing food security and even empowering women. Many of the workers in such farms are women as it requires less physical labour. Little investment is required as country boats are used for collection of water hyacinth, carrying of produce and other inputs. It is also an environmentally friendly agricultural practice as it requires no pesticides and uses fewer chemical fertilizers. Vidanage et al., (2022, Chap. 15 of this volume) and Kattel and Nepal (2021, Chap. 11 of this volume) have also showed examples of communities-level mobilization to promote adoption of climate resilient technologies in rural areas.

12.5 Conclusions

This study gives three examples from three communities in rural Bangladesh. In the first case, a local ecosystem was threatened by an influx of refugees from Myanmar where the community was outnumbered by 4:1 because of their entry. The community welcomed them in open arms and provided shelters in their locality. However, the pressure on the ecosystem was enormous. It led to depletion of forest cover, exposed the community to storms and cyclones, dried up the streams and increased human–wildlife conflicts. Recognizing the threat, the international community which provided support to settle the refugees has developed alternative fuel supply. As a result, a supply chain of LPG was created and led to 80% reduction in use of fuel-wood compared to the pre-influx level. This happened through a market mechanism and shows how markets can be used to develop resilience in communities and also to protect ecosystems.

The second case is about riverbank erosion. This has been a threat for communities who are living on riverbanks. With climate change the threat is likely to increase. Governments are spending millions to build structures to stop erosion. However, communities in remote areas often live outside the radar of government policymakers. This case shows how a small community used local knowledge to build simple structures to reduce bank erosion. The knowledge that existed in the local communities was centuries old and yet very effective but required community participation. It is a shining example of low cost, low carbon and a green solution to a huge problem.

The third case is about floating agriculture. It shows how local farmers, using a small amount of support from government and NGOs developed the knowhow to overcome the challenge of waterlogging. The small technical support from NGOs and the government worked like magic to build resilience in a community which was suffering from long-term waterlogging.

References

Anik, S. I., & Khan, M. A. S. A. (2012). Climate change adaptation through local knowledge in the north eastern region of Bangladesh. *Mitigation and Adaptation Strategies for Global Change, 17*(8), 879–896.

Bangladesh Bureau Statistics. (2015). *District statistics 2011—Cox's bazar. Census 2011.* Bangladesh Bureau of Statistics.

Bangladesh Bureau of Statistics. (2016). *Bangladesh disaster-related statistics 2015: Climate change and natural disaster perspectives.* Ministry of Planning. Government of People's Republic of Bangladesh.

Banholzer, S., Kossin, J., & Donner, S. (2014). The impact of climate change on natural disasters. In A. Singh, & Z. Zommers (Eds.), *Reducing disaster: Early warning systems for climate change.* Springer, Dordrecht. https://doi.org/10.1007/978-94-017-8598-3_2. https://link.springer.com/chapter/10.1007/978-94-017-8598-3_2

Bhuiyan, M. A. H., Islam, S. M. D. U., & Azam, G. (2017). Exploring impacts and livelihood vulnerability of riverbank erosion hazard among rural household along the river Padma of Bangladesh. *Environmental Systems Research, 6*(1), 1–15.

Chowdhury, R. B., & Moore, G. A. (2017). Floating agriculture: A potential cleaner production technique for climate change adaptation and sustainable community development in Bangladesh. *Journal of Cleaner Production., 150*, 371–389.

Dasgupta, S., Kamal, F. A., Khan, Z. H., Choudhury, S., & Nishat, A. (2014). *River salinity and climate change: Evidence from Coastal Bangladesh.* Policy Research Working Papers: The World Bank Group.

Dev, P. K. (2013). Waterlogging through soil-less agriculture as a climate resilient adaptation option. In W. Leal Filho (Ed.), *Climate change and disaster risk management. Climate change management.* Springer.

Forest Department of Bangladesh. (2017). *District-wise forest land. forest department.* Government of the People's Republic of Bangladesh.

Haque, C. E. (1988). Human adjustments to riverbank erosion hazard in the Jamuna floodplain, Bangladesh. *Human Ecology, 16*(4), 421–437.

Hasan, S., Mohammad, A., Ghosh, M., & Khalil, M. (2017). Assessing of farmers' opinion towards floating agriculture as a means of cleaner production: A case of Barisal District, Bangladesh. *British Journal of Applied Science and Technology., 20*(6), 1–14.

Huq, S., Ali, S. I., & Rahman, A. A. (1995). Sea-level rise and Bangladesh: A preliminary analysis. *Journal of Coastal Research* 44–53.

Iftekhar, M. S. (2006). Forestry in Bangladesh: An overview. *Journal of Forestry, 104*(3), 148–153.

Iftekhar, M. S., & Hoque, A. K. F. (2005). Causes of forest encroachment: An analysis of Bangladesh. *Geo Journal, 62*(1–2), 95–106.

Inter Sector Coordination Group, Energy and Environment, & Government of Bangladesh. (2018). *LPG project: RRRC requirements.* Humanitarian Response. Cox's Bazar.

Inter Sector Coordination Group. (2019). *2019 joint response plan for Rohingya humanitarian crisis—January to December.* Inter Sector Coordination Group: Cox's Bazar.

Intergovernmental Panel on Climate Change. (2007). *Climate change 2007: Synthesis report.* In R. K. Pachauri & A. Reisinger (Eds.), *Contribution of working groups I, II and III to the fourth assessment report of the intergovernmental panel on climate change. Core writing team.* IPCC.

International Union for Conservation of Nature and Natural Resources. (2019). Impact of LPG distribution among the Rohingya and Host communities of Cox's Bazar South Forest division on forest resources. UNHCR Assessment Report 73248. United Nations High Commissioner for Refugees.

Irfanullah, H. M. (2013). Floating gardening: A local lad becoming a climate celebrity? *Policy Brief: Practical Action.* https://infohub.practicalaction.org/bitstream/handle/11283/366224/51c01544-ce1c-498a-9c6c-111c0a000075.pdf?sequence=1. Accessed May 21, 2020.

Irfanullah, H. M., Adrika, A., Ghani, A., Khan, Z. A., & Rashid, M. A. (2008). Introduction of floating gardening in the north-eastern wetlands of Bangladesh for nutritional security and sustainable livelihood. *Renewable Agriculture and Food Systems, 23*(2), 89–96.

Irfanullah, H. M., Azad, M. A. K., Kamruzzaman, M., & Wahed, M. A. (2011). Floating gardening in Bangladesh: A means to rebuild lives after devastating flood. *Indian Journal of Traditional Knowledge, 10*(1), 31–38.

Islam, T., & Atkins, P. (2007). Indigenous floating cultivation: A sustainable agricultural practice in the wetlands of Bangladesh. *Development in Practice, 17*(1), 130–136.

Kattel, R. R., & Nepal, M. (2021). Rainwater harvesting and rural livelihoods in Nepal. In A. K. E. Haque, P. Mukhopadhyay, M. Nepal, & M. R. Shammin (Eds.), *Climate change and community resilience: Insights from South Asia* (pp. 159–173). Springer.

Kharin, V. V., Zwiers, F. W., Zhang, X., & Hegerl, G. C. (2007). Changes in temperature and precipitation extremes in the IPCC ensemble of global coupled model simulations. *Journal of Climate, 20*(8), 1419–1444.

Ministry of Agriculture. (2017). *Floating garden agricultural practices in Bangladesh: A proposal for globally important agricultural heritage systems (GIAHS).* Food and Agriculture Organization. http://www.fao.org/publications/card/en/c/4c57fc69-e538-46e0-969c-d197be 845a5f/. Accessed May 21, 2020.

Nakagawa, H., Hiroshi, T., Kenji, K., Yasuyuki, B., & Hao, Z. (2011). Analysis of bed variation around bandal-like structures. *Annuals of Disaster Prevention Research Institute, Kyoto University, 54(B)*, 497–510.

Noss, R. F. (2001). Beyond Kyoto: Forest management in a time of rapid climate change. *Conservation Biology, 15*(3), 578–590.

Rahman, M. L. (2019). Study of Bamboo bandalling structures in the tidal river for riverbank erosion. In K. Murali, V. Sriram, A. Samad, & N. Saha (Eds.), *Proceedings of the fourth international conference in ocean engineering. Lecture notes in civil engineering* (vol. 23, pp. 49–57). Springer.

Rahman, A. (1995). *Beel Dakatia: The environmental consequences of a development disaster.* University Press.

Rahman, L., & Osman, S. (2015). Riverbank erosion protection using bamboo bandalling structure: A case study. *Journal of Civil Engineering, 43*(1), 1–8.

River Research Institute. (2018). *Annual Report 2017–18* (Serial no. 45/17–18). River Research Institute.

Rounsevell, M. D. A., Evans, S. P., & Bullock, P. (1999). Climate change and agricultural soils: Impacts and adaptation. *Climatic Change, 43*(4), 683–709.

Salam, M. A., Noguchi, T., & Koike, M. (1999). The causes of forest cover loss in the hill forests in Bangladesh. *Geo Journal, 47*(4), 539–549.

Shishir, B. H. (2020, July 30). Bandal can transform the life of river and its people. *Oxfam Blogs.* August 9, 2020. https://oxfamblogs.org/bangladesh/bandal-trasnfrom-life-river-people/19

Siddique, A. (2020). Locals use nature-based solution to tame river in Bangladesh. *The Third Pole.* June 30, 2020. https://www.thethirdpole.net/2020/06/30/nature-based-solutions-bangladesh/. Accessed December 19, 2020.

Smith, K., & Ward, R. (1998). *Floods—Physical processes and human impacts.* Wiley.

Sultana, P., Johnson, C., & Thompson, P. (2008). The impact of major floods on flood risk policy evolution: Insights from Bangladesh. *International Journal of River Basin Management, 6*(4), 339–348.

Thomas, L., Balakrishna, R., Chaturvedi, R., Mukhopadhyay, P., & Ghate, R. (2022). What influences rural poor in india to refill their LPG? In A. K. E. Haque, P. Mukhopadhyay, M. Nepal, & M. R. Shammin (Eds.), *Climate change and community resilience: Insights from South Asia* (pp. 191–203). Springer.

United Nations High Commissioner for Refugees. (2020). *Population Arrivals after 25 August 2017.* Operational Portal of Refugee Situations.

Vidanage, S. P., Kotagama, H. B., & Dunusinghe, P. M. (2022). Sri Lanka's small tank cascade systems: Building agricultural resilience in the dry zone. In A. K. E. Haque, P. Mukhopadhyay, M. Nepal, & M. R. Shammin (Eds.), *Climate change and community resilience: Insights from South Asia* (pp. 225–235). Springer.

Younus, M. A. F., & Harvey, N. (2013). Community-based flood vulnerability and adaptation assessment: A case study from Bangladesh. *Journal of Environmental Assessment Policy and Management, 15*(03), 1350010. https://www.worldscientific.com/doi/abs/10.1142/S1464333213500105

Chapter 13
What Influences Rural Poor in India to Refill Their LPG?

Liya Thomas, Raksha Balakrishna, Rahul Chaturvedi, Pranab Mukhopadhyay, and Rucha Ghate

Key Messages

- Rural income generation schemes, female literacy, positively influence LPG refills.
- While male work force participation increases LPG refills, female workforce does not.
- Vicinity to forest has heterogeneous effects depending on type of forest.

13.1 Introduction

Under the Nationally Determined Contributions (NDC) in the Paris Agreement (2015) India has committed to reduce emission intensity by 33–35%; increase the share of non-fossil-based energy to 40%; and improve its forest and tree cover to create an additional carbon sink of 2.5–3 GT-CO_2e (UNFCCC, 2018). Meeting these

L. Thomas · R. Balakrishna · R. Chaturvedi · R. Ghate
Foundation for Ecological Security, Anand, India

L. Thomas
e-mail: liyabensy@gmail.com

R. Balakrishna
e-mail: raksha.balakrishna@gmail.com

R. Chaturvedi
e-mail: rahul.chaturvedi@fes.org.in

R. Ghate
e-mail: rucha@fes.org.in

P. Mukhopadhyay (✉)
Goa Business School, Goa University, Taleigao Plateau, Goa, India
e-mail: pm@unigoa.ac.in

© The Author(s) 2022
Haque et al. (eds.), *Climate Change and Community Resilience*,
https://doi.org/10.1007/978-981-16-0680-9_13

carbon-mitigation commitments requires the adoption of cleaner and more efficient alternatives. Liquefied petroleum gas (LPG) and natural gas have been internationally recommended as a mitigation measure to reduce black carbon emissions (IPCC, 2018). A push towards cleaner cooking technologies like LPG would help in achieving targets under five of the 17 SDGs, namely SDG 3—Good health and well-being; SDG 5—Gender equality; SDG 7—Affordable and clean energy; SDG 13—Climate action and SDG 15—Life on land (Rosenthal et al., 2018). This chapter examines the impact of rural employment generation programmes along with various socio-economic and local environmental factors on LPG use.

LPG is a naturally occurring, unavoidable by-product of oil and natural gas extraction and crude oil refining. Earlier, LPG was vented or flared at sites, wasting valuable fuel and spewing black carbon into the atmosphere (Van Leeuwen et al., 2017). Utilizing it instead has been recognized as beneficial for both environment and human health in comparison with alternatives such as solid biomass fuels as it releases lower levels of black carbon and methane (Bruce et al., 2017).

A 2016 report states that as many as 819 million people (nearly 60% of the population) in India use traditional biomass such as fuelwood, cow dung, and coal, for their daily cooking needs, sourced primarily from nearby forests and wooded areas (IEA, 2016). Widespread use of these fuels poses serious risk to both human and environmental health (Junaid et al., 2018). Incomplete combustion of the fuels on inefficient stoves, and other devices used for cooking, lighting and heating, leads to household air pollution (HAP). High levels of HAP include health-damaging pollutants such as fine particles and carbon monoxide and contribute to about 4–6% of the burden of disease in India (Smith, 2000). Since women and children spend most time at home, they are the most adversely affected (Kankaria et al., 2014; Smith & Sagar, 2014). Mitigating the ill-effects of HAP is crucial not just to achieve targets of improved health (SDG 3) but also gender equality (SDG 5). In addition, shifting to cleaner fuels like LPG reduces the burden of fuel wood collection and reduces cooking time, thus allowing for empowerment of women (Rosenthal et al., 2018). Studies have estimated that HAP contributes to between 22 and 52% of ambient PM2.5 exposure in India also adding to the climate crisis (Conibear et al., 2018).

Burning fuelwood emits climate pollutants such as black carbon, methane, carbon monoxide and other ozone-depleting gases. In South Asia, over half of black carbon comes from cook stoves, disrupting the monsoon and expediting the Himalayan–Tibetan glacier melting (Chung et al., 2012). In rural areas of developing countries, emission from biomass-based cooking alone was 49.0 $GtCO_2$-eq (recorded in 2004) (IPCC, 2007). Though LPG has been criticized as a fossil fuel, till such time as there are renewable alternatives, LPG could be promoted as the available cleaner solution with the potential of reducing emissions from 49.0 $GtCO_2$-eq to 0.70. This would directly help meet NDC commitments of reduced emission and targets under SDG 3 (Good health and well-being) and SDG 13 (Climate action).

In addition, fuelwood extraction for fuel and energy is also a major contributor to deforestation and threatens the health of forests and other wooded areas. Global estimates indicate that about 30% of wood fuel harvesting is unsustainable (Bailis et al., 2015). In 2010–2011, the annual fuelwood consumption by India was 216.4

million tonnes per year (FSI, 2011). By protecting forests from fuelwood and charcoal extraction, LPG use could reduce the pressure on local resources and thereby enable carbon sequestration.

13.1.1 Policy Evolution Towards Cleaner Cooking: LPG

The rural poor in South Asia are heavily dependent on natural resources and thus directly influenced by extreme weather events (IPCC, 2014). In the wake of the warming temperature and decreasing precipitation, studies have projected an increased risk of climate disasters in India (Bisht et al., 2019). At the household level, this would translate to reduced availability of food, fodder, water and fuelwood in the short term and ecological and socio-economic consequences in the long term. When faced with such shortages, disadvantaged groups are likely to be most affected. In this context, adapting and promoting innovative cleaner energy sources such as LPG could potentially increase the resilience of rural communities to changing climate. Shamin and Haque (2022, Chap. 14 this volume) examine a similar question with respect to the adoption of solar systems in Bangladesh.

Realizing this, India has made many attempts to introduce improved cooking technologies that provide "triple benefits"—reduction in HAP and time-saving for households (health benefit), reduction in forest dependence (local environmental benefit) and reduced emission of carbon (global benefit) (Bhojvaid et al., 2014; Jeuland & Pattanayak, 2012). Since 2009, the government has attempted to promote the use of LPG as a fuel choice for households in remote and rural areas.

Starting with the Rajiv Gandhi Gramin LPG Vitarak Yojana (RGGLVY) (Sankhyayan & Dasgupta, 2019), the scheme evolved into the Pradhan Mantri Ujjwala Yojana (PMUY) in May 2016. This intervention aimed at bringing the benefit of efficient and low-emission fuel options to households that could not afford it because of their income status (Dabadge et al., 2018). The initial aim was to provide 50 million women belonging to poor (below the poverty line, BPL) families with gas (liquefied petroleum gas, LPG) connections. The scheme aimed to provide financial support for new LPG connection (installation).

Apart from RGGLVY and PMUY, the government has introduced other schemes like Pahal and complementing campaigns like "Give it Up" that have been crucial in ensuring that subsidies for LPG reach those who need them most (Gould & Urpelainen, 2018). While Pahal Consumers Scheme, launched in June 2013 aimed at directly transferring LPG subsidies to the bank accounts of consumers, the 2015 "Give it Up" scheme focused on motivating LPG consumers who can afford to pay full price for the cylinders to give up the LPG subsidy voluntarily.

Over the past decade, there has been steady progress towards the adoption of clean fuels in India. The number of LPG connections in the country has more than doubled, from 106 million households in 2009 to 263 million in 2018; total household consumption of LPG has increased from 10.6 million tonnes to 20.4 million tonnes during the same period (PPAC, 2018). With a push towards the adoption and use of cleaner cooking fuels, nearly 90% of Indian households now have LPG connections, making it the world's second-largest consumer of LPG (PPAC, 2019). However, sustained use of this fuel remains a challenge (Kar et al., 2019).

13.1.2 Factors Limiting Sustained Use of LPG

While there is general acceptance that the adoption of cleaner fuels like LPG has the potential to deliver health, social and environmental benefits including positive climate impacts in the short term, there has been mixed success on their sustained use despite state-subsidized efforts (Bruce et al., 2017; Rosenthal et al., 2018). Earlier studies suggest that there is a wide heterogeneity of factors influencing its use (Jain et al., 2018; Kumar et al., 2017; Singh et al., 2017). This includes; price (Sankhyayan & Dasgupta, 2019), women's participation in household decision-making (Gould & Urpelainen, 2018), seasonality (Kar et al., 2019) and household characteristics like house type and household size, and ease of access (Giri & Aadil, 2018).

Households with irregular income and easy accessibility to biomass fuel are less likely to use LPG for all their cooking needs (Mani et al., 2020). Forested areas and shared land resources in and around villages have been the primary source of this fuelwood (Pandey, 2002). Households that have traditionally depended on fuelwood for cooking purposes continue to do so, especially for heating water and large-scale cooking. In rural areas, the annual average fuelwood consumption per capita was estimated at 796 kg (Pandey, 2002). With continued population growth, demand for fuelwood is only likely to grow in the future resulting in the degradation of the forests in the vicinity of villages and the formation of barren lands. With improved access to LPG connections, households have started practising fuel stacking, wherein they stack both traditional biomasses such as fuelwood along with LPG, to meet requirements. However, in Bangladesh, Bari, Haque and Khan (2022, Chap. 14 of this volume) found that better supply of LPG reduced forest dependence of rural migrant communities.

There is a recognized need for a policy push to offset the use of biomass fuel by cleaner cooking technologies such as LPG. This shift could help India to meet NDC commitments as well as five of the Sustainable Development Goals, 2030.

13.1.3 MGNREGA a "Window of Opportunity" to Improve LPG Use?

Affordability has been recognized as one of the crucial barriers in LPG use (Khandker et al., 2012). This can be ensured either by making money available to rural households through more work and better wages or by extending higher subsidies. In the long run, increasing the disposable income of rural households to buy refills is more sustainable than providing subsidies. We examine the potential of the Mahatma Gandhi National Rural Employment Guarantee Act (MGNREGA), which guarantees 100 days of wage employment per year to rural households, in influencing LPG use in India. We expect that a district that has a higher per capita MGNREGA expenditure presumably has employed more people and/or for longer days and hence gives the rural poor of that district a better income status. It is contended that by utilizing the otherwise untapped labour potential in rural areas, the programme effectively increases the purchasing power of rural households.

While testing this expectation, the chapter also examines other socio-economic and environmental factors that could influence LPG use in India. LPG use can be inferred not from the number of connections but from the frequency of refills. We, therefore, test the relationship between the frequency of refills and various socio-economic and environmental factors.

In rural areas, households still primarily depend on rain-fed agriculture. Therefore, rainfall in a district would strongly predict the agricultural income of a region, *ceteris paribus* (Gadgil & Gadgil, 2006; Krishna Kumar et al., 2004; University of East Anglia Climatic Research Unit (CRU) et al., 2019). Supply and cost drive fuel choice, i.e. village communities who live in the proximity of forests are likely to choose fuelwood over LPG as the relative shadow price of fuelwood is much lower than LPG. The economic status of households would be reflected by the extent of poverty in the district. While poverty rates are a direct way to understand the income distribution of a region, the economic well-being can also be gauged by the participation of the population in the workforce. This would directly indicate income generation opportunities—we expect that the higher the workforce participation rate in a district, the better off the households of that district due to available income from employment.

Given the demographic structure of Indian societies, women's empowerment through education could have significant implications for family decision-making (Sen, 2000) which includes decisions on expenditure on fuel and women's health. Education is a known tool for empowerment within and outside the household (Walker, 2005). People (especially women) of a more literate society are likely to choose cleaner fuel even if it costs more as they would value their health and make more informed choices. Economic deprivation in India is closely linked to social categories. Scheduled Tribes (ST) and Scheduled Castes (SC) have, for long, been known to be historically deprived (Deshpande, 2011). We, therefore, use SC and ST proportions in rural populations to understand the extent of deprivation at the district level.

13.2   Material and Methods

We have used data available from government sources on LPG connections and refills (PMUY, 2018); MGNREGA expenditure for the year 2017–2018; forest survey data (FSI, 2019); district-level rainfall data (IMD, 2015, 2016, 2017, 2018; University Of East Anglia Climatic Research Unit (CRU) et al., 2019); poverty data (Chaudhuri &

Gupta, 2009); and demographic data (Census, 2011). After matching LPG data with all the above data, we were left with complete data for 582 districts across 29 states and three union territories.

We use a formal regression model for our analysis. Our dependent variable is the proportion of LPG refills to the number of LPG connections registered under the PMUY scheme in each district. This we have treated as an indicator of LPG adoption. We anticipate, as stated above, that this would be dependent on multiple factors.

$$
\begin{aligned}
\text{LPG refills in 2019} = f\ (\text{amount of per capita expenditure per capita} \\
\text{on MGNREGA in 2018 [preceding year], the extent} \\
\text{of rainfall in 2018 [preceding year], rural female} \\
\text{literacy rate, the proportion of SC and ST} \\
\text{in rural areas, the proportion of the rural} \\
\text{population in the workforce, percentage of poor} \\
\text{in rural districts and extent of different types of forests})
\end{aligned}
\tag{13.1}
$$

The specific model using the ordinary least squares (OLS) multiple regression method is discussed below.

$$
\begin{aligned}
Y_i = \beta_0 + \beta_1 X_{1i} + \beta_2 X_{2i} + \beta_3 X_{3i} + \beta_4 X_{4i} + \beta_5 X_{5i} \\
+ \beta_6 X_{6i} + \beta_7 X_{7i} + \beta_8 X_{8i} + \beta_9 X_{9i} + \beta_{10} X_{10i} \\
+ \beta_{11} X_{11i} + \beta_{12} X_{12i} + \beta_{13} X_{13i} + \beta_{14} X_{14i} + \varepsilon_i
\end{aligned}
\tag{13.2}
$$

where

Y = Proportion of refills four times from among those who got LPG connection under PMUY.

X_1 = MGNREGA expenditure per capita (ratio of MGNREGA expenditure to state population).

X_2 = Square of MGNREGA expenditure per capita (X_1).

X_3 = Total rainfall in 2018 (in millimetre).

X_4 = Square of total rainfall in 2018.

X_5 = Rural female literacy rate.

X_6 = Proportion of ST in rural population.

X_7 = Proportion of SC in the rural population.

X_8 = Female workforce participation rate.

X_9 = Male workforce participation rate.

X_{10} = Per cent of rural population under the poverty line.

X_{11} = Area under very dense forest (in hectare).

X_{12} = Area under moderate dense forest (in hectare).

X_{13} = Area of open forest (in hectare).

X_{14} = Area under scrub (in hectare).

ε_1 = Stochastic error.

We use Stata 15.1 *"regress"* command to estimate OLS results (see Table 13.2) and the post-estimation commands to confirm that the data fulfils the OLS assumptions to validate our estimated coefficients. We conducted three post-estimation tests for (1) normality, (2) heteroskedasticity and (3) influential observations. We found that for all three tests, the null hypothesis of normality, homoscedasticity and non-influential observations holds.

(1) Heteroskedasticity: We did a Breusch–Pagan test which has a chi-square value of 1.879 (with p-value: 0.170).

(2) Normality of residuals: We did a Shapiro–Wilk W normality test which has a "z" value of 1.242 (with a p-value: 0.107).

(3) Influential observations: We did a Cook's distance test, which is less than 1.00, and there is no distance which is above the cut-off.

While most of these variables are used commonly as independent variables, the case of forests is not self-evident. LPG adoption is expected to reduce demand for wood fuel and therefore forest dependence. There are two points to be noted here. First, many researchers have noted that fuelwood use does not reduce the density and canopy cover of trees. The fuelwood demand for forest-dependent communities is met by loping of lower branches and dry wood. Second, the impact of fuel wood collection on forest quality is not necessarily dependent on access to the forest or the density of forest-dependent population but on the availability of wage labour and local markets (Davidar et al., 2010). Third, LPG adoption is unlikely to show results in the very short run and is more of a long-term intervention.

13.3 Results

The summary statistics of the above variables is presented (in Table 13.1) below. Our findings indicate that around 48% of all those who got an LPG connection reported refilling the LPG four times a year ranging from a low of 6% to a high of 92%. The distribution nearly approximates a normal distribution (see Supplementary information, Graph S.G1). The average per capita expenditure on MGNREGA in 2017 was INR754. The reported average rainfall in 2018 in India was 1103 mm, ranging from 804 to a high of 5065 mm. The wide range in rainfall is a reminder of the 15 agro-climatic zones in the country. The heterogeneity is not just in geography, but also in social characteristics. Female literacy on average was 55% and varied between a low of 12% to a high of 89%. The districts differed in terms of marginalized groups (SC and ST populations). While the average ST population was 19% (minimum 0 to a maximum of 99%), the SC population on average was 16%, with a smaller range of 1–53%. The female workforce participation, which is an indicator of the presence of women in the paid workforce, had a national average of 32% (from a low of 5% to a high of 65%).

Table. 13.1 Summary statistics of variables

Variable	Unit	Obs	Mean	Std. dev	Min	Max
Proportion of PMUY beneficiaries who refilled four times	Number	610	47.74	17.67	5.92	91.53
MNREGA expenditure per capita	Rs. lakh (INR)	597	0.008	0.01	0	0.11
Rainfall in 2018	Mm	577	1103.35	804.67	0	5065.9
Female literacy rate	%	633	0.55	0.12	0.24	0.89
Proportion of ST in rural population	Number	624	0.19	0.28	0	0.99
Proportion of SC in rural population	Number	624	0.16	0.1	0	0.53
Female workforce participation rate (rural)	%	624	32	13	5	65
Rural poverty rate	%	509	28.25	19.71	0	88.4
Very dense forest	Ha	634	156.31	407.51	0	4699.29
Moderately dense forest	Ha	634	483.44	742.04	0	5881.18
Scrub forest	Ha	633	72.52	159.91	0.26	1520.19
Open forest	Ha	634	475.23	522.35	0	3538.63

Source Authors' calculations from multiple sources

Variables influencing these refills were—MGNREGA expenditure, rainfall, female literacy, the proportion of SC and ST populations, female workforce participation rate, percentage of poor in a district, extent of density of forest, prevalence of open and scrub forests (see Table 13.2). Female literacy rate, the proportion of SC population, as well as prevalence of very dense forest and scrub forest, influence the refills positively. On the other hand, the proportion of the ST population, female workforce participation and percentage of poor and open forest negatively impact refills. The negative relation with female workforce participation may seem odd because the greater this value the more likely it is to have greater family income, and therefore potentially a cause for LPG adaption. However, we are aware that the official statistics on female workforce participation may be underreporting the value. A large proportion in the female rural workforce may not be part of the paid workforce but participate in productive activity. This could be a possible reason for this result.

There is a nonlinear relationship between refills and its two determinants, MGNREGA and rainfall (U-shaped). Both of these variables influence the income of rural households. An initial increase in MGNREGA expenditure or rainfall reduces refills. However, as these values—MGNREGA expenditure or rainfall (below a calamity level)—rise, the increased household income positively impacts on refills after a threshold level. It comes as no surprise that refills are low in areas of high poverty.

Table. 13.2 Results of ordinary least squares regression (robust standard errors)

Dependent Variable: Proportion of PMUY beneficiaries who refilled four times	
Independent variables	*Coefficient (t-value)*
MNREGA expenditure per capita	-2677.2 *** (-7.03)
Square of MNREGA expenditure per capita	103,518.7 *** (-5.76)
Total rainfall in 2018	-0.0094 *** (-3.66)
Square of total rainfall in 2018	0.0000017** (2.75)
Female literacy rate (rural)	13.34* (1.85)
Scheduled Caste population (rural)	12.75* (1.65)
Scheduled Tribe population (rural)	-20.79*** (-4.84)
Workforce participation (male, rural)	1.17 (0.07)
Workforce participation (female, rural)	-13.23** (-1.99)
Very dense forest in 2019	0.005* (2.41)
Moderately dense forest in 2019	-0.0004 (-0.29)
Open forest in 2019	-0.004** (-2.24)
Scrub area in 2019	0.008** (1.95)
Constant	65.2*** (9.67)
N	445
R-square	0.4939
adjusted *R*-square	0.4775
F (14, 430)	29.98
Prob > *F*	0.0000
*p < 0.05, **p < 0.01, ***p < 0.001	

Source Authors' calculations

We have four measures of forest types—very dense, moderately dense, scrub and open. Refills are higher in areas with very dense forest and scrub areas. The reasons for this could be that in very dense forests fuelwood collection would be difficult and regulated by the forest department. Therefore, there is a higher adoption of alternate fuels. In scrub areas, there is a lower availability of fuelwood which, again, leads to a higher number of refills. However, in open forests, there is scope for fuelwood availability and so refills are less frequently observed.

Information on forest category indicates higher instances of refills in dense and scrub forests for reasons given above, the relationship with moderately dense shows as insignificant. In moderately dense forests, state monitoring is relatively less. There is also relatively greater availability of fallen branches and dry wood. Harvesting of fuelwood from this category of forests can be significant. However, this relationship needs to be more closely studied. Vidanage et al., (2022, Chap. 15 of this volume) and Devi et al., (2022, Chap. 8 of this volume) have shown that state support for programmes could elevate outcomes to being more sustainable.

13.4 Conclusion and Policy Implications

Today, about three-fifths of India's households rely on fuelwood and other solid fuels. Continuing these consumption patterns could lead to significant environmental impacts, especially considering India's high population growth and increasing fuelwood extraction. India will need to move away from fossil fuels gradually to meet sustainable development targets and carbon-mitigation targets. Increasing household LPG use is one of the several pathways to achieve this.

However, projections of the International Energy Outlook report suggest that in 2030, 580 million people in India will still be using traditional fuels and India would then fall short of its target under SDG 7 (Affordable and Clean Energy) (IEA, 2017). This is despite government efforts to improve access to subsidized connections through various schemes. The main reasons cited for this gap in meeting targets are poor implementation, supply shortage and lower affordability. While two of these issues need to be addressed from the supply side, this chapter focused on the push needed from the demand side to improve LPG uptake.

Our analysis indicates that poorer households are more likely to switch fuels if their disposable income increases through employment generation schemes. The expenditure on MGNREGA is a policy-determined variable, and the decision-makers could ensure a win–win situation of triggering the triple benefits of reduced household air pollution, reduction in forest dependence and reduced emission of carbon, in turn promoting affordable and clean energy (SDG 7) for rural households.

Switching to cleaner cooking fuels such as LPG has the potential to deliver extensive health, social and environmental benefits, including positively affecting climate in the short term (Bruce et al., 2017; Rosenthal et al., 2018; Singh et al., 2017). It can further support achieving a few of the targets under SDG. Since India is committed to achieving the Sustainable Development Goals by balancing economic, social and environmental goals, the wide use of LPG would be a small but sure step towards achieving these objectives.

References

Bailis, R., Drigo, R., Ghilardi, A., & Masera, O. (2015). The carbon footprint of traditional woodfuels. *Nature Climate Change, 5*(3), 266–272. https://doi.org/10.1038/nclimate2491

Bari, E., Haque, A. K. E., & Khan, Z. K. (2022). Local strategies to build climate resilient communities in Bangladesh. In A. K. E. Haque, P. Mukhopadhyay, M. Nepal, & M. R. Shammin (Eds.), *Climate change and community resilience: Insights from South Asia* (pp. 175–189). Springer.

Bhojvaid, V., Jeuland, M., Kar, A., Lewis, J., Pattanayak, S., Ramanathan, N., Ramanathan, V., & Rehman, I. (2014). How do people in Rural India perceive improved stoves and clean fuel? Evidence from Uttar Pradesh and Uttarakhand. *International Journal of Environmental Research and Public Health, 11*(2), 1341–1358. https://doi.org/10.3390/ijerph110201341

Bisht, D. S., Sridhar, V., Mishra, A., Chatterjee, C., & Raghuwanshi, N. S. (2019). Drought characterization over India under projected climate scenario. *International Journal of Climatology, 39*(4), 1889–1911. https://doi.org/10.1002/joc.5922

Bruce, N. G., Aunan, K., & Rehfuess, E. A. (2017). *Liquefied petroleum gas as a clean cooking fuel for developing countries: Implications for climate, forests, and affordability* (Materials on development financing No. 7, pp. 44). KfW Development Bank. https://static1.squarespace.com/static/53856e1ee4b00c6f1fc1f602/t/5b16ec08352f538a85f57d7c/1528228877332/2017_Liquid-Petroleum-Clean-Cooking_KfW.pdf

Census (2011). Primary census abstracts, Registrar General of India, Ministry of Home Affairs, *Government of India.* http://www.censusindia.gov.in

Chaudhuri, S., & Gupta, N. (2009). Levels of living and poverty patterns: A district-wise analysis for India. *Economic and Political Weekly, XLIV*(9), 94–110.

Chung, C. E., Ramanathan, V., & Decremer, D. (2012). Observationally constrained estimates of carbonaceous aerosol radiative forcing. *Proceedings of the National Academy of Sciences, 109*(29), 11624–11629. https://doi.org/10.1073/pnas.1203707109

Conibear, L., Butt, E. W., Knote, C., Arnold, S. R., & Spracklen, D. V. (2018). Residential energy use emissions dominate health impacts from exposure to ambient particulate matter in India. *Nature Communications, 9*(1). https://doi.org/10.1038/s41467-018-02986-7

Dabadge, A., Sreenivas, A., & Josey, A. (2018). What has the Pradhan Mantri Ujjwala Yojana achieved so far? *Economic and Political Weekly, 53*(20), 7–8. https://www.epw.in/journal/2018/20/notes/what-has-pradhan-mantri-ujjwala-yojana-achieved-so-far.html

Davidar, P., Sahoo, S., Mammen, P. C., Acharya, P., Puyravaud, J.-P., Arjunan, M., Garrigues, J. P., & Roessingh, K. (2010). Assessing the extent and causes of forest degradation in India: Where do we stand? *Biological Conservation, 143*(12), 2937–2944. https://doi.org/10.1016/j.biocon.2010.04.032

Deshpande, A. (2011). *The grammar of caste: Economic discrimination in contemporary India.* Oxford University Press.

Devi, P. I., Sam, A. S., & Archana Raghavan Sathyan, A. R. (2022). Resilience to climate stresses in South India: Conservation responses and exploitative reactions. In A. K. E. Haque, P. Mukhopadhyay, M. Nepal, & M. R. Shammin (Eds.), *Climate change and community resilience: Insights from South Asia* (pp. 113–127). Springer.

FSI. (2011). *Carbon stock in India's Forests.* Forest Survey of India, Ministry of Environment and Forest. http://fsi.nic.in/carbon_stock/

FSI. (2019). *State of forest report 2019.* Forest Survey of India, Ministry of Environment, Forests and Climate Change.

Giri, A., & Aadil, A. (2018). *Pradhan Mantri UjjwalaYojana:A demand-side diagnosticstudy of LPG refills (Policy Brief).* Microsave Consulting Services. http://www.microsave.net/wp-content/uploads/2018/11/Pradhan_Mantri_Ujjwala_Yojana_A_demand_side_diagnostic.pdf

Gadgil, S., & Gadgil, S. (2006). The Indian Monsoon, GDP and Agriculture. *Economic and Political Weekly, 41*(47), 4887–4895. JSTOR. https://www.jstor.org/stable/4418949

Gould, C. F., & Urpelainen, J. (2018). LPG as a clean cooking fuel: Adoption, use, and impact in rural India. *Energy Policy, 122*, 395–408. https://doi.org/10.1016/j.enpol.2018.07.042

IEA. (2016). *World energy outlook 2016*. IEA. https://www.iea.org/reports/world-energy-outlook-2016

IEA. (2017). *World Energy Outlook 2017*. IEA, Paris. https://www.iea.org/reports/world-energy-outlook-2017

IMD. (2015). *All India district rainfall statistics: India meteorological department*. Government of India. https://mausam.imd.gov.in/imd_latest/contents/rainfallinformation.php

IMD. (2016). *All India District Rainfall Statistics: India Meteorological Department*. Government of India. https://mausam.imd.gov.in/imd_latest/contents/rainfallinformation.php

IMD. (2017). *All India District Rainfall Statistics: India Meteorological Department*. Government of India. https://mausam.imd.gov.in/imd_latest/contents/rainfallinformation.php

IMD. (2018). *All India district rainfall statistics: India meteorological department*. Government of India. https://mausam.imd.gov.in/imd_latest/contents/rainfallinformation.php

IPCC. (2007). *Climate change 2007: Synthesis report*. Intergovernmental Panel on Climate Change.

IPCC. (2014). *The IPCC's fifth assessment report what's in it for South Asia?* Overseas Development Institute and Climate and Development Knowledge Network. https://cdkn.org/wp-content/uploads/2014/04/IPCC_AR5_CDKN_Whats_in_it_for_South_Asia_FULL.pdf

IPCC. (2018). *Global warming of 1.5°C* (E-edition). Intergovernmental Panel on Climate Change. http://www.ipcc.ch/report/sr15/

Jain, A., Tripathi, S., Mani, S., Patnaik, S., Shahidi, T., & Ganesan, K. (2018). *Access to clean cooking energy and electricity: Survey of states 2018*. CEEW, Council on Energy, Environment and Water. https://www.ceew.in/sites/default/files/CEEW-Access-to-Clean-Cooking-Energy-and-Electricity-11Jan19_0.pdf

Jeuland, M. A., & Pattanayak, S. K. (2012). Benefits and costs of improved cookstoves: Assessing the implications of variability in health, forest and climate impacts. *PLoS ONE, 7*(2), e30338. https://doi.org/10.1371/journal.pone.0030338

Kankaria, A., Nongkynrih, B., & Gupta, S. K. (2014). Indoor air pollution in India: Implications on health and its control. *Indian Journal of Community Medicine, 39*(4), 203. https://doi.org/10.4103/0970-0218.143019

Kar, A., Pachauri, S., Bailis, R., & Zerriffi, H. (2019). Using sales data to assess cooking gas adoption and the impact of India's Ujjwala programme in rural Karnataka. *Nature Energy, 4*(9), 806–814. https://doi.org/10.1038/s41560-019-0429-8

Khandker, S. R., Barnes, D. F., & Samad, H. A. (2012). Are the energy poor also income poor? Evidence from India. *Energy Policy, 47*, 1–12. https://doi.org/10.1016/j.enpol.2012.02.028

Krishna Kumar, K., Rupa Kumar, K., Ashrit, R. G., Deshpande, N. R., & Hansen, J. W. (2004). Climate impacts on Indian agriculture. *International Journal of Climatology, 24*(11), 1375–1393. https://doi.org/10.1002/joc.1081

Kumar, P., Dhand, A., Tabak, R. G., Brownson, R. C., & Yadama, G. N. (2017). Adoption and sustained use of cleaner cooking fuels in rural India: A case control study protocol to understand household, network, and organizational drivers. *Archives of Public Health, 75*(1). https://doi.org/10.1186/s13690-017-0244-2

Mani, S., Jain, A., Tripathi, S., & Gould, C. F. (2020). Sustained LPG use requires progress on broader development outcomes. *Nature Energy, 5*(6), 430–431. https://doi.org/10.1038/s41560-020-0635-4

Pandey, D. (2002). *Fuelwood studies in India: Myth and reality*. Center for International Forestry Research.

PMUY. (2018). *State-wise PMUY refill profile for the connections installed till 31.12.2018 since beginning (May, 2016) and refill upto 03.06.2019*. Pradhan Mantri Ujjwala Yojana, Ministry of Petroleum and Natural Gas Government of India. https://pmuy.gov.in/registereduser.html

PPAC. (2018). *Consumption of petroleum products*. Petroleum Planning and Analysis Cell, Ministry of Petroleum and Natural Gas, Government of India. https://www.ppac.gov.in/content/147_1_ConsumptionPetroleum.aspx

PPAC. (2019). *Energizing and empowering India: Annual report 2018–19*. Petroleum Planning and Analysis Cell, Ministry of Petroleum and Natural Gas, Government of India. http://petroleum. nic.in/sites/default/files/AR_2018-19.pdf

Rosenthal, J., Quinn, A., Grieshop, A. P., Pillarisetti, A., & Glass, R. I. (2018). Clean cooking and the SDGs: Integrated analytical approaches to guide energy interventions for health and environment goals. *Energy for Sustainable Development, 42*, 152–159. https://doi.org/10.1016/ j.esd.2017.11.003

Sankhyayan, P., & Dasgupta, S. (2019). 'Availability' and/or 'affordability': What matters in household energy access in India? *Energy Policy, 131*, 131–143. https://doi.org/10.1016/j.enpol.2019. 04.019

Sen, A. (2000). *Development as freedom*. Anchor Books.

Shammin, M. R., & Haque, A. K. E. (2022). Small-scale solar solutions for energy resilience in Bangladesh. In A. K. E. Haque, P. Mukhopadhyay, M. Nepal, & M. R. Shammin (Eds.), *Climate change and community resilience: Insights from South Asia* (pp. 205–224). Springer.

Singh, D., Pachauri, S., & Zerriffi, H. (2017). Environmental payoffs of LPG cooking in India. *Environmental Research Letters, 12*(11), 115003. https://doi.org/10.1088/1748-9326/aa909d

Smith, K. R. (2000). National burden of disease in India from indoor air pollution. *Proceedings of the National Academy of Sciences, 97*(24), 13286–13293. https://doi.org/10.1073/pnas.97.24. 13286

Smith, K. R., & Sagar, A. (2014). Making the clean available: Escaping India's Chulha trap. *Energy Policy, 75*, 410–414. https://doi.org/10.1016/j.enpol.2014.09.024

UNFCCC. (2018). *INDC—Submissions*. United Nations. http://www4.unfccc.int/Submissions/ INDC/Submission%20Pages/submissions.aspx

University Of East Anglia Climatic Research Unit (CRU), Harris, I. C., & Jones, P. D. (2019). *CRU TS4.03: Climatic research unit (CRU) time-series (TS) version 4.03 of high-resolution gridded data of month-by-month variation in climate (Jan. 1901–Dec. 2018)* [Application/xml]. Centre for Environmental Data Analysis (CEDA). https://doi.org/10.5285/10D3E3640F004C5784034 19AAC167D82

Van Leeuwen, R., Evans, A., & Hyseni, B. (2017). *Increasing the use of liquefied petroleum gas in cooking in developing countries* (World Bank Other Operational Studies No. 26569). The World Bank. https://econpapers.repec.org/paper/wbkwboper/26569.htm

Vidanage, S. P., Kotagama, H. B., & Dunusinghe, P. M. (2022). Sri Lanka's small tank cascade systems: Building agricultural resilience in the Dry Zone. In A. K. E. Haque, P. Mukhopadhyay, M. Nepal, & M. R. Shammin (Eds.), *Climate change and community resilience: Insights from South Asia* (pp. 225–235). Springer.

Walker, M. (2005). Amartya Sen's capability approach and education. *Educational Action Research, 13*(1), 103–110. https://doi.org/10.1080/09650790500200279

Chapter 14
Small-Scale Solar Solutions for Energy Resilience in Bangladesh

Md Rumi Shammin and A. K. Enamul Haque

Key Messages

- Solar home systems (SHSs) offer a cost-effective, climate-friendly alternative power source in off-grid communities.
- SHS serve both climate adaptation and mitigation as a win–win solution.
- There are opportunities for SHS to accomplish multiple sustainable development goals (SDGs) as co-benefits.
- Innovative strategies can be developed to make SHS more accessible and equitable in rural communities.

M. R. Shammin (✉)
Environmental Studies Program, Oberlin College, Oberlin, OH, USA
e-mail: rumi.shammin@oberlin.edu

A. K. E. Haque
Department of Economics, East West University, Dhaka, Bangladesh
e-mail: akehaque@gmail.com

© The Author(s) 2022
Haque et al. (eds.), *Climate Change and Community Resilience*,
https://doi.org/10.1007/978-981-16-0680-9_14

14.1 Introduction

> We are like tenant farmers chopping down the fence around our house for fuel when we
> should be using Nature's inexhaustible sources of energy – sun, wind and tide … I'd put my
> money on the sun and solar energy.
> — Thomas Edison.[1]

Solar power is a key piece of the puzzle as humanity confronts climate change and strives to transition to a just, sustainable, and decarbonized future. The technology has improved, costs have become competitive, and implementation rates have grown exponentially. Global solar photovoltaic capacity increased from 15 GW in 2008 to 505 GW in 2018—generating ~ 640 TWh or 2.4% of global electricity annually (REN 21, 2019). However, there is still a long way to go for solar and other climate-friendly renewables, which currently account for 11% of total primary energy supply in the world, to displace the 85% coming from carbon intensive fossil fuel sources (BP, 2019). This shift will require humanity to make bold commitments and find creative ways of accelerating the adoption of renewable energy across a wide spectrum of demographic, economic, social, geopolitical, and environmental circumstances. While China, the United States, and the European Union lead solar power generation as of 2019, distributed solar installations in homes, commercial buildings and industrial facilities are expected to double by 2024 accounting for 50% of total growth in solar power (IEA, 2019). Expansion of solar powered electricity across the globe is not homogeneous. In countries like China and the USA, it is mostly driven by commercial interests to supply electricity to the grid. Europe has additionally focused on the household level using net-metering. In South Asia, most installations are for at-home consumption using rooftop solar systems.

Deconstruction of the recent upsurge of solar power exposes very different contexts in which the technology is manifesting—revealing innovative opportunities to leverage solar solutions to achieve climate mitigation, support climate adaptation, build community resilience, and help accomplish the United Nations Sustainable Development Goals (United Nations, n.d.). This chapter documents lessons learned from small-scale solar solutions in remote rural climate vulnerable communities in Bangladesh.

14.2 Powering the Poor in a Changing Climate

Globally, 789 million people live without electricity and hundreds of millions more live with insufficient or unreliable access to it (The World Bank, 2020) More than 1.2 billion people including 40% of the world's rural population living in off-grid

---

[1] Attributed in: Newton, J. D. (1989). Uncommon friends: life with Thomas Edison, Henry Ford, Harvey Firestone, Alexis Carrel, and Charles Lindbergh. Mariner Books.

rural areas in developing and less developed countries do not have reliable access to electricity (IEA: World Energy Outlook, 2016).

Since energy-related carbon emissions are primarily responsible for global climate change, grid expansion to deliver electric power to off-grid communities may lead to continued dependence on existing fossil fuel-based electricity sources and potentially incremental greenhouse gas (GHG) emissions (Komatsu et al., 2011). This presents a philosophical and practical dilemma: on the one hand, these communities have near-zero contribution to the current causes of climate change and, therefore, are least obligated to make major compromises to mitigate GHG emissions; on the other hand, many of these rural communities are in areas that are highly vulnerable to rising sea levels and the effect of more frequent extreme weather conditions caused by global climate change. Any initiative, local or global that reduces the impact of climate change would therefore be beneficial to them. Hence, SHS has the potential for local mitigation initiatives to reduce future local impacts.

In Bangladesh as of 2018, ~ 15% of 166 million residents do not have access to electricity, a decrease from ~ 39% in 2014 (The World Bank, n.d.). This is most likely due to increased production capacity and grid expansion, as well as growth of solar home systems (SHSs) . The SHS installations have focused on off-grid rural communities and served predominantly poor and marginalized communities.

14.3 Homes Powered by the Sun

Solar home systems (SHSs), installed on rooftops of individual households, offer a win–win solution for rural electrification and climate mitigation. SHS have been evolving worldwide since the late 70s and early 80s. The world's first solar electric neighborhood in Gardner, Massachusetts, consists of 30 solar homes—each fitted with a 2 kW grid connected system. The Pal Town Solar City in Japan has 550 homes—each fitted with a 4 kW system (Kamal, 2011).

The number of small-scale SHS projects has been steadily increasing in Asia, South America, and Africa since the 90s with nearly a million SHS installed by the year 2000. This growth has accelerated in the new millennium—with significant momentum documented in several South Asian countries including Sri Lanka, India, and Bangladesh (Komatsu et al., 2011). With annual solar radiation of more than 1900 kWh/m^2 and average daily solar radiation of 4–6.5 kWh/m^2, SHS are particularly attractive for Bangladesh (Khanam et al., 2018).

To bridge the financing gap for developing medium and large-scale infrastructure and renewable energy projects in Bangladesh, the Infrastructure Development Company Limited (IDCOL) was established in 1997 by the Government of Bangladesh—licensed as a non-bank based financial institution. IDCOL spearheads the dissemination of SHS in Bangladesh through its solar energy program with financial support from the World Bank, Global Environment Facility, Kreditanstalt fürWiederaufbau (KfW), German Agency for International Cooperation (GIZ) formerly known as Gesellschaft für Technische Zusammenarbeit (GTZ), Asian

Development Bank, and Islamic Development Bank. IDCOL started this program in January 2003 with an initial target of financing 50,000 SHS by the end of June 2008. The target was achieved in September 2005, three years ahead of schedule, and two million dollars below budget. IDCOL then revised its target and decided to finance 200,000 SHS by the end of 2009. This was also achieved seven months ahead of schedule. In subsequent years, IDCOL consistently overshot its SHS implementation goals—making it one of the fastest growing renewable energy programs in the world. As of 2019, over 4 million solar home systems (SHS) have been installed in rural off-grid communities in Bangladesh—creating over 70,000 jobs and bringing electric power to more than 18 million people or 11% of the country's population (IDCOL, n.d.). This is about 12.2% of all connected users in Bangladesh (GoB, 2019). Most of these users are low-income and consume a very small amount of electricity at their homes.

The SHS installed in developing countries represent significantly different technologies and scales. While most systems installed in developed countries are grid connected and operate on AC power through an inverter, implementation in off-grid communities in developing countries rely on energy storage in batteries running DC powered appliances through a charge controller. The typical sizes of solar home systems in OECD countries range from 1000 W (1 kW) to roughly 6000 W (6 kW) per household. The size of the SHS installed in rural homes in Bangladesh are orders of magnitude smaller—typically ranging between 20 and 100 W (see Fig. 14.1a, b for a visual comparison).

(a)   (b)

Fig. 14.1 **a** Trail magic in Oberlin, Ohio, USA with a 5.2 kW solar system. **b** Solar home in Batiaghata, Bangladesh with a 40 W solar system. *Photo credit* **a** Carl McDaniel, **b** Md Rumi Shammin

14.4   SHS in Bangladesh: A Closer Look

To evaluate potential co-benefits of SHS in the context of climate change, a study was conducted in climate vulnerable coastal communities in the coastal districts of Khulna and Bagherhat in south-eastern Bangladesh in 2012–13. The SHS surveyed were installed by Bangladesh Rural Integrated Development for Grub-Street Economy (BRIDGE)—a partner organization (PO) of IDCOL. Systematic random samples of 1000 households were drawn from ten different BRIDGE project locations. 50% of the surveys ($n = 500$) were carried out with households currently using SHS and the remaining 50% ($n = 500$) of surveys were carried out with SHS non-users randomly selected from the same general area. The purpose of surveying the two groups was to have a control population to analyse changes achieved by SHS and explore untapped future opportunities.

The SHS user survey included questions on demographics (name, age, income, occupation, family size, education, etc.), satisfaction, user-friendliness, cost, maintenance, past energy use, present energy use, quality of life (before/after), idea of alternatives, perception of opportunity cost, distribution of benefits, barriers, equity, etc. The SHS non-user survey included the same questions with appropriate modifications and additional questions on unmet energy needs and willingness to pay for SHS. When available, the head of households was surveyed. In their absence, the spouse of the head of household was surveyed. The final sample count and locations are shown in Fig. 14.2.

Statistical tests indicate that SHS users and non-users are different in terms of their demographic characteristics (Table 14.1). It appears that households that are relatively well-off within these communities adopted the SHS. A similar result was also observed in a study by the World Bank on SHS users in Bangladesh (Asaduzzaman et al., 2013). Finally, informal conversations were carried out with selected public officials, private sector entrepreneurs, representatives of non-government organizations and local people to distill contextual and anecdotal information about the opportunities and challenges of SHS.

14.5   SHS and SDGs: Grassroots Lessons

SHS provide a range of benefits associated with and beyond providing an alternate source of power. These benefits include access to new income generating activities, reduced travel cost (or opportunity cost of time) to buy kerosene, educational benefits for children, increased security, reduced indoor air pollution, access to information through television and access to cell phone service (Komatsu et al., 2011; Urmee et al., 2009). These experiences indicate that SHS have the potential to address multiple SDGs and improve community resilience. Shammin et al., (2022, Chap. 2 of this volume) has developed an integrative framework for climate resilient communities that connects climate adaptation, resilience, and SDGs.

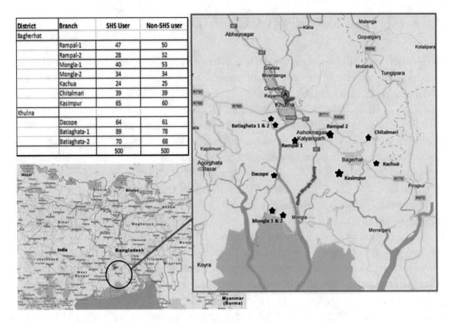

District	Branch	SHS User	Non-SHS user
Bagherhat			
	Rampal-1	47	50
	Rampal-2	28	32
	Mongla-1	40	53
	Mongla-2	34	34
	Kachua	24	25
	Chitalmari	39	39
	Kasimpur	65	60
Khulna			
	Dacope	64	61
	Batiaghata-1	89	78
	Batiaghata-2	70	68
		500	500

Fig. 14.2 Sample count and distribution of field surveys. *Source* Authors' creation based on Google Maps

Table 14.1 Key characteristics of the respondents and their families

Description	SHS users	SHS non-users	All	Significance
Female (respondent)	21.00	26.85	23.92	**
Male (respondent)	79.00	73.15	76.08	**
Married (respondent)	91.67	92.77	76.08	**
Age (respondent)	40.34	39.13	39.74	
Household size	4.88	4.70	4.79	*

Note * means 10%, ** means 5% and *** means 1% level of significance
Source Field data 2012–13

The results of this study have been organized below under primary benefits and secondary benefits with their relevance to SDGs depicted using corresponding icons. Additionally, the outcomes of SHS have been assessed based on the way they contribute to building community resilience.

14.5.1 Primary Benefits of SHS

More than 90% of the households sampled in this study report using kerosene lamps as their current or previous source of lighting energy (for non-SHS users and SHS users, respectively). When SHS users were asked a question about the benefits of solar power as part of this study, the majority report improvements in quality of life in terms of comfort and convenience (~ 83% of users surveyed). In addition to lights, they use mobile phones, televisions, fans and other small appliances (Fig. 14.3).

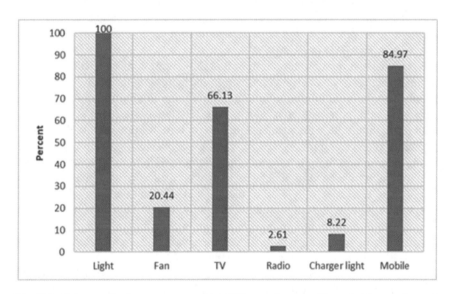

Fig. 14.3 End uses of solar power reported by SHS users (percent of households reporting use). *Source* Field data 2012–13

Lighting remains the most important use of solar power as about 70% of the SHS users mention 'insufficient light' as a major problem with their previous energy source. They report higher quality of light from solar power and ~ 25% extension of time of use of lighting source at night. Using a paired sample test, this study finds that solar users use electricity significantly longer with solar than with their previous light source (t (496) $= -$ 19.57, $p <$ 0.001). 99.4% of the users state that the solar lights are better than their previous energy sources with 91.5% characterizing the improvements as moderate to extraordinary. These are consistent with the findings by previous studies that SHS users clearly prefer the quality of light provided by the SHS compared to the kerosene lamps they used before. They like the ability to power additional equipment and also experience better night-time security (Asaduzzaman et al., 2013; Biswas, 2004; Komatsu et al, 2011; Urmee et al, 2009). Another study of two off-grid communities in Africa found that SHS reduce the use of disposable batteries (Stojanovskia et al., 2017).

Kerosene lamps produce black carbon and CO_2 during combustion that can affect the lungs, increase risks of asthma and cancer, and increase vulnerability to infectious diseases (Apple et al., 2010; Lam et al., 2012; Tedsen, 2013). If materials in lamps contain lead, this poses additional health risks (Lakshmi et al., 2013). Asaduzzaman et al. (2013) found that the incidence of several types of preventable illness such as general ailment, respiratory diseases, and gastrointestinal problems were lower among the members of the households that purchased SHS in Bangladesh. However, when non-users were asked about problems with their current source of lighting energy (e.g., kerosene), only 11% of the respondents identified air pollution from lighting as a concern and only some SHS users mentioned the pollution-free indoor environment that contributes to good health and well-being. It is possible that lack of awareness of indoor air pollution, and its impact is the reason for the limited acknowledgment of this benefit of solar lights.

This study finds that the question of affordability is more nuanced than a simple cost–benefit analysis and payback estimates. Currently, the IDCOL program receives approximately 10% of the cost through grant money. Partner organizations receive an additional, smaller percent as support for institutional development. The remaining 90% cost of the system is micro-financed by the households at interest rates between 6 and 12% with a 15% down payment. The households surveyed were charged 6% interest rate for a three-year loan—resulting in an average monthly payment of Tk. 816 which is about 4 times higher than the average monthly cost of kerosene lamps. Beyond the payment period, however, the monthly cost of solar power would be reduced to zero for the life of the panel, excluding the costs of system mainte-nance, troubleshooting or battery replacement. If future benefits are not discounted, the all-inclusive average monthly cost of solar power over ten years would be 282 Taka/month (for a three-year financing scheme). Results from this study show that the average cost of lighting energy from kerosene is approximately Taka 210/month. Therefore, at the very least, solar power is about 35% more expensive in the long run than their previous energy source for households who make the switch. While this may raise questions about the affordability of SHS, it should be noted that the SHS provides better quality lighting and additional benefits compared to their existing source of lighting. While we do not have data on the actual longevity of these solar panels, the typical life of solar photovoltaic (PV) panels in developed countries is 25–30 years (Solar Reviews, 2020).

When asked about maintenance issues, 80% of the users reported that they had no maintenance issues with their systems. This is primarily because the maintenance warranty was included in the cost of the systems for the duration of the loan and all households surveyed were still within that period. 91% of the users state that the community representative of the PO that installed the systems, visited them monthly to conduct inspections. During informal conversations, SHS users appeared to be comfortable with the technology and generally satisfied with the performance and maintenance of the systems. Hence, the households are spending more money, but availing of more services, with greater benefits and a reliable source of energy.

Diesel generator is another possible energy source for rural off-grid communities. Biswas et al. (2004) reported that solar electricity is significantly cheaper than small diesel generators in rural, off-grid areas in Bangladesh. This is due to the poor economy of scale at the local level for producing and distributing generators, where the low-electricity consumption is not worth the associated infrastructure costs. There are also noise pollution, air pollution, and GHG emissions associated with generators. This survey reveals that less than 5% of the non-SHS users currently use generators and less than 1% of the SHS users consider it as an alternative to solar power.

Overall, SHS deliver a clean energy source for people in rural off-grid climate vulnerable communities and ensure better indoor air quality for improved health and well-being; but their affordability remains questionable and nuanced.

Technological leapfrogging toward renewable energy has been a centerpiece of developing country participation in climate mitigation under the Paris Agreement. The GHG reduction potential of displacing kerosene lamps is known to be relatively small per household (Baurzhan & Jenkins, 2016), but the accumulated global CO_2

reductions by SHS can be significant if millions of people in developing countries adopt this technology. The SHS are thus a part of Bangladesh's national contribution to global greenhouse gas reduction goals. The 4.1 million SHS installed in Bangladesh as of early 2019 are expected to displace 3.6 million tons of kerosene over the next 15 years (IDCOL, n.d.) and prevent emission of 10 billion tons of CO_2. This translates to about 163 kg CO_2 per household per year. This appears to be a low estimate as Hoque and Das (2013) report about four times higher GHG emission reductions by SHS in Bangladesh, comparing laboratory test results of the specific type of kerosene lamps used with 50 W solar panels.

Observations made during field work also revealed interesting ways the SHS are contributing to climate change adaptation at the community level. Solar power has made cell phones accessible in these communities, in turn providing them with new tools for early warning systems and disaster response during floods, storms, and cyclones – which occur frequently in these communities due to climate change.

14.5.2 Secondary Benefits of SHS

Urmee (2009) reported new income generating activities as one of the co-benefits of SHS. This is confirmed through survey results, field observation, and anecdotal evidence from this study. Access to electricity makes better quality lighting available for longer hours for SHS users and creates opportunities for household crafts, tutoring of children, and other productive activities. More productive hours in the evening frees up time during the day to engage in farming, business, and other enterprises.

Several new business enterprizes were observed in the rural marketplaces of the communities surveyed in this study. Solar power made it possible for SHS users to have access to electronic equipment such as cell phones, radio, television, etc. To support the maintenance of this new equipment, the marketplaces now feature services for charging cell phones and repairing electronics as well as sale of voice and data packages. With cell phone access, money transfer services such as *bKash* also became accessible to these communities. While the SHS project primarily targeted households, it turned out that businesses also adopted solar power. Barber shops and convenience stores were able to use solar lights to extend their operating hours into

Fig. 14.4 Solar powered convenience store, barbershop, and electronics repair service from left to right. *Photo credit* Md Rumi Shammin

the evening. Halder (2016) found similar benefits of SHS in local small businesses in two randomly selected villages in Sirajgonj and Jessore districts in Bangladesh.

Field observations in the communities surveyed in this study illustrate creative and innovative ways of using solar power (Fig. 14.4). In one village, multiple stores were sharing one solar system. In another instance, a solar panel was installed on a box frame that could be moved around throughout the day to maximize exposure to sunlight and then stored inside the store at night for security.

About 82% of the SHS users expressed that their previous light source affected their children's education and a nationwide study found statistically significant differences in the study habits of children with solar lights. (Asaduzzaman et al., 2013).

The majority of non-users reported that their current source of light does not meet their needs. The extended use of lights by SHS users was statistically significant when compared with non-users (see Table 14.2).

Table 14.2 Differences in lighting use duration between SHS users and non-users

Hours of lighting	obs	Mean	SE	SD	Significance
Prior to SHS connection	497	3.97	0.041	0.923	
After SHS connection	497	4.83	0.041	0.906	
Full sample	994	4.40	0.032	1.009	
Difference in mean		-0.85	0.058	-0.967	***

Note *** significance at 1% level
Source Field data 2012–13

Table 14.2 indicates that after SHS connections, households increased their total hours of lighting by nearly an hour (0.85 * 60 = 51 min). The increase is statistically significant, and it can be argued that their standard of living increased due to the solar home systems. Since it is the women who do the chores every day, ease of lighting or convenience meant that it is women who enjoyed the benefits. It probably provided them with some extra time to do other things of their choice or enjoy the time in leisure.

Anecdotal evidence documented during field surveys reveals specific ways SHS are empowering women. Salma,[2] a high school educated divorced self-employed woman, reported a doubling of her household income from tutoring children after hours. She described how having more productive hours in the evening freed up time during the day for her to engage in other income generating activities. She also indicated how better lighting in the kitchen reduces the risk of cooking related accidents (e.g., fire hazards). This is consistent with findings by Asaduzzaman et al. (2013) that women are directly impacted by SHS by having access to the longer study time for their children and better lighting for cooking at night, while maintaining the nutritional quality of food. While this study did not investigate the impact of SHS on women's health, a study from India found that the health of women improved after replacing kerosene lamps with lighting powered by SHS – evidenced by reduced incidences of issues such as eye problems, headache, and coughing (Barman et al., 2017).

Equitable distribution of development opportunities is a key component of ensuring a just and sustainable future. The role of SHS in this regard can be investigated at multiple scales: community, national, and international.

At the community level, this study finds several key differences in household demographics between the randomly surveyed SHS users and non-users from the same communities. First, the household income of SHS users is significantly higher than non-users ($F = 41.05$, $p < 0.001$). Second, there is a significant gap in the level of education between SHS users and non-users ($F = 50.22$, $p < 0.001$). As evident in Table 14.1, about 57% of the non-SHS users in the sample have no education or just elementary education, whereas about 61% of the SHS users in the sample have at least middle/high school education (in Bangladesh, middle school and high school are not separately distinguished beyond fifth grade). Third, significant differences were found in the occupation between SHS users and non-users (Pearson Chi square

---

[2] Real name has not been reported in order to maintain confidentiality.

Table 14.3 Differences in household income, level of education and nature of occupation between SHS users and non-users

	Mean	SE	SD	Significance
Monthly income				
SHS users	8078.067	198.977	4449.260	
SHS non-users	6477.567	151.036	3377.256	
Differences in mean	− 1600.500	249.807	− 2090.707	***
Highest level of family Education				
SHS users	3.706	0.079	1.755	
SHS non-users	3.190	0.080	1.781	
Differences in mean	− 0.516	0.112	− 0.735	***

Source Field data 2012–13

(5) $= 78.46$, $p < 0.001$). Head of households that adopted the SHS were more likely to be businesspersons, government workers, or non-governmental workers, while those who did not are more likely to be engaged in daily labor, farming, or fisheries. Both SHS users and non-users were asked about their perception of who benefits and who are left behind from the SHS initiatives in their communities. The results indicated are consistent with inferences made from the statistical analysis presented in Table 14.3 that SHS technology favors professionals, shop owners, and the more affluent (middle and high income) households.

Table 14.3 reveals that the much talked about SHS program missed the poorest strata in the society. Differences in income and education between users and non-users are statistically significant at 1% level of significance. One reason for this exclusion is that the 15% down payment and monthly installments for the microfinance loan are cost prohibitive for lower-income households as discussed below.

When asked about the barriers to access SHS, 63% of the non-users indicate cost as the main factor that prevented them from adopting this technology. Since the cost of the SHS is considerably higher than current energy sources for non-users in the short run, those who are willing and able to pay for the additional services provided by solar power become adopters. Mondal (2010) argues that in such cases, the social and environmental benefits of SHS need to be internalized into their financing metrics by leveraging national and international initiatives to improve accessibility and make the implementation of this technology more equitable.

At the national level, governments in developing countries with an obligation to bring electric power to all residents are faced with particular challenges when it comes to rural off-grid communities often located in climate vulnerable areas. Three aspects of the rural households are noteworthy: (1) they are generally more widely dispersed over the landscape; (2) they are often located in remote areas sometimes separated by a river or other natural barrier from their nearest electricity grid; and (3) they have very low-power needs per household. These aspects make it difficult and expensive to expand the existing electricity grid to incorporate them. To do so would require elaborate infrastructural additions such as power stations, substations, long transmission lines across unfavorable terrains, etc. (Komatsu et al., 2011; Zerriffi, 2011). Pode (2013) argues that due to the remoteness, isolation, low-electricity demands, and high-investment costs of grid expansion, these communities are unlikely to be reached by simply extending the power grid. These villages are therefore ideal for small-scale, distributed power solutions that deliver electricity in a convenient and cost-effective manner. SHS provide clean energy access to previously underserved communities, reduce inequalities on the national scale and advance social and environmental justice efforts where implemented.

At the international level, SHS are making a positive contribution to climate change mitigation and should receive financial assistance from international climate change funds to reduce costs and improve access. This would be ethically just given that these communities have historically contributed very little toward climate change and yet find themselves more vulnerable to the impacts of climate due to their low-lying coastal locations. Even though international agencies are featured as partners in IDCOL's SHS program in Bangladesh, most of the costs are borne by the end users through micro-credit programs. Figure 14.5 illustrates that the total value of external grants per system is less than the interest paid by the average SHS users on their micro loan and constitutes about 12% of the final cost to consumers.

A fair and equitable measure would be to leverage the international climate funds earmarked for developing countries under the Paris Agreement and find creative ways to participate in international carbon markets to generate additional foreign aid alongside earnings. IDCOL has already recognized that the avoided GHG emissions of SHS is a global commodity and is working with the World Bank to participate in the global carbon market (Asaduzzaman et al., 2013). The international community can thus play an important role in expanding the deployment of SHS and make them more

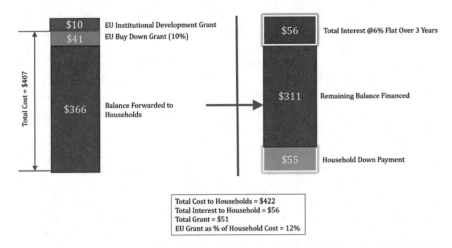

Fig. 14.5 Distribution of cost, funding, and financing of a 50-W SHS. *Note* Conversion rate used: 1 US Dollar = 80 Taka. *Source* Authors' creation based on information from IDCOL and POs

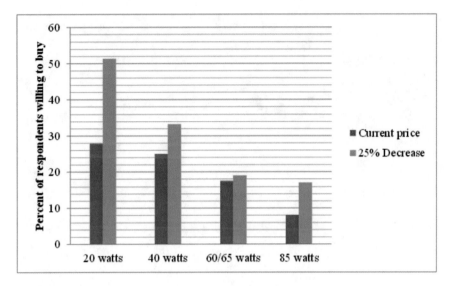

Fig. 14.6 Willingness to purchase SHS systems of different capacities under different price scenarios by households without solar power. *Source* Field data 2012–13

accessible and equitable (Chaureya & Kandpal, 2009). When non-SHS users were randomly asked about their willingness to buy solar systems of various capacities under different price scenarios, the results show that there is significant additional demand for solar systems in the communities surveyed under all price scenarios. Specifically, a 25% decrease in price and expanding distribution of smaller panels (20 and 40 W) have the potential for significantly increasing the number of SHS users (see Fig. 14.6).

Despite the uncertainties due to past political shifts in the US and the ongoing Covid-19 pandemic, any financial assistance that might still be available through international climate funds can reduce the cost of the SHS. This will make the system more affordable and lessen the financial burden of current SHS users and allow additional disposable income for them to improve their economic well-being.

14.6 Emerging Solar Solutions

While SHS remain the flagship solar project in Bangladesh, several other emerging technologies and models have been initiated by government, non-government and private initiatives in recent years. In addition to traditional rooftop solar systems, IDCOL has started two other solar projects for off-grid communities in Bangladesh: Solar irrigation and solar mini-grids. The solar irrigation project aims to install 50,000 solar PV-based irrigation pumps by 2025 in areas with three annual cropping seasons. With support from the World Bank and several other international aid agencies, over

Fig. 14.7 SOLshare grid in a marketplace in Bangladesh. *Photo credit* SOLshare

1300 pumps went into operation by fall 2019. Solar mini-grid projects are intended to provide grid quality electricity to households and small commercial users. Seven PV-based mini-grids are already operational and serve about 5000 rural households (IDCOL, n.d.). Both projects compliment the SHS project.

Another interesting private market-based initiative is SOLshare—a new approach to expand affordable solar electricity accessibility using smart peer-to-peer grids (Fig. 14.7). SOLshare has developed a marketplace called SOL bazaar, a trading platform which enables people to trade the excess solar energy generated by SHS. SHS users can sell their excess energy to non-SHS-users. This creates a win–win solution where SHS users have additional income generation opportunities and the buyer gets access to electricity (SOLshare, n.d.).

There is ongoing research and development in the design and delivery of the SHS. Zubi et al. (2019) proposed a modified layout of the SHS that integrates a lithium-ion battery-pack and is complemented with LED lamps and an energy efficient multi-cooker. Coupled with creative financing mechanisms, they hope to accomplish a more efficient and affordable alternative to the current system. Similarly, the design of LED lamps using solar-based batteries (like the NiCd, NiMH, Lithium-ion, or Lithium Polymer batteries typically used in mobile phones) instead of using lead acid batteries is also a major game changer in the uptake of SHS across the world.

14.7 Resilient Communities that Run on Sunshine

SHS deliver significant benefits to rural off-grid populations in developing countries and contribute to improving the quality of life for some of the poorest and most marginalized people in the world. This study focuses specifically on the interactions between SHS, climate mitigation, and community resilience by focusing on remote off-grid climate vulnerable communities in Bangladesh. The life cycle analysis of upstream and downstream GHG emissions from solar PV systems is similar to that of other renewables and nuclear energy, and ten times lower than coal (NREL, 2012).

SHS have the potential for delivering outcomes related to multiple SDGs. In addition to providing clean renewable energy and achieving climate mitigation, these systems create economic opportunities, empower women, advance children's education and improve living environment. While the technology appears to be reliable and user-friendly, the system remains cost prohibitive for low-income households—particularly farmers, fishermen, and landless day-laborers. These barriers may be removed by lowering the cost of the systems and running more targeted awareness campaigns. Revenues from international climate funds and carbon markets, if used to reduce the price of SHS, will increase access to this technology by aiding households who are currently left behind. Proactive initiatives that explicitly connect SHS and other solar solutions with SDGs at early stages of program development have the potential for greater attainment of the goals. This will make the use of development aid and climate financing more coordinated and efficient.

This study also reveals that solar solutions have the potential to advance community resilience. Hopkins (2009) identified modularity and tightness of feedback as two important properties of resilient systems. The SHS are distributed and modular and hence immune to a whole-system failure in case any individual unit malfunctions. They are not subject to supply disruptions of kerosene or diesel—allowing individual households to retain their energy access when communication routes are cut-off during natural disasters such as floods or cyclones. The SHS are local sources of energy, and their maintenance is also serviced by a trained local workforce. Hence, the feedback loop between the systems and their end-use is short, transparent, and accessible. Additionally, access to cell phones during natural disasters facilitates quick dissemination of early warning systems and coordination during disaster response. Cell phones also provide access to communication technology during other emergencies (e.g., medical). Bangladesh has nationwide coverage of internet data service which is now available to these communities—thus opening up access to global databases of knowledge and information. Since these are climate vulnerable communities, new enterprises made possible by solar solutions provide alternate economic opportunities for farmers and fishermen whose livelihood might be threatened by climate change—thus reducing the risk of displacement and climate migration.

Finally, the study of SHS demonstrates how solar solutions offer communities the ability to take ownership of and participate in climate change mitigation and adaptation. They allow for technological penetration without technological imperialism. However, the most important point to note is that some of the poorest and most marginalized populations in the world are participating in a carbon-free energy future, while many developed countries are yet to act boldly. As long as the sun shines, communities across the developing world can use the power of the sun to survive the wrath of the rising seas and roaring skies—striving toward a just, equitable, sustainable, and resilient future.

Acknowledgements Partial funding for this research was provided by Arthur M. Blank Fund and H.H. Powers Travel Grant with additional support from the faculty research portfolio at Oberlin College. The authors would like to acknowledge the contributions of the following people: Mr. Zohurul Haque, Executive Director of BRIDGE, for logistical support; Mr. Akhtarujjaman Shohel for assisting with survey administration; Eashan Ahmed and Abu Mosayeb—undergraduate research assistants at United International University, Bangladesh; Ethan Abelman, Courtney Koletar, Amanda Jacir and Savitri Sedlacek—undergraduate research assistants at Oberlin College, USA; and student assistants from local colleges in the Khulna/Bagherhat area of Bangladesh who were part of the field survey team: Md. Tanvir Rahman, Md. Suez Khan, Md. Kamruzzaman, Md. Robiul Islam, and Md. Arif Hossain. Finally, we thank Salma Islam for information and photo about SOLshare. Preliminary results of the survey reported in this chapter was presented in Shammin and Haque (2012) as a conference paper.

References

Apple, J., Vicente, R., Yarberry, A., Lohse, N., Mills, E., Jacobson, A., & Poppendieck, D. (2010). Characterization of particulate matter size distributions and indoor concentrations from kerosene and diesel lamps: Indoor particulate matter concentrations from kerosene lamps. *Indoor Air, 20,* 399–411.

Asaduzzaman, M., Yunus, M., Haque, A. K. E., Azad, A. K. M. A. M., Neelormi, S., & Hossain, M. A. (2013). *An evaluation of institutional effectiveness and impact of solar home systems in Bangladesh.* Bangladesh Institute of Development Studies (BIDS).

Barman, M., Mahapatra, S., Palit, D., & Chaudhury, M. K. (2017). Performance and impact evaluation of solar home lighting systems on the rural livelihood in Assam, India. *Energy for Sustainable Development, 38,* 10–20.

Baurzhan, S., & Jenkins, G. P. (2016). Off-grid solar PV: Is it an affordable or appropriate solution for rural electrification in Sub-Saharan African countries? *Renewable and Sustainable Energy Reviews, 60,* 1405–1418.

Biswas, W., Diesendorf, M. & Bryce, P. (2004). Can photovoltaic technologies help attain sustainable rural development in Bangladesh? *Energy Policy, 32*(10), 1199–1207.

BP Statistical Review of World Energy. (2019). 68th edition. BP.

Chaureya, A., & Kandpal, T. (2009). Carbon abatement potential of solar home systems in India and their cost reduction due to carbon finance. *Energy Policy, 37*(1), 115–125.

Government of Bangladesh. (2019). *Bangladesh economic review 2019.* Ministry of Finance.

Halder, P. K. (2016). Potential and economic feasibility of solar home systems implementation in Bangladesh. *Renewable and Sustainable Energy Reviews, 65,* 568–576.

Hopkins, R. (2009). *Resilience thinking.* Resurgence No. 257, November/December 2009.

Hoque, S. M. N., & Das, B. K. (2013). Analysis of cost, energy and CO2 emission of solar home systems in Bangladesh. *International Journal of Renewable Energy Research, 3*(2), 2013.

Infrastructure Development Company Limited. (n.d.). http://idcol.org/home/solar. Last accessed April 2020.

International Energy Agency. (2016). *World energy outlook 2016*. IEA, Paris. https://www.iea.org/reports/world-energy-outlook-2016. Last accessed 11-10-2020

International Energy Agency (2019). *Renewables 2019*. IEA. https://www.iea.org/reports/renewables-2019. Last accessed 10/10/2020.

Kamal, S. (2011). *The renewable revolution*. Earthscan.

Khanam, M., Hasan, M. F., Miyazaki, T., Saha, B. B., & Koyama, S. (2018). Key factors of solar energy progress in Bangladesh until 2017. *EVERGREEN Joint Journal of Novel Carbon Resource Sciences and Green Asia Strategy, 5*(2), 78–85.

Komatsu, S., Kaneko, S., & Ghosh, P. (2011). Are micro-benefits negligible? The implications of the rapid expansion of solar home systems (SHS) in rural Bangladesh for sustainable development. *Energy Policy, 39*(7), 4022–4031.

Lakshmi, P. V. M., Virdi, N. K., Sharma, A., Tripathy, J. P., Smith, K. R., Bates, M. N., & Kumar, R. (2013). Household air pollution and stillbirths in India: Analysis of the DLHS-II National Survey. *Environmental Research, 121*, 17–22.

Lam, N. L., Smith, K. R., Gauthier, A., & Bates, M. N. (2012). Kerosene: A review of household uses and their hazards in low- and middle-income countries. *Journal of Toxicology and Environmental Health Part b: Critical Reviews, 15*, 396–432.

Mondal, M. (2010). Economic viability of solar home systems: Case study of Bangladesh. *Renewable Energy, 35*(6), 1125–1129.

National Renewable Energy Laboratory. (2012). Life cycle greenhouse gas emissions from solar photovoltaics. *Journal of Industrial Ecology*.

Pode, R. (2013). Financing LED solar home systems in developing countries. *Renewable and Sustainable Energy Reviews, 25*, 596–629.

REN21. (2019). *Renewables 2019 global status report*. REN21 Secretariat.

Solar Reviews. (2020). *How long do solar panels actually last?* https://www.solarreviews.com/blog/how-long-do-solar-panels-last. Last accessed 10/10/2020.

Shammin, M. R., & Haque, A. K. E. (2012). The economics and ethics of solar home systems in remote rural areas of Bangladesh. Paper presented at the Meeting of the International Society for Ecological Economics (ISEE), June 16–19 in Rio De Janeiro, Brazil.

Shammin, M. R., Haque, A. K. E., & Faisal, I. M. (2022). A framework for climate resilient community-based adaptation. In A. K. E. Haque, P. Mukhopadhyay, M. Nepal, & M. R. Shammin (Eds.), *Climate change and community resilience: Insights from South Asia* (pp. 11–30). Springer.

SOLshare. (n.d.). https://www.me-solshare.com/. Last accessed November 11, 2020

Stojanovskia, O., Thurber, M. and Wolaka, F. (2017). Rural energy access through solar home systems: Use patterns and opportunities for improvement. *Energy for Sustainable Development, 37*, 33–50.

Tedsen, E. (2013). *Black carbon emissions from kerosene lamps*. Ecological Institute of Berlin.

The World Bank. (2020). *Understanding poverty*. https://www.worldbank.org/en/topic/energy/overview. Last accessed 10/10/2020.

The World Bank. (n.d.). *Access to electricity (% of population)—Bangladesh*. https://data.worldbank.org/indicator/EG.ELC.ACCS.ZS?locations=BD. Last accessed 11/10/2020.

United Nations. (n.d.). *Sustainable development goals*. Knowledge Platform. https://sustainabledevelopment.un.org/sdgs. Last accessed 11/10/2020.

Urmee, T., Harries, D., & Schlapfer, A. (2009). Issues related to rural electrification using renewable energy in developing countries of Asia and Pacific. *Renewable Energy, 34*(2), 354–357.

Zerriffi, H. (2011). *Rural electrification: Strategies for distributed generation*. Springer.

Zubi, G., Fracastoro, G. V., Lujano-Rojas, J. M., Bakari, K. E. & Andrews, D. (2019). The unlocked potential of solar home systems; an effective way to overcome domestic energy poverty in developing regions. *Renewable Energy, 132*, 1425–1435.

Chapter 15
Sri Lanka's Small Tank Cascade Systems: Building Agricultural Resilience in the Dry Zone

Shamen P. Vidanage, Hemasiri B. Kotagama, and Priyanga M. Dunusinghe

Key Messages

- Small tank cascade systems in the Dry Zone of Sri Lanka harvest rainwater and mitigate floods and drought.
- These are traditional systems of building famer resilience to climate variability.
- Farmers have expressed willingness to pay to restore the degraded small tank cascades, and the government has identified restoration and sustainable management of STCS as a priority adaptation action in irrigation and agriculture.

15.1 Introduction

In keeping with the requirements of addressing the threat of climate change, Sri Lanka has formulated an extensive national policy framework to meet the challenges. The national climate change policy, climate change strategy, sector vulnerability assessments for key sectors, national adaptation plan, technology needs assessment and Nationally Determined Contributions (NDCs) are the major components of the adaptation strategy. The Climate Change Secretariat of the Ministry of Environment is spearheading the national action on climate change including liaison with the United Nations Framework Convention on Climate Change (UNFCCC), and the country

S. P. Vidanage (✉)
University of Kelaniya, Kelaniya, Sri Lanka
e-mail: shamenpv@kln.ac.lk

H. B. Kotagama
Sultan Qaboos University, Muscat, Oman
e-mail: hemkot@squ.edu.om

P. M. Dunusinghe
University of Colombo, Colombo, Sri Lanka
e-mail: dunusinghe@econ.cmb.ac.lk

© The Author(s) 2022
Haque et al. (eds.), *Climate Change and Community Resilience*,
https://doi.org/10.1007/978-981-16-0680-9_15

is in the process of finalising the Third National Communication to the UNFCCC based on a review of the national strategy. As part of its adaptation strategy, Sri Lanka has identified traditional systems such as village tanks (small reservoirs constructed during ancient times) as a time-tested adaptation mechanism in helping Dry Zone agrarian communities in coping with climate variability. This chapter on Small Tank Cascade Systems discusses their role in the Dry Zone of Sri Lanka as a mechanism for climate change adaptation with the local community willing to contribute to restoration and sustainable management.

15.2 Small Tank Cascade System

The irrigation systems of Sri Lanka are broadly categorised into minor, medium, large and those designated as special projects—such as Mahaweli and Walawe systems (Murray & Little, 2000). This chapter focuses on tanks with command areas of 80 ha or less, classified as minor irrigation systems or village or minor irrigation works. Work by Panabokke et al. (2002) indicates that these village tanks are not situated randomly, but organised to collect rainwater from well-defined micro catchments. These individual tanks are components of large systems or units called 'cascades', defined as 'a connected series of village irrigation tanks organised within a micro— (or meso-) catchment of the Dry Zone landscape, storing, conveying, and utilising water from an ephemeral rivulet' (Madduma Bandara, 1985).

Cascade systems comprise of a number of components including many types of tanks; *mahawewa,*[1] *olagamwewa,*[2] *kuluwewa,*[3] *godawala*[4] and *pinwewa*[5] (Tennakoon, 2004). Multiple uses are generated from irrigation water, rather than just crop irrigation (Renwick, 2001). Some of the direct and indirect agricultural benefits of cascade systems are provisioning water for fisheries and livestock, control of soil erosion, flood prevention, water quality control, storage of water for irrigation, reducing vulnerability to drought, and retaining the health of the soil.

Small tank cascade systems are unique soil–water conservation systems prevalent in the Dry Zone of Sri Lanka. As explained by Madduma Bandara (2007), these tank cascade systems are linked with a diverse ecological and socio-economic subsystem within which they have evolved, covering the following:

---

[1] *Mahawewa* is the larger tank in villages used for irrigation and other domestic purposes.

[2] *Olagamwewa* means the tank without a village, cultivation from the tank is done by villages from an adjoining village.

[3] *Kuluwewa* is a small tank constructed upstream for the purpose of trapping silt brought down by runoff water. Usually, there are no settlements or paddy field for these tanks.

[4] *Godawala* are the water holes in forest areas above tanks, mainly for the wild animals around which lush vegetation prevails.

[5] *Pinwewa* are constructed closer to the temples to meet their water needs. These tanks were not used for irrigation purposes earlier, but paddy cultivation has been seen in recent times.

i.   The ecological system with catchment forests, aquatic habitats, and the commons
ii. Land use zoning systems
iii. Various crop combination systems
iv. Elaborate water management systems including, sluices, spills, water control weirs (*Karahankota*) with rotational water distribution systems, and
v.   Management systems such as *Gamarala* (Village Headman) system that dates back to pre-colonial times.

Madduma Bandara (2007) has further elaborated that the village tank cascade systems emerged as a response to ensure sustainable agriculture given the challenge of recurrent water shortages and drought conditions in a seasonally dry environment. Despite the adverse effects of various socio-economic, institutional and political changes experienced over centuries, these systems still exist and continue to operate mainly on the biophysical, socio-economic and ecological principles on which they were created, albeit at a reduced level of efficacy. STCS have been recognised as Globally Important Agricultural Heritage System (GIAHS) by the Food and Agriculture Organisation (FAO).

Scholars have different estimates on the total number of small tanks in Sri Lanka. Dharmasena (2004) suggested that over 30,000 small tanks are in existence in Sri Lanka, whereas others like Panabokke (2004) estimated it as around 18,000 small tanks, both operational and abandoned, distributed across 70 well-defined river basins in the Dry Zone of Sri Lanka. Nearly, 90% of these small tanks were found to be organised as clusters or cascades (Madduma Bandara, 1985; Panabokke et al., 2002) in the Dry Zone. The total number of remaining cascades in Sri Lanka were estimated at 1166 (DAD, 2005), out of which 457 cascades are found in the North Central Province of Sri Lanka (Panabokke et al., 2002).

As described by Panabokke et al. (2001) the micro-morphological features of Dry Zone such as *heennas*[6] and *mudunnas* have had a great influence in the distribution, density, alignment, size, shape and use of small tanks within cascades. They further explained that though the small tanks within a cascade differ physically from one another, these eco-friendly pools of water have a hydrological and socially determined pattern that is economically and socially beneficial. About 2 millennia after their invention, the STCS are still an integral part of the Dry Zone of Sri Lanka for the economic, socio-cultural, and ecological needs of the present generation and possibly also the future if well managed.

---

[6] *Heennas* and *mudunnas* are micro-morphological characteristics of the areas where cascades are constructed. Inside low ridges within sub watershed boundaries are called *heennas* whilst summits within those sub watersheds are called *mudunnas* (Tennnakoon, 2001).

15.3   Evolution of STCS

According to Shannon and Manawadu (2007), the first planned settlement in Sri Lanka was recorded in 1000 BC in the Dry Zone. Due to the seasonality of watercourses, the settlements suffered from water scarcity and tanks were built to conserve rainwater to overcome the same (Shannon & Manawadu, 2007). Systematic study of the irrigation works of Sri Lanka is believed to have started during the British period by irrigation engineers employed by the government. Brohier (1934) in his monumental work on ancient irrigation works mentions the belief that small tank technology in Sri Lanka dates back to the pre-Aryan settlements (fifth century BC) but that its further technical development and wider usage is evident mostly from the Anuradhapura period (from *circa* twelve Century AD) onwards. He referred to these amazing clusters of tanks [cascades] in the following words:

> So careful were the inhabitants in husbanding the liquid resources on which their very existence depended that even the surplus waters from one tank would spill to the next, when water was plentiful, were not allowed to escape. The tanks were built in an orderly method, at slightly varying elevations so that there often was a series of reservoirs to take the overflow from one above it … (Brohier, 1934, p. 2.)

The evolution of minor tank systems through long periods of history spanning several millennia has resulted in the accumulation of a considerable wealth of indigenous knowledge in the field of irrigation and agriculture (Plan Sri Lanka, 2012). At the same time, the evolution of tanks had its own vicissitudes with times of recession and desolation. It is believed the bigger tank systems often collapsed due to numerous factors, ultimately leading to the collapse of the entire 'hydraulic civilization' in the Dry Zone. In contrast, the number of small tanks within their cascades expanded and contracted to adapt to changes in the population. This resilience of the minor tank systems made them more sustainable than the bigger ones, enabling them to survive through the centuries (Plan Sri Lanka, 2012).

Shah et al. (2013) conclude that "Sri Lanka's Dry Zone is the only ancient irrigation culture that can boast of an unbroken history of local management of village tanks for rice irrigation over millennia". Despite the changes in political dynasties, the social organisation around small and large tanks have remained intact, according to several twentieth-century scholars. Whilst the small tank systems continue to provide many ecological, cultural, spiritual, aesthetic and economic benefits, their functionality has been reduced due to unplanned development activities. As the political and administrative boundaries do not overlap with hydrological boundaries, there has been a dissection of cascades into different administration units, disturbing the ecological cohesiveness of the cascades and this too may have contributed to their degradation. The subject requires further research.

15.4 Environmental Conditions of STCS

STCS are man-made socio-ecological systems and can be regarded as early applications of a landscape approach to spatial planning. This is an agricultural production system that functions in harmony with the ecology within a hydrological boundary. Work by Dharmasena (2004) on small tanks indicates that the village tank systems have been developed to cater to diverse micro as well as macro land uses such as *gangoda* (hamlet), *chena* (shifting cultivation areas), *welyaya* (paddy field), *gasgommana* (forest), *godawala* (waterhole in forest areas for wild animals), *perahana* (filter), *iswetiya* (soil conservation bund), *kattakaduwa* (trees for trapping salts), and *kivulela* (drainage canal). The different interacting components of the STCS provide the habitat for many species of flora and fauna. According to Somasiri (1991), small tanks are probably the most important source of water that recharges the shallow groundwater aquifers of hard rock areas of the Dry Zone that supply potable water to the inhabitants. Restoration of degraded cascades has been identified as a key climate change adaptation mechanism in the National Adaptation Plan for Sri Lanka (Ministry of Mahaweli Development and Environment [MMDE],).

15.5 Social and Institutional Setting of STCS

Though the cascades have been explained in hydrological context, there is no literature available on the existence of cascade level governance mechanism. As Madduma Bandara noted (1985, 2007), this may be because the early management systems of small tanks buried in the historical past still remain poorly understood.

During the pre-colonial era, the village irrigation systems were farmer-managed systems regulated through customary laws and traditions and well managed under '*Rajakariya*' system which saw mandatory community labour being used in the maintenance of commons. As highlighted by Aheeyar (2001), these management actions were governed by '*Gamsabawa*' (village council) headed by '*Gamarala*' (village headman). The British viewed the *Rajakariya* system as forced labour, and with its abolishment in 1832 the customary regulations and traditions of community management of small tanks began to collapse. The vacuum in the management of the small irrigation systems created by the British (Aheeyar, 2001) contributed to the degradation of the STCS. Subsequent interventions such as the Paddy Lands Irrigation Ordinance No. 9 of 1856 introducing '*Velvidane*' (irrigation headman) in place of *Gamarala* and establishment of the Irrigation Department in 1900 led to greater centralisation and bureaucratised what was once a community-managed system. In the post-colonial era, the subject of managing small tanks changed hands between the Irrigation Department and the Department of Agrarian Development, established in 1958 under the *Paddy Lands Act*. With this act the position of the *Velvidane* was abolished and a *Govi Karaka Sabha* (farmer committee council) with a *PalakaLekam* (administrative secretary) was appointed. With this, the mode of compulsory labour maintenance virtually collapsed, but in most places the voluntary *Velvidane* system was able to continue. After several misguided attempts, the Farmer Organisation (FO) system came into operation in the late 1980s, and this has continued up to now with the FOs functioning with the help of the former *Velvidane* (Panbokke et al., 2002). Over time, the institutional arrangements for small irrigation systems changed from community-owned well-managed small irrigation systems to a government institution led dual management system today with no clear community ownership.

The small tank cascades have evolved over a very long period of time incorporating the principles of Integrated Water Resources Management (IWRM), and landscape approaches into planning and possibly in governance. Due to various historical and socio-cultural reasons, these systems were neglected over a long period of time. Cascades were 'rediscovered' as an interconnected system of tanks by Madduma Bandara (1985); however, there are no records on the governance of these systems. Small tanks, the individual components of STCS, are easier to study than cascades and hence, over time, we have been managing individual tanks instead of treating them as a part of a complex system of cascades. These unique soil and water management systems based on hydrological and ecological principles have received more attention in recent times in view of their potential to address climate change vulnerability in agricultural communities in the Dry Zone of Sri Lanka. During the restoration of *Kapiriggama* cascade tank system in Anuradhapura by the International Union for Conservation of Nature (IUCN) together with the Department of Agrarian Development, the local community contributed nearly 10% of the restoration cost through providing their labour free of charge.

15.6 Economics of STCS

Farmers in South Asia can benefit from accessible and affordable technologies for provisioning water for agriculture. Kattel and Nepal (2021, Chap. 11 of this volume) reported significant impact of the adoption of rainwater harvesting on farm income and profitability. For small tank associated agrarian societies, the tank is the most important asset as it provides numerous services in addition to supplying water for irrigation. The tanks provide water for other purposes such as drinking, washing bathing, livestock and wildlife, and help maintain groundwater and the micro-environment. In addition, tank fish has been the main source of protein for small tank associated communities. There are many more food items (Non-Timber Forest Products—NTFP) such as lotus root, seeds, *kekatiya* and other edible aquatic plants that they get from small tanks.

As highlighted by Wijekoon et al. (2016) based on the data from the Department of Census and Statistics, village tank systems contributed to 26% of the 2014/15 *Maha* season paddy extent cultivated (203,836 ha out of total extent of 772,626 ha of paddy) in Sri Lanka. Similarly, they reported that the village tanks' contribution to 2015 *Yala* season was 25% covering 123,375 ha out of the total 480,662 ha paddy land cultivated. Wijekoon et al. (2016), further estimated that the minor irrigation systems represent 28% of 2014/15 total *Maha* season production and 24% of the total *Yala* paddy production in 2015. It is noted that small tank related values capturing their environmental aspects were at individual tank level, cascade level such values are yet to be estimated.

It is also now being increasingly recognised that the use of water for several other essential purposes such as inland fisheries, livestock needs during the dry season, replenishment of groundwater conditions, domestic bathing needs and environment amelioration during the enhanced dry months from July–September should all collectively be assigned an economic and social value.

The rehabilitation or reconstruction of a minor tank is often beyond the capacity of poor communities inhabiting tank villages even though they are cognisant of the importance of the tanks for survival and improvement of their own living conditions. In recent decades, increasing uncertainties of rainfall and water availability associated with on-going climate change have further discouraged the farming communities from investing their meagre resources towards tank rehabilitation. It is in this context that both the governmental as well as several international agencies have come forward to undertake minor irrigation tank rehabilitation during the last few decades (Aheeyar, 2013; Plan Sri Lanka, 2012). However, in addition to giving their labour for restoration as in-kind contribution, a recent study by Vidanage (2019) estimated the local community's willingness to pay for restoration and sustainable management of the *Pihimbiyagollawa* STCS where the non-market values of STCS were estimated using the Choice Experiment method (Table 15.1). These values were subsequently used in an extended cost benefit analysis that determined that the ecological restoration and sustainable management of the STCS are financially justified.

Table 15.1 Willingness to pay for non-market values of STCS

Attribute	WTP/household/year	
	LKR	US$
Water for paddy	25,109	134
Water for other uses	16,366	88
Cascade ecology components	1,973	11
Cascade biodiversity	5,880	31
Cascade wide total value	49,328	264

Source Vidanage (2019)

15.7 Sustainable Management of the STCS

As highlighted in previous sections, maintenance of small tank cascade systems has been neglected over a long period of time. One of the fundamental reasons for this is a lack of knowledge about the ecosystem values of the STCS. People disregard the benefits of the ecological components of the system limiting the calculation of benefits to just the productive components (i.e., individual tanks). This is further aggravated when there is no governance mechanism in place to take care of the entire cascade. Tanks are managed individually by Farmer Organisations devoid of the linkages between them and other ecological components.

As ecosystem goods and services are not provided and valued through the market system their monetary value is not reflected and they have not been recognised by formal economic investment analysis tools nor informal political decision-making processes. This coupled with loss of social values which would view the STCS as communal property has led individuals try to maximise private benefits by engaging in activities which are detrimental to sustaining the STCS as a functional unit.

The lack of understanding of cascade dynamics, ad hoc restoration of individual tanks (part of the system) without looking at a cascade as a functional system/unit also have given rise to negative results. Various poorly planned development activities such as catchment deforestation, and land alienation also disturb these systems. Poor understanding of the values that small tank cascades provide as a multipurpose system is arguably the main issue for not drawing the attention of the policy makers on these systems.

Even though these systems are meant to generate multiple benefits to the local economy, and actually do so, the planners/decision-makers tend to take only the irrigation benefits of these systems for assessing the economic feasibility of restoration of degraded systems, as these are the only market values available. They often conclude that restoration of the entire STCS is not financially or economically feasible. Hence, the tendency is to rehabilitate/restore one or two tanks in a given system ignoring what it means to the cascade system of tanks as a whole. However, as they were designed to work as interconnected systems of tanks, the rehabilitation of a few tanks in isolation do not bring about the expected results and may even contribute to destabilise the systems.

Vidanage (2019) demonstrated the economic feasibility of restoration of degraded STCS looking at their multiple benefits. Such restoration will increase the climate resilience of the tank associated communities as the well-functioning restored systems would prove to be an adaptation measure for climate change vulnerability. The National Adaptation Plan for Climate Change Impacts in Sri Lanka 2016–2025 has identified participatory cascade management programmes in selected village tank catchments as one of the sector actions in its water resources action plan (MMDE, 2016). Further, the government of Sri Lanka is expecting funds under the Green Climate Fund for 'Strengthening the resilience of smallholder farmers in the Dry Zone' to climate variability and extreme events through an integrated approach to water management. This will restore 17 cascades consisting of 320 small tanks

Fig. 15.1 Areal view of *Konakumbukwewa*—a small tank in the *Kapiriggama* small tank cascade system. *Photo Credit* Gayan Pradeep, IUCN

amongst other things covered in the Project (Green Climate Fund, 2017). There are other initiatives such as feeding cascades in the North Central Province, whilst conveying water from *Moragahakanda and Kaluganga* Reservoirs through the Upper *Elahera* Canal of the North Central Province Canal Project (NCPCP) to the North (Mahaweli Authority of Sri Lanka, 2015).

Long neglected STCS are now being revived looking at their climate resilience and restoration of degraded STCS is being recognised as a viable option for adaptation in the National Action Plan for climate change in Sri Lanka as well as articulated in the Nationally Determined Contributions (NDCs) under the Paris Agreement as adaptation actions. Community participation by way of contributing free labour in recent projects for cascade-wide restoration has demonstrated greater farmer willingness to restore and sustainably manage the STCS (IUCN, 2016). This has also contributed towards the economic feasibility of STCS restoration. Findings of Vidanage (2019) indicate that there will be greater interest in reviving these systems for their multiple benefits including their climate change adaptation ability for small scale farming communities in the Dry Zone of Sri Lanka.

Fig. 15.2 Typical degraded small tank in a cascade. *Photo credit* Shamen Vidanage

References

Aheeyar, M. M. (2001). Socio-economic and institutional aspects of small tank system in relation to food security. In H. P. M. Gunasena (Eds.), *Food security and small tank systems in Sri Lanka: Workshop proceedings*. National Science Foundation.

Aheeyar, M. M. (2013). *Alternative approaches to small tank/cascade rehabilitation: Socio-economic and institutional perspective*. Hector Kobbekaduwa Agrarian Research and Training Institute.

Brohier, R. L. (1934). *Ancient irrigation works in Ceylon*. Ceylon Government Press.

Department of Agrarian Development. (2005). The map of main watersheds, sub watersheds, village tank cascades and anicut clusters of Sri Lanka.

Dharmasena, P. B. (2004). Small tank heritage and current problems. In *Small tank settlements in Sri Lanka* (pp. 31–39). Hector Kobbekaduwa Agrarian Research and Training Institute.

Green Climate Fund. (2017). Retrieved from http://www.greenclimate.fund/what-we-do/projects-programmes.

International Union for Conservation of Nature. (2016). *Ecological restoration of kapiriggama cascade*. IUCN.

Kattel, R. R., & Nepal, M. (2021). Rainwater harvesting and rural livelihoods in Nepal . In A. K. E. Haque, P. Mukhopadhyay, M. Nepal, & M. R. Shammin (Eds.), *Climate change and community resilience: Insights from South Asia* (pp. 159–173). Springer.

Madduma Bandara, C. M. (1985). Catchment ecosystems and village tank cascades in the dry zone of Sri Lanka. In U. L. Lungquist, & M. Halknmark (Eds.), *Strategies for river basin development*. Riedel Publishing Co.

Madduma Bandara, C. M. (2007). *Village tank cascade systems of Sri Lanka: A traditional technology of water and drought management*. EDM-NIED, Kobe.

Mahaweli Authority of Sri Lanka (MASL). (2015). *EIA report of the upper Elehara Canal, canal from Mannakkattiya Tank to Mahakanadarawa tank and Kaluganga-Moragahakanda Link Canal project*. MASL, Colombo.

Ministry of Environment. (2010). *National climate change adaptation strategy for Sri Lanka*. Ministry of Environment.

Ministry of Mahaweli Development and Environment. (2016). *National adaptation plan for climate change in Sri Lanka: 2016–2025*. Climate Change Secretariat, MMDE.

Murray, F. J., & Little, D. C. (2000). *The nature of small-scale farmer managed irrigation systems in North West Province, Sri Lanka and potential for aquaculture.* Working Paper SL1.3. Project R7064.

Panabokke, C. R., Tennakoon, M. U. A., & Ariyabandu, R. De. S. (2001). Small tank system in Sri Lanka: Summary of issues and considerations. In H. P. M. Gunasena (Eds.), *Food security and small tank systems in Sri Lanka: Workshop proceedings.* National Science Foundation.

Panabokke, C. R., Sakthivadivel, R., & Weerasinghe, A. (2002). *Small tanks in Sri Lanka: Evolution, present status and issues.* International Water Management Institute (IWMI).

Panabokke, C. R. (2004). *Small tank settlements in Sri Lanka* (pp. 56–68). Hector Kobbekaduwa Agrarian Research and Training Institute.

Plan Sri Lanka. (2012). *Cascade irrigation systems for rural sustainability: Experience of plan Sri Lanka's cascade systems development project in the North Central Province of Sri Lanka 2004–2010.* Plan Sri Lanka.

Renwick, M. (2001). *Valuing water in irrigated agriculture and reservoir fisheries: A multiple-use irrigation system in Sri Lanka.* International Water Management Institute (IWMI). Research Report 51.

Shah, T., Samad, M., Ariyaratne, R., & Jinapala, K. (2013). Ancient small-tank irrigation in Sri Lanka: Continuity and challenge. *Economic and Political Weekly* (XLVIII), 11.

Shannon, K., & Manawadu, S. (2007). Indigenous landscape urbanism: Sri Lanka's reservoir and tank system. *Journal of Landscape Architecture, 2*(2), 6–17. https://doi.org/10.1080/18626033.2007.9723384

Somasiri, S. (1991). Irrigation potential of minor tanks and their agricultural stability. *Tropical Agriculturist, 147,* 41–58.

Tennakoon, M. U. A. (2001). Evolution and role of small tank cascade systems in relation to the traditional settlement of the Rajarata. In *Food Security and Small Tank Systems in Sri Lanka* (pp. 13–31). National Science Foundation.

Tennakoon, M. U. A. (2004). Tanks are not mono functional: They are multifunctional. In M. M. Aheeyar (Ed.), *Small tank settlements in Sri Lanka.* Hector Kobbekaduwa Agrarian Research and Training Institute.

Vidanage, S. P. (2019). Economic Value of an Ancient Small Tank Cascade, System in Sri Lanka, Doctoral Research Series of the Department of Economics, University of Colombo.

Wijekoon, W. M. S. M., Gunawardena, E. R. N., & Aheeyar, M. M. M. (2016). Institutional reforms in minor (village tank) irrigation sector of Sri Lanka towards sustainable development. In *Proceedings of the 7th International Conference on Sustainable Built Environment 2016,* Kandy.

Part IV
Disaster Risk Reduction

Chapter 16
Frameworks, Stories and Learnings from Disaster Management in Bangladesh

Md Rumi Shammin, Remeen Firoz, and Rashadul Hasan

Key Messages

- Bangladesh has developed a robust framework of disaster management institutions, plans and rules to deal with climate vulnerability over three decades.
- The framework includes vertical integration of institutions across various levels of government and horizontal integration of stakeholders within each level.
- Participatory processes and community engagement at the grassroots level have emerged as successful examples of efficient and cost-effective solutions that leveraged traditional knowledge and reduced loss of lives and livelihoods.

16.1 Introduction

The specter of climate change threatens worsening natural disasters, rapid urbanization, forced migration, and economic hardship for the most vulnerable—Tedros Adhanom, Director-General, World Health Organization (Adhanom, 2017).

As the world warms, the frequency and severity of climate change-related extreme events such as heatwaves, extreme precipitation, and coastal flooding are increasing globally (IPCC, 2014). The link between climate change and natural disasters is

M. R. Shammin (✉)
Environmental Studies Program, Oberlin College, Oberlin, OH, USA
e-mail: rumi.shammin@oberlin.edu

R. Firoz
Dhaka, Bangladesh
e-mail: firozremeen@gmail.com

R. Hasan
Bangladesh Disaster Preparedness Centre, Dhaka, Bangladesh
e-mail: rashadul.hasan@gmail.com

© The Author(s) 2022
Haque et al. (eds.), *Climate Change and Community Resilience*,
https://doi.org/10.1007/978-981-16-0680-9_16

becoming clearer—especially through developments in impact attribution research. High-resolution datasets and more sophisticated models have allowed researchers to find the fingerprint of climate change in individual weather events (Ornes, 2018). A report published in the Bulletin of the American Meteorological Society included studies examining 168 specific weather events in 2018, of which 122 were found to be influenced by climate change (Herring et al., 2020). IPCC defines climate-induced extreme weather events as climate extremes (IPCC, 2012). The impacts of observed climate extremes have been particularly intense for people and communities in vulnerable and disaster-prone regions of the world (Eckstein et al., 2019).

While Bangladesh accounts for ~ 0.5% of global GHG emissions (Ge & Friedrich, 2020), it has been experiencing the impact of climate extremes for over three decades and thus has had long experience with disaster risk reduction and emergency preparedness programmes. These initiatives are often cited as successful examples that have greatly minimized the loss of life and ensured livelihood and food security for millions of people annually (Molla, 2019).

This chapter summarizes the disaster management framework in Bangladesh. We have deconstructed the vulnerability, impacts and responses to natural disasters and unpacked the national and sub-national institutional mechanisms for disaster risk reduction. We describe how the disaster management process works at the grassroots level from the perspectives of the operatives on the ground through site visits and conversations with stakeholders. We also tell stories of unique initiatives undertaken by the government, a local NGO, and an international organization. We highlight the lessons learned for wider application.

16.2 Climate and Disaster Vulnerability

The character and severity of impact from climate extremes depend on exposure and vulnerability (IPCC, 2012). Bangladesh is geographically, geologically, hydrologically, and meteorologically vulnerable to natural disasters. It is the largest delta on earth facing the Bay of Bengal formed at the confluence of three major rivers: Brahmaputra, Ganges, and Meghna. This unique landscape creates ideal circumstances for cyclones, floods, and storm surges. Meandering rivers change course due to riverbank erosion with silted islands (*chars*) forming and disappearing in a matter of years. Bangladesh has always wrestled with abundance and scarcity of water: too much water floods a large swath of the country in the wet season, while too little water causes droughts during the dry season. The plate tectonics of the Himalayan region also brings moderate to severe earthquakes to the country (GoB, 2009a).

The demography of Bangladesh also adds to its vulnerability to natural disasters. It is the most densely populated country in the world (except a few smaller nations/states with population < 10 million) with a total population of ~ 165 million and a population density of ~ 3,277 per square mile (United Nations, 2019). About 63% of the population living in rural communities are predominantly engaged in farming and fishing, often poor and without access to adequate infrastructure (The

World Bank, 2018). Many of these communities are situated in disaster-prone areas and are exposed to a multitude of natural disasters that exacerbate poverty, landlessness, and loss of livelihood. A study published in Nature Communications states that more than 70% of the total number of people worldwide currently living on climate implicated land are in eight Asian countries, including Bangladesh and India (Kulp & Strauss, 2019).

From the north of the sub-continent, climate-induced glacial melt is bringing more water with heavy sediment loads from the eroded soil of the Himalayas down to the riverine delta. Monsoon rain floods large parts of the country and dry season droughts triggered by extreme heat evaporates the water and deposits silt in riverbeds. This creates the perfect storm for catastrophic annual floods that erode riverbanks and inundate farmlands, villages, and towns. Normal flooding (*barsha*) affects about 25% of Bangladesh each year, but land use and settlements are well adapted to it. Severe flooding (*bonya*) or flash floods (*baan*) can submerge more than half of the landmass—damaging crops and property, disrupting economic activities, and causing injury and loss of life.

From the south, a rising ocean aided by stronger tidal and wind movements is forcing volumes of water inland causing coastal flooding, erosion, and saltwater intrusion. Formerly fertile agricultural lands have become unfit for cultivation and freshwater fisheries are rendered unfeasible. Before and after the monsoon season, cyclones with wind speeds up to 280 kmph develop in the Bay of Bengal with predictable regularity. Storm surges and associated tidal waves rise to 10 m in height resulting in catastrophic consequences for the coastal inhabitants (Schmuck, 2003). Cyclones bring multiple threats of severe winds, storm surges, and heavy rainfall that result in both surface and riverine flooding, with bigger storm surges often inundating areas protected by embankments. The frequency of floods and storms (including cyclones) in Bangladesh shows a clear increase in the total number of events in the three most recent decades compared to the previous two (Fig. 16.1).

Over the centuries, the people of Bangladesh have learned to develop ways of life adapted to the inevitable natural calamities of the area while still aspiring to improve the socio-economic circumstances of the poor and the vulnerable—especially since the nation's independence in 1971. However, climate change has become the straw that threatens to break the camel's back for this naturally and demographically vulnerable country. It is predicted that with a 0.5-m sea-level rise, Bangladesh could lose approximately 11% of its land, affecting an estimated 15 million people living in its low-lying coastal region by 2050 (EJF, 2020). A 1.0-m sea-level rise could submerge 17% of the coast, inundate 20% of the country, and potentially displace more than 30 million people (Glennon, 2017). It is thus not surprising that Bangladesh has often been labelled as "ground zero for climate change or the poster child of climate vulnerabilities and impacts" (Szczepanski et al., 2018).

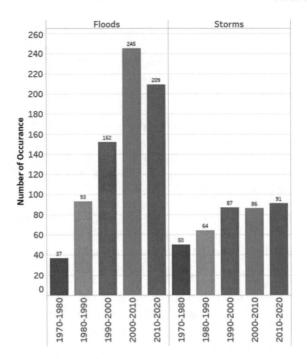

Fig. 16.1 Occurrence of floods and storms by decade in Bangladesh since 1970. *Source* Authors' calculations based on data from the Emergency Events Database, Centre for Research on the Epidemiology of Disasters, Université catholique de Louvain

16.3 Anatomy of Disaster Management in Bangladesh

Major developments in disaster management in Bangladesh began in the 1990s. At the national level, the Ministry of Disaster Management and Relief (MoDMR) was responsible for inter-ministerial planning, coordination, and disaster response. Under the MoDMR, there are two agencies: the Disaster Management Bureau (DMB) and the Directorate of Relief and Rehabilitation (DRR). The DMB, established in 1993, was a small professional unit at the national level that performed specialist functions at both national and local levels. The government also set up a national council and various committees at national, district, Upazila (sub-district) and union (local council) levels for disaster preparedness. The main focus of disaster management was relief and rehabilitation guided by the first Standing Orders on Disaster (SOD) published in 1997—later modified and translated in English in 1999 (Haque & Uddin, 2013).

The 2000s saw a global shift in disaster management towards a more holistic model that prioritized prevention, preparedness, response, and recovery. The aim was to address multiple types of hazards, harness the power and resources of community-based initiatives, and involve multiple stakeholders. The Government of Bangladesh developed the Comprehensive Disaster Management: A Framework for Action (2005–2009). Bangladesh also signed the *Hyogo Framework for Action 2005–2015*, adopted at the 2005 World Conference on Disaster Reduction (held in Kobe, Japan), and later, hosted a summit of South Asian countries who collectively adopted *Disaster Management in South Asia: A Comprehensive Regional Framework for Action 2006–2015*. These developments, coupled with two major floods in 2004 and 2007, and the catastrophic cyclone Sidr in 2007 followed by Aila in 2009, prompted revisions to the SOD 1997, catalyzed the creation of national plans and policies, and shifted the paradigm of disaster management in Bangladesh from *relief and rehabilitation* to *disaster risk reduction*.

16.3.1 National Regulatory and Institutional Mechanisms

Under the new paradigm, a National Plan for Disaster Management (NPDM) 2010–2015 was created (and subsequently updated for 2016–2020) and updated Standing Orders on Disaster (SOD) were published (GoB,). These umbrella documents promoted two main categories of activities: disaster risk reduction (DRR) and emergency response management (ERM). These activities are mandated by a set of regulatory and institutional mechanisms at the national level. The current regulatory framework and institutional mechanisms in Bangladesh are shown in Table 16.1.

According to the 2010 SOD, the Department of Disaster Management or DDM (formerly Disaster Management and Relief Division or DM&RD) under the auspices of the Ministry of Disaster Management and Relief or MoDMR (formerly Ministry of Food and Disaster Management or MoFDM) is responsible for administering all disaster management issues. The Disaster Management Information Centre (DMIC) within the DMB is mandated to assist the Ministry with all necessary information during normal time, warning and activation, emergency response, relief, and recovery. The Ministry provides information to the NDMC, IMDMCC, the relevant cabinet committee and the National Disaster Response Coordination Group (NDRCG) and assists them in the decision-making process. The Secretary of the Disaster Management and Relief Division of MoFDM coordinates the activities of all officials engaged either directly or indirectly in emergency response, relief, and recovery activities.

The SOD is the most comprehensive framework for disaster management in Bangladesh. Taking a holistic approach to disaster risk reduction, the SOD created structures for inclusive management and implementation of DRR and ERM. The subsequent development of acts and rules provided the necessary legal authority for local level disaster management institutions to leverage cross-sectoral resources and deliver prompt and well-coordinated response in the event of a disaster. The local level operatives that we have spoken to all mentioned this document as the primary reference for their workflow. It is, therefore, not just a comprehensive document on paper, but an applied document that is actively used for disaster management and response on the ground. The updated SOD released in 2019 is expected to continue the tradition.

Table 16.1 Regulatory framework and institutional mechanisms for disaster management in Bangladesh 2010 onwards

Regulatory framework	Institutional mechanisms
National Plan for Disaster Management (2016–20, 2010–15)	National Disaster Management Council (NDMC)
Standing Orders on Disaster (SOD), 2010 (*updated SOD has just been released in 2019*)	Inter-Ministerial Disaster Management Coordination Committee (IMDMCC)
Disaster Management Act 2012	National Disaster Management Advisory Committee
National Disaster Management Policy (2015)	Earthquake Preparedness and Awareness Committee (EPAC)
Guidelines for Government at all Levels	National Platform for Disaster Risk Reduction (NPDRR)
	National Disaster Response Coordination Group (NDRCG)
	Cyclone Preparedness Programme (CPP) Policy Committee
	Committee for Focal Points Operational Co-ordination Group
	Co-ordination Committee of NGOs relating to Disaster Management
	Disaster Management Training and Public Awareness Task Force (DMTPATF)

Source Authors' adaptation

The National Disaster Management Policy of 2015 was formulated to define the national perspective on disaster risk reduction and emergency management, describe the strategic framework, and outline the national principles of disaster management in Bangladesh. It is strategic in nature and describes the broad national objectives and strategies in disaster management. The NPDM 2016–20, published in 2017 offers several enhancements to SOD 2010 and NPDM 2010–15. It takes inspiration from the four priorities of the *Sendai Framework for Disaster Risk Reduction 2015–2030* adopted at the Third UN World Conference on Disaster Risk Reduction in Sendai, Japan, on March 18, 2015: (i) Understanding disaster risk; (ii) Strengthening disaster risk governance to manage disaster risk; (iii) Investing in disaster reduction for resilience and; (iv) Enhancing disaster preparedness for effective response, and to "Build Back Better" in recovery, rehabilitation and reconstruction. The Sendai Framework aims to achieve substantial reduction of disaster risk and losses in lives, livelihoods, and health and in the economic, physical, social, cultural, and environmental assets of persons, businesses, communities, and countries (United Nations, 2015). NPDM 2016–20 also incorporates the UN Sustainable Development Goals (SDGs) which became available in 2015 as part of the 2030 Agenda for Sustainable Development (United Nations, 2020). Notable additions in the updated NPDM were: resilience as a central theme; gender equality by ensuring participation of women in committees; synergy between DRR and climate change adaptation plans; and connecting DRR with sustainable development. Shammin et al. (2022, Chap. 2 of this volume) has developed an integrative framework for climate-resilient communities that incorporates DRR and SDGs.

The national-level framework described above is intended to foster cross-sectoral coordination, streamline collaboration and information sharing to develop and update policy and management guidelines, leverage necessary resources and support local-level organizations in developing and implementing long-term DRR programs and event-specific emergency response measures.

16.3.2 Local-Level Institutional Mechanisms

SOD 2010 outlines institutional mechanisms at the local level (district and below) to facilitate coordination, institutional capacity and logistics of prevention, mitigation, preparedness and response, and relief (Table 16.2). A larger disaster management committee brings together representatives from all government departments including police and armed forces, non-government organizations, key religious and educational establishments, and civil society. A smaller group under the auspices of Local Disaster Response Coordination Group (LDRCG) is created with fewer key stakeholders to ensure coordination during and after disasters at all levels.

While the committees and coordination groups are designed to be inclusive, the lower tiers at the municipal and union levels offer the greatest potential for community engagement, place-based solutions and cost savings.

Table 16.2 Local-level Institutional Mechanisms for Disaster Management in Bangladesh 2010 onwards

Local Level institutional mechanisms
District Disaster Management Committee (DDMC) and Response Coordination Group (DDRCG)
City Corporation Disaster Management Committee (CCDMC) and Response Coordination Group (CCDRCG)
Upazila Disaster Management Committee (UzDMC) and Response Coordination Group (UDRCG)
Pourashava (Municipal) Disaster Management Committee (PDMC)and Response Coordination Group (PDRCG)
Pourashava Ward Disaster Management Committee (PWDMC) and Response Coordination Group (PWDRCG)[a]
Union Disaster Management Committee (UDMC) and Response Coordination Group (UDRCG)
Union Ward Disaster Management Committee (UWDMC) and Response Coordination Group (UWDRCG)[a]

Source Authors' adaptation
[a] Added in SOD 2019

16.4   Economics of Disaster Management

Natural disasters have had significant economic costs for Bangladesh. Between 1998 and 2011, five major disasters incurred damages amounting to roughly 15% of GDP—an average of 2.7% per event (Ahmed et al., 2015). A significant part of the losses can be attributed to climate change (GoB, 2012). From 2010–2015, there were 17 high/medium impact natural disasters in Bangladesh and many localized hazard events, with economic losses ranging from 0.8 to 1.1% of GDP (GoB, 2017). According to a study on disaster-related public spending, the Government of Bangladesh allocated Tk. 15,097 crore (~ US$19 billion) for 164 projects between 2011 and 2015. Out of this allocation, 68.5% was spent for disaster preparedness and risk reduction. Bangladesh contributed 63.4% of the cost and the remaining was funded by external donors (GoB, 2016). During the fiscal year 2020–2021, the Government of Bangladesh has allocated Tk 9,836 crore (~ US$12.5 billion) for disaster management and relief (Dhaka Tribune, 2020).

Mechler (2016) reviewed 39 CBA case studies from around the world, including three from Bangladesh, and determined that the benefits of investing in DRR in terms of avoided and reduced losses outweigh the costs of implementing these programs by a factor of four. Das (2021, Chap. 17 of this volume) determined that the annual storm protection value of mangrove forests is more than twenty times higher than the return from alternative land uses. Taking advantage of indigenous knowledge, local resources and community participation often result in solutions that build on past practices and produce low-cost alternatives to top-down infrastructure-oriented interventions. Valerie Amos, the Under-Secretary-General for Humanitarian Affairs of the United Nations, recognised the DRR initiatives of Bangladesh as examples of low-cost approaches that save lives (UN News, 2012).

16.5   Disaster Management on the Ground

We visited several climate-vulnerable, disaster-prone areas in the Cox's Bazar district of south-western Bangladesh to better understand how the disaster management framework was operating on the ground. We had detailed conversations with the District Disaster Risk and Rehabilitation Office (DRRO), Executive Engineer of the Local Government Engineering Department (LGED), local people (e.g. school teachers, volunteers) and representatives of Upazila Porishod (sub-district administration), village police, and non-government organizations. Through these conversations, we documented the local-level DRR and ERM implementation process from their perspectives.[1,2]

---

[1] Field notes by Md Rumi Shammin from Khurushkul and Cox's Bazar, Bangladesh in 2019.

[2] Field notes by Md Rumi Shammin from Dhaka and Cox's Bazar, Bangladesh in 2020.

The DRRO referred to the SOD 2010, NPDM 2016–20 and Disaster Management Act 2012 as primary reference documents for planning, organization, and implementation at the district level and below. The DRRO talked about the SDGs as an overarching mission and coordination with climate change adaptation as an important operational strategy. Climate change and natural disasters regularly negate progress towards the SDGs goals in vulnerable communities making disaster risk reduction an essential prerequisite. The DRRO emphasized the importance of allocating 15–20% of the SDG budget towards DRR and ERM.

The 2012 Disaster Management Act provides legal mandates to local level administrations to secure logistical support from stakeholders during disasters and ensure action and accountability of all involved. This allows committees at the district level and below to engage with government, private, and non-governmental entities efficiently and authoritatively. For example, transport unions are called to provide transportation support, merchant unions are asked to help with emergency food supply and post-disaster price stabilization, and industries are asked to share warehouses for shelter and/or food storage. Fire service and civil defense forces are leveraged as needed. About 36 local and international NGOs are involved in various stages of DRR and ERM in the Cox's Bazar area including the World Food Program (WFP), Bangladesh Rural Advancement Committee (BRAC), Mukti and the Bangladesh Red Crescent Society (BDRCS).

The BDRCS has been instrumental in Bangladesh for nearly five decades working on disaster management, emergency recovery and relief. One of their flagships programs is the Cyclone Preparedness Program (CPP). The CPP is unique in the sense that it engages volunteers such as students and teachers from the local community (CPP, n.d.).

In Cox's Bazar, the CPP operation is administered by a deputy director who supervises 8 officers, one in each Upazila. Each officer oversees the union level volunteer leaders. Each volunteer group consists of about 20 volunteers. At the time of our field visits in 2018 and 2019, there were 538 cyclone shelters in place and 19 more under construction. Communication infrastructure includes UHF/VHF radios and mobile phones managed through the District Disaster Management Control Room and Emergency Operation Centre (EOC) with generator support for power outages. These are open year-round with 24-hour phone service for people to report events or seek support.

Under normal circumstances, disaster management committee meetings are held once every two months. Anytime there is a warning signal of level 4 or above, the committees at all levels meet on the same day and begin to mobilize personnel and resources. Bangladesh Meteorological Department issues early warnings of impending disasters which are communicated to both national and local level institutions. The control room at the district level then communicates the warning and subsequent development of the event to all local level committees, CPP representatives and other stakeholders down to the union level. Each union consists of nine wards. In addition to cell phones and radios, the ward level action is often handled through volunteers walking through neighborhoods with hand mikes and megaphones. In case of major disasters, people are guided to nearby designated shelters with provisions

Table 16.3 Deaths due to floods and storms by decade in Bangladesh since 1970

	Decade	Total deaths	Total impacted
Floods	1970–1980	29,375	52,614,127
	1980–1990	7,064	123,268,327
	1990–2000	2,944	58,165,806
	2000–2010	2,404	58,649,763
	2010–2020	658	26,156,421
Storms	1970–1980	307,144	3,740,174
	1980–1990	18,977	17,275,998
	1990–2000	142,594	23,658,485
	2000–2010	5,588	13,282,990
	2010–2020	592	9,404,929

Source Authors' calculations based on data from the Emergency Events Database, Centre for Research on the Epidemiology of Disasters, Université catholique de Louvain

for dry food and ramps to allow people to bring their livestock with them. The CPP program volunteers work alongside the government program participants/volunteers to execute this emergency management process. The participants and volunteers are trained beforehand through programs developed in collaboration between public disaster management entities, BDRCS and local fire service personnel. Appropriate chain of command is maintained at all levels of implementation.

The hallmark of this entire process is a broad collaboration between public, private, and non-profit organizations supplemented by representative public participation and engagement of local volunteers—ensuring financial, resource, and program efficiencies in reducing the impacts of natural disasters in Bangladesh on lives and livelihoods (BDRCS, n.d.; UN News, 2012). Table 16.3 demonstrates that despite the increased frequency of floods and storms shown in Fig. 16.1, the number of impacted people and fatalities have decreased in recent decades.

16.6   Stories of Innovation in Disaster Management

16.6.1   Empowering Women Through Community-Based Adaptation in Teknaf

The Department of Environment (DoE) in Bangladesh has designated ecosystems such as forests, islands, or wetlands in the country as "Ecologically Critical Areas" (ECAs) that require immediate conservation action to halt further degradation. In Cox's Bazar, the DoE manages the ECAs through Village Community Groups (VCGs) and ECA Management Committees at the union, Upazila and district levels. In Cox's Bazar and Teknaf areas the DoE has implemented several initiatives in recent decades and one noteworthy example is the Community-based Adaptation for Ecologically Critical Areas (CBA-ECA) project. In addition to several anthropogenic threats such as land conversion and encroachment, climate change-induced disasters pose a major threat to the conservation of ECAs and the lives and livelihoods of the adjoining communities. This project aims to build resilient communities that are better able to cope with disasters and withstand the effects of cyclones and tidal surges (DOE, 2015).

The CBA-ECA project uses skills and technology to empower women through various training programs and introduction of alternative livelihood options. It introduced the use of improved cooking stoves that drastically reduced indoor air pollution and made kitchens a safer place for women. Solar irrigation and desalination plants installed in selected sites in Cox's Bazar and Shah Porir Dweep (island) added significantly to the safety and well-being of women who are primarily in charge of collecting water and fuel.

During a Focus Group Discussion held with a VCG in Tulatoli, Cox's Bazar, several interesting findings were revealed from the consultation. The VCG has 30 members—all of them are women. These women members are vulnerable coastal mothers and daughters, who collect molluscs and shells, hunt crabs, and cut small branches of trees to make a living. Ever since they were associated with the project, they have been trained and engaged in making handicrafts such as beaded necklaces, hand-woven lace caps, sewing and embroidery, etc. They are also conversant in maintaining books and accounts and aware of conservation and disaster-related issues. These women contribute a meagre amount of 50 Taka (BDT) every month to the community fund and accumulated a handsome capital; over time, they have started a system of "rotational loans" within the group. When asked how their lives have changed by being members of the VCG, one of the women said *"taka jaar haathe, khomota taar kache"* meaning that the one with money is the one with power. Resilient and economically solvent communities in a gender-balanced society are better equipped to tackle disasters and climate change, as witnessed in the case of the women of Tulatoli.[3]

16.6.2   Ward-Level Organization in Cyclone Fani Response in Goroikhali

Goroikhali is a Union Parishad (rural level administrative unit) under Paikgacha Upazila on the bank of Shibsha River near the Bay of Bengal. An embankment was constructed on the bank of the river after cyclone Sidr of 2007 to protect the land from storm surges and saline water intrusion. The *Panii Jibon* (Water is Life) project of Bangladesh Disaster Preparedness Centre (BDPC) has been working to strengthen disaster preparedness and livelihoods in the south-western coastal Upazilas of Paikgacha and Koyra since January 2018. In the Goroikhali union, BPDC facilitated

---

[3] Field Notes by Remeen Firoz from Tulatoli, Teknaf, and Cox's Bazar, Bangladesh in 2015.

the formation of the Union Ward Disaster Management Committee (UWDMC)[4]—a level below the UDMCs that existed in the government framework—to strengthen community-based disaster risk reduction. The WDMC consists of 11 members from different strata of the local society under the leadership of local UP member of Ward#4. BDPC organized training for the UWDMC members on disaster risk reduction, climate change adaptation, community risk assessment, and early warning systems. The training was followed by the development of a Risk Reduction Action Plan (RRAP).

When tropical cyclone Fani developed in the Bay of Bengal on 2 May 2019, the water level in the Shibsha River started to increase—causing a part of the embankment to breach near Ward#4. A complete embankment failure could submerge the nearby four Unions. Community members observed the embankment breach and the president of the WDMC immediately informed the Union level committee (UDMC) of Goroikhali. The UDMC immediately called an emergency meeting and informed both the Upazila level committee (UzDMC) and the Bangladesh Water Development Board (BWDB). The UDMC promptly employed 100 labourers to repair the embankment (using an earmarked fund: The 100-day work budget). In addition, 50 volunteers, including the WDMC members of Ward nos. 2, 3 and 4 participated in the repair work. They finished the primary repair in three days—sparing the people of the four Unions from the potential disaster.

While the repair work was ongoing, the local BWDB officials visited the damaged embankment area and initiated the process of permanently repairing the embankment with proper design standards administered by their engineers. The WDMC of Ward#4 played a critical role in reducing the risk of embankment failure throughout the process: from the initial communication and information dissemination to the evacuation of people and completion of the emergency repair work. Cyclone Fani caused widespread damage elsewhere including the death of 81 people in India and Bangladesh and USD 8.1 billion in economic losses (AON, 2020). The President of the WDMC said: "As a local community member it is our responsibility to do the immediate response, the WDMC gave us a platform and idea about formal process of response through the appropriate authority. Our quick response has saved a large area of Paikgacha and Koyra from being submerged by saline water."[5]

16.6.3 "Let's Hear Rana Bhai": A Climate Education Program in Noakhali

In association with the Char Development and Settlement Project (CDSP) III, the Bangladesh branch of the International Union for Conservation of Nature (IUCN) implemented an education and awareness project in the coastal areas of Noakhali,

[4] Even though the official name has the 'Union' prefix to distinguish it from the 'Pouroshova' level committee, these are referred to as WDMC in practice.

[5] Field notes by Rashadul Hasan from Goroikhali, Paikgacha, Bangladesh in 2019.

Fig. 16.2 Educational material showing Rana Bhai talking about climate change. *Source* IUCN (2010) (with authors' translations)

with a special focus on the children of 15 different schools and madrasa (religious school). The activities consist of essay and writing workshops, art camps, participatory drama, and distribution of awareness materials like posters and stickers and creation of a climate change mascot named "Rana Bhai".

Inspired by the success of 'entertainment education' of UNICEF's *Meena Communications Initiative*, the mascot 'Rana Bhai' was created, with close consultation of children from the coastal communities (see Fig. 16.2). Rana Bhai is an iconic character who radiates knowledge and wisdom. His very name Rana has been derived from the scientific name for bullfrogs, *Rana tigerina*. As a Bangladeshi national, he takes the appearance of a *shona bang*, a frog species indigenous to Bangladesh. Being an indicator species, the amphibious bright yellow Rana Bhai is very susceptible to climate change. Children related to him better because Rana is a common Bangla (Bengali) name and the suffix 'bhai' literally means 'brother.' The participatory drama 'Let's hear Rana Bhai' was performed and filmed to advocate climate change messages and encourage school children to discuss adaptation measures with friends and families. Rana Bhai, the climate change mascot, is also the star of his own documentary on climate change targeted at school children (Adrika et al., 2011) (Fig. 16.2).

Although there is no documented evidence of the impact of the communication materials of the project, we had consulted the school children and local communities following cyclone Sidr and Aila. They informed us that because of the enhanced awareness, there were fewer casualties as most of them had taken shelter in the nearby school buildings. The project also went on to become one of the 100 most impactful climate change projects in 2012, selected by the British Broadcasting Corporation (BBC). Some of this project's activities were replicated and scaled up for other coastal regions in Bangladesh like Barisal, Potuakhali and Borguna through the Climate Livelihood Adaptation Project (CLAP) funded by the Deutsche Gesellschaft für Internationale Zusammenarbeit (GIZ).[6]

16.7   Lessons Learned and Recommendations

An overview of the disaster risk reduction and emergency response mechanisms employed in Bangladesh to deal with ongoing and worsening exposure to natural disasters reveal the importance of coordination at multiple levels and broad engagements of stakeholders. These systems have greatly reduced the scale of loss of lives, livelihood and property in the country (see Table 16.3). The takeaway lessons from the DRR and ERM systems in place in Bangladesh are summarized below.

Government Commitment and Institutional Framework

Disaster preparedness, management and recovery are high priority activities for Bangladesh. The government has taken a holistic, multi-level, multi-sectoral and multi-stakeholder approach to develop policies, guidelines, management and action plans, legal frameworks, rules, and implementation plans to address these issues. The Prime Minister of the country heads the National Disaster Management Council (NDMC). Legal mandate afforded by the Disaster Management Act of 2012 empowers the disaster management institutions to coordinate across various levels of the government and leverage resources from a wide range of stakeholders.

Stakeholder Engagement

The national level committees have representation from all relevant government units and ministries as well as inclusion of international agencies, non-governmental agencies, and private entities as appropriate. At local levels, government representatives interface with local organizations (religious and educational institutions), local dignitaries and volunteer networks. Vertical integration of management bodies between all levels ensures chain of command, information flow, clear designation of roles and responsibilities and appropriate legal and logistical support. This institutional setup also facilitates the development of joint implementation programs by government entities, international organizations, NGOs, and the civil society.

Community-Based Programs

NGO initiatives parallel to government programs are creating innovative new programs at the grassroots level on education, awareness building, infrastructure solutions and grassroots-level disaster management. The *Let's Hear Rana Bhai*

---

[6] Field Notes by Remeen Firoz from the coastal areas of Bangladesh in 2013.

example illustrates one such initiative. These community-based approaches coordinate local response with the engagement of local institutions and people through trained volunteers, prior installation of critical infrastructure (e.g. cyclone shelters) and logistical support during and after disasters. Access to local knowledge, resources and volunteers facilitate the development of cost-effective alternatives to top-down infrastructure solutions.

Livelihood Solutions

Forest conservation programs in several coastal communities in Bangladesh are achieving disaster risk reduction while protecting community access to subsistence resources. All along the coastal belt, there are thousands of multi-purpose cyclone shelters which function as schools during normal times and act as a refuge during disasters. Many of these shelters in the Cox's Bazar area have been retrofitted with ramps for movement of livestock during disasters—thus saving lives. Livelihood and income-generating opportunities empower vulnerable communities in tandem with DRR efforts.

Gender Inclusive Approaches

Women are often disproportionately burdened by natural disasters. For example, water scarcity due to salinity intrusion often results in women having to travel longer distances to collect drinking water. The institutional mechanisms in Bangladesh require representation and participation of women in all committees at the local levels. The DRR program development process also mandates consideration of gender-specific issues and solutions. The CBA-ECA program in Teknaf demonstrates how this can be accomplished. Ensuring women's health and well-being make their communities more resilient.

Resilience Building and Sustainable Development Goals

NPDM 2016–20 is organized around the theme of building resilience for sustainable human development which recognizes how building resilient communities go hand in hand with DRR and ERM. NPDM 2016–20 had also introduced attainment of SDGs as an umbrella mission of disaster management. Both infrastructure solutions and community-based programs of DRR and ERM have potential for reducing setbacks towards SDGs. However, there are untapped opportunities for proactive planning and program development that are designed to enhance SDG attainment in the process of DRR and ERM.

Feedback and Learning

The existing disaster management mechanisms in Bangladesh are receptive to feedback, innovation, and new ideas. NGO initiatives to pilot WDMCs like the *Panii Jibon* project by BPDC in Goroikhali have led to formal adoption of the WDMC model in the updated 2019 Standing Orders on Disaster (SOD). The DRRO of Cox's Bazar, the lead government official at the district level, lauded the idea of WDMCs promoted by NGOs and acknowledged that this model would fill a gap at the grassroots level and make DRR and ERM more effective. This positive and collaborative culture of cooperation is not a happenstance, but the result of a multi-level institutional framework that fosters relationship building in the service of an essential humanitarian cause.

Finally, we note that this study is an effort to document the framework of disaster management in Bangladesh and tease out important lessons and best practices. A critical analysis of the limitations and opportunities for improvements of disaster management systems in Bangladesh is also important but has not been addressed in this chapter.

16.8 Concluding Remarks

Disaster management and emergency response will be part of the new reality as the world experiences a rapidly changing climate with uncertainties and inadequate global response to reverse the course of human activities that fuelled this crisis. Unexpected events such as the Covid-19 pandemic will further endanger communities, exacerbate vulnerabilities and alter the traditional ways of life. We hope that the detailed description of the institutional mechanisms, perspectives on implementation of the programs at the ground level and stories of innovation from Bangladesh presented here will offer useful lessons on disaster management for other South Asian countries, developing nations around the world and vulnerable communities of western countries. As for Bangladesh, while the country has managed to significantly reduce human casualty, many more are likely to succumb to future disasters. There is a long road ahead with opportunities for further innovation, but no room for complacency.

Acknowledgements Partial support for this research was provided by an implementation grant from the Henry Luce Foundation's Luce Initiative on Asian Studies and the Environment (LIASE) for faculty research support at Oberlin College. We thank Leo Lasdun, student research assistant at Oberlin College, for assistance with data on natural disasters. We also thank all the government officials, agency representatives and community members who helped us better understand grassroots DRR operations in Bangladesh.

References

Adhanom, T. (2017). *Peace, prosperity and global health security.* Contributed article. Huffpost. April 15, 2017. https://www.huffpost.com/entry/peace-prosperity-and-glob_b_9694778. Accessed April 10, 2020

Adrika, A., Firoz, R., & Khan, N. A. (2011). Coping with climate change at community level. In *Moving coastlines, emergence and use of land in the Ganges-Brahmaputra-Meghna Estuary.* University Press Limited.

Ahmed, A. U., Haq, S., Nasreen, M., & Hassan, A. W. R. (2015, January). *Climate change and disaster management: Sectoral inputs towards the formulation of Seventh Five Year Plan (2016–2021).* Government of Bangladesh.

AON. (2020). *Weather, climate & catastrophe insight*. 2019 Annual Report. AON.

Bangladesh Red Crescent Society. (n.d.). http://www.bdrcs.org/. Accessed May 10, 2020

Cyclone Preparedness Programme. (n.d.). http://www.cpp.gov.bd. Accessed June 2020

Das, S. (2021). Valuing the role of mangroves in storm damage reduction in coastal areas of Odisha. In A. K. E. Haque, P. Mukhopadhyay, M. Nepal, & M. R. Shammin (Eds), *Climate change and community resilience: Insights from South Asia*. Springer.

Dhaka Tribune. (2020). *Budget FY21: Allocation for Disaster Ministry reduced*. Tribune Desk. June 11, 2020. https://www.dhakatribune.com/business/economy/2020/06/11/budget-fy21-allocation-for-disaster-ministry-reduced. Accessed October 17, 2020

Department of Environment. (2015). *Community based adaptation in ecologically critical areas of Bangladesh: Responding to changing climate and nature*. Department of Environment, Government of Bangladesh.

Eckstein, D., Hutfils, M., & Winges, M. (2019). *GLOBAL CLIMATE RISK INDEX 2019: Who suffers most from extreme weather events? Weather-related loss events in 2017 and 1998 to 2017*. Briefing Report. Germanwatche.V.

Environmental Justice Foundation. (2020). *Climate displacement in Bangladesh*. https://ejfoundation.org/reports/climate-displacement-in-bangladesh. Accessed June 14, 2020

Ge, M., & Friedrich, J. (2020, February 06). *4 charts explain greenhouse gas emissions by countries and sectors*. World Resource Institute.

Glennon, R. (2017, April 21). *The unfolding tragedy of climate change in Bangladesh*. Guest Blog. Scientific American.

Government of Bangladesh. (2009a). *Bangladesh climate change strategy and action plan (BCCSAP)*. Ministry of Environment and Forests.

Government of Bangladesh. (2009b). *National plan for disaster management 2010–2015*. Ministry of Environment and Forests, Government of People's Republic of Bangladesh.

Government of Bangladesh. (2010), *Standing orders on disaster*. Ministry of Food and Disaster Management, Disaster Management & Relief Division, Disaster Management Bureau.

Government of Bangladesh. (2012, November). *Second national communication: Adaptation, contribution to Second National Communication (SNC) of GOB*. Ministry of Environment and Forest (MOEF), Government of Bangladesh (GOB).

Government of Bangladesh. (2016). *Trends of disaster related public fund allocation in Bangladesh: An analysis of ADPs during 6th Five Year Plan period (FY 2011–FY 2015)*. Programming Division, Planning Commission & NARRI Consortium.

Government of Bangladesh. (2017). *National plan for disaster management (2016–2020): Building resilience for sustainable human development*. Ministry of Disaster Management and Relief. Government of the People's Republic of Bangladesh (GoB).

Haque C. E., & Uddin, M. S. (2013). Disaster management discourse in Bangladesh: A shift from post-event response to the preparedness and mitigation approach through institutional partnerships. In J. P. Tiefenbacher (Ed.), *Approaches to disaster management: Examining the implications of hazards, emergencies and disasters*. IntechOpen.

Herring, S. C., Christidis, N., Hoell, A., Hoerling, M. P., & Stott, P. A. (Eds.). (2020). Explaining extreme events of 2018 from a climate perspective. *Bulletin of American Meteorological Society, 101*(1), S1–S128.

Intergovernmental Panel for Climate Change. (2012). *Managing the risks of extreme events and disasters to advance climate change adaptation. A special report of working Groups I and II of the Intergovernmental Panel on Climate Change* (582 pp.). Cambridge University Press.

Intergovernmental Panel for Climate Change. (2014). Summary for policymakers. In *Climate change 2014: Impacts, adaptation, and vulnerability. Part A: Global and sectoral aspects. Contribution of Working Group II to the Fifth Assessment Report*. IPCC.

International Union for the Conservation of Nature. (2010). *Addressing Climate Change: Issues and solutions from around the world*. IUCN.

Kulp, S. A., & Strauss, B. H. (2019). New elevation data triple estimates of global vulnerability to sea-level rise and coastal flooding. *Nature Communications, 10*.

Mechler, R. (2016). Reviewing estimates of the economic efficiency of disaster risk management: Opportunities and limitations of using risk-based cost–benefit analysis. *Natural Hazards, 81*, 2121–2147.

Ornes, S. (2018). Core concept: How does climate change influence extreme weather? Impact attribution research seeks answers. *PNAS, 115*(33), 8232–8235. https://doi.org/10.1073/pnas.1811393115.

Schmuck, H. (2003). Living with cyclones. Strategies for disaster preparedness in Cox's Bazar district, Bangladesh. *GeographischeRundschau, 55*(11), 34–39.

Shammin, M. R., Haque, A. K. E., & Faisal, I. M. (2022). A framework for climate resilient community-based adaptation. In A. K. E. Haque, P. Mukhopadhyay, M. Nepal, & M. R. Shammin (Eds.), *Climate change and community resilience: Insights from South Asia.* Springer.

Szczepanski, M., Sedlar, F., & Shalant, J. (2018). *Bangladesh: A country underwater, a culture on the move.* onEarth. Natural Resources Defense Council (NRDC).

The World Bank. (2018). *The World Bank data.* https://data.worldbank.org/indicator/SP.RUR.TOTL.ZS?locations=BD. Accessed October 18, 2020

United Nations. (2015). *Sendai framework for disaster risk reduction 2015–2030.* United Nations Office for Disaster Risk Reduction (UNISDR).

United Nations. (2019). *World population prospects 2019* (Online Edition, Rev. 1). Department of Economic and Social Affairs, Population Division.

UN News. (2012). *In Bangladesh, UN humanitarian chief commends country's disaster preparedness.* News release following the visit of Valerie Amos, the Under-Secretary-General for Humanitarian Affairs of the United Nations.

United Nations. (2020). *United Nations Sustainable Development Goals (UNSDGs).* https://sustainabledevelopment.un.org/. Accessed June 17, 2020

Chapter 17
Valuing the Role of Mangroves in Storm Damage Reduction in Coastal Areas of Odisha

Saudamini Das

Key Messages

- Storm protection service of mangroves is very high for cyclone prone regions.
- During 1999 super cyclone in Odisha, every hectare of mangroves provided storm protection in the range of USD 4335 to USD 43,352 to the Kendrapada district, which is 25–249 times the 1999 per capita income of the district (USD 174).
- The annualized storm protection value of a mangrove hectare is more than two times the land price of cleared forests and more than twenty times the annual return from alternative land uses clearly justifying mangrove conservation to receive storm protection.

17.1 Introduction

In disaster management, resilience has been defined as the "*ability of an entity (individuals, communities, organizations, states) to recover from the effects of exogenous shocks, such as natural hazards, without compromising the long-term prospects of growth*" (Kousky & Shabnam, 2015). This is possible if damage from natural disasters is low (static resilience) or people recover quickly (dynamic resilience).

Disclaimer: The presentation of material and details in maps used in this book does not imply the expression of any opinion whatsoever on the part of the Publisher or Author concerning the legal status of any country, area or territory or of its authorities, or concerning the delimitation of its borders. The depiction and use of boundaries, geographic names and related data shown on maps and included in lists, tables, documents, and databases in this book are not warranted to be error-free nor do they necessarily imply official endorsement or acceptance by the Publisher, Editor(s), or Author(s).

S. Das (✉)
Institute of Economic Growth, Delhi, India
e-mail: saudamini@iegindia.org

© The Author(s) 2022
Haque et al. (eds.), *Climate Change and Community Resilience*,
https://doi.org/10.1007/978-981-16-0680-9_17

With climate change and increased threats from tropical storms to coastal dwellers, resilience building is an urgent need and the conservation of coastal vegetation provides both static and dynamic resilience from storms to people (Das & D' Souza, 2019). This chapter examines whether mangroves should be conserved for building coastal resilience.

Mangrove wetlands are one of the most important tropical and sub-tropical coastal wetlands and provide a range of provisioning, supporting, regulating, and cultural services to humans (MA, 2003). However, mangroves are threatened by change of land use to settlement, agriculture, aquaculture, or industrial uses (Field et al., 1998). This is because most of the important services of mangroves are indirect, invisible and occur off-site, whereas when these wetlands are converted to other land use like aquaculture or coastal development, the returns are visible, instantaneous, direct, and commercially very significant. Population pressure has resulted in high demand for land for different economic activities. Unless the benefits of the ecosystem services are explicitly measured, these benefits would be ignored in decisions on land use and result in underconservation of the mangroves. Ecosystem service valuation is therefore essential for sustainable land-use planning.

This research examines and quantifies the storm protection services of mangroves based on the October 1999 super cyclone damage data related to human lives, residential houses, and livestock loss in Kendrapada district of the eastern Indian state of Odisha.[1] Mangroves are seen to provide static resilience to coastal people by reducing loss of lives and damage to property during this storm and the storm protection value of mangroves is used to examine whether mangrove conservation is economically viable or not. In the coastal zones of Bangladesh which is also affected by frequent cyclones, Mahmud et al., (2021, Chap. 20 of this volume) describe local level learning effects by those affected. While in Indian Sunderbans, Ghosh and Roy (2022, Chap. 26 of this volume) find that younger educated residents and migrating as an adaptation strategy.

17.2   Why Use Averted Damage Approach to Measure Storm Protection Services

The measurement of storm protection value of mangroves, which was earlier equated to only that of constructing a sea wall at the coastline (Chan et al., 1993), has undergone tremendous methodological innovations in course of time. Both stated and revealed preference methods have been used to measure storm protection, the former being less advised due to the fear that people usually overestimate risks (Spanink & Beukering, 1997). Use of surrogate market-based methods like defensive expenditure and hedonic prices are also discouraged as they either overestimate or underestimate the storm protection value of mangroves because of high maintenance cost of substitutable structures or imperfect property markets (Bann, 1997). Researchers have also

---

[1] Called Orissa before the 113th amendment to the Indian Constitution on 24 March 2011.

used avoided expenditures and replacement costs methods to value this service (Sathi-rathai, 1998; Tri et al., 1996). However, all such methods measure storm protection indirectly and produce a proxy value. In comparison, the avoided damage approach takes into account the actual damage suffered in mangrove protected areas compared to damage in areas not protected by mangroves and provides a more realistic measure. It follows the production function approach where the storm damage as a function of storm features, location, and socio-economic factors including mangroves is esti-mated in step 1 and in step 2; the damage averted due to mangrove presence is quan-tified. It was pioneered by Farber (1987) and has been used to measure the protection provided by mangroves from storm (Costanza et al., 2008) as well as tsunami damage (Kathiresan & Rajendran, 2005). The expected damage function (EDF) has been suggested as an alternative method to measure the protection services of mangroves (Barbier, 2007). Presence of wetlands in some areas will reduce damage, and thus, the amount of compensation to be paid to the household and this change can measure the storm protection value of the wetland. However, the estimation technique as devel-oped by Barbier (2007) is a variant of avoided damage (Costanza et al., 2008, pp 246).

Though the averted damage approach has the advantage of being based on the actual damage, it can estimate the protective service of mangroves accurately provided one controls for the impact of other factors that influence the occurrence of storm damage (Das, 2007). Otherwise, it can generate either a spurious or a highly inflated protection value due to omitted variable biases. The present paper follows this methodology and takes into account a wide range of socio-economic, geo-physical, and meteorological variables as controls to separate the impact of mangroves from those of other factors on storm damage. I arrive at a comparatively lower but possibly more accurate estimate of the storm protection value of the mangroves.

17.3   Study Area and the Mangroves

This study is based on village and gram panchayat level damage data from the *Kendrapada district* in Odisha (Fig. 17.1). This district is one of the most vulnerable districts in India having a high annual probability (nearly equal to one) of being hit by cyclones (Das, 2009) and was severely impacted by a super cyclone in Oct 1999. The cyclone had its landfall at a place called *Ersama*, 20 km southwest of Kendrapada. The district was the ideal choice to measure the storm protection services of mangroves as (1) it was situated north of the eye of the cyclone and path of the cyclone throughout,[2]

---

[2] In northern hemisphere, the direction of the cyclonic wind is anticlockwise and thus the wind direction in Kendrapada was from sea to land through the mangrove forest.

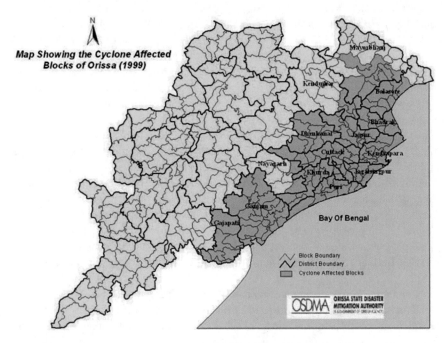

Fig. 17.1 Kendrapada district in cyclone hit Orissa. *Source* Orissa State Disaster Management Authority, Government of Odisha

(2) has mainly mangrove forest[3] and barren areas on the coast line, and (3) is devoid of highlands, the average elevation being less than 10 m everywhere (NATMO, 2000).

Kendrapada was an economically backward district with nearly 50% of the population living below the poverty line, 94% living in rural areas and around 2% of the rural houses having concrete structures when the storm struck in 1999.

17.3.1 The Mangroves of Kendrapada

The State of Orissa has 480 km of coastline covering seven coastal districts and 5133.60 km^2 of coastal wetlands. The state was endowed with rich mangrove cover historically; with nearly 500 km^2 in 1944, which was destroyed over time leaving it with 227 km^2 of mangrove forests, most of which (88%) is located in the Kendrapada district.

---

[3] The main forests were the mangroves though a few patches of Casuarina plantations were also to be found in the coastal areas before the cyclone. But the width of these plantations everywhere was between 200 and 400 m.

Mangrove Forest Cover in 1999 and the Cyclone path

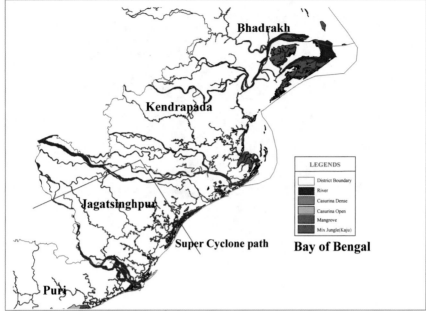

Fig. 17.2 Mangrove and other coastal forests of Kendrapada and Jagatsinghpur districts in October 1999. *Source* Das (2011)

Though both Jagatsinghpur and Kendrapada were the major mangrove districts of the state and witnessed mangrove loss, the loss was nearly 100% for Jagatsinghpur district (from 177.27 km^2 in 1944 to 5 km in 2001), whereas it was around 37% for Kendrapada (from 306.7 in 1944 to 192 km in 2001). In Kendrapada district, the mangroves are found in two patches as seen from Fig. 17.2 that shows the mangrove cover in Jagatsinghpur and Kendrapada districts as it existed on 11 Oct. 1999. In Kendrapada, 89 villages have been established after cutting down the mangroves, which are labelled as mangrove *habitat villages* and have been accounted for separately in the analysis.

17.3.2 Drivers of Mangroves Loss in Orissa

Figure 17.3 shows the mangrove forest map of the districts Jagatsinghpur and Kendrapada as it existed in the year 1944. As evident from the figure, more than 80% of the coastline from the mouth of the river Devi to the mouth of the river Dhamra was covered by mangrove forests of more than 10 km width as these areas are crisscrossed by river channels and their tributaries and rivulets (seen from the figure also). The mangrove forest of Jagatsinghpur district and the Mahanadi delta mangroves of

Mangroves of 1950, Rivers and cyclone path

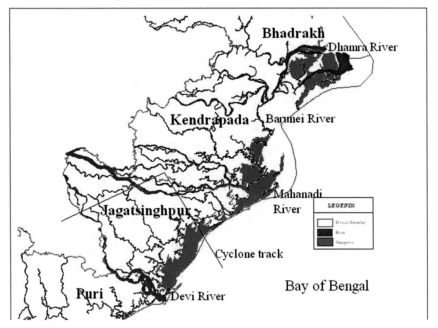

Fig. 17.3 Mangroves of Kendrapada and Jagatsinghpur districts as in 1944. *Source* Das (2011)

Kendrapada district were known historically as the Kujang Forest, and the mangroves of Bhitarkanika region, Bhadrakh, and Balasore districts were known as the Kanika forest after the name of the princely states that used to rule over these areas. Though there is less research on the drivers of mangrove loss in the State of Orissa, local vernacular publications and independent studies done by researchers and NGOs overwhelmingly link the loss of mangrove forests to the political economy of the state. The maximum mangrove destruction occurred during the 1960s and 1970s for various reasons including the lack of proper jurisdiction during the period following the abolition of Zamindari in 1957 till the formation of the Wild Life Division in 1980. The creation of the Paradeep port, rehabilitation of refugees from then East Pakistan (present-day Bangladesh), lack of knowledge of mangrove values, and conversion of mangrove land for betel vine, agriculture, and aquaculture farms, etc. are some of the main reasons for the destruction of the mangroves (Choudhaury, 1990; Das, 2009; Mohanty, 1992). The Ministry of Environment and Forest, Government of India, had listed the existence of 15 different types of threats to mangroves of the region (Das, 2009), the maximum being anthropogenic in nature, with the clearing of the forest due to the subsistence requirements of the people being the most prominent one. Another interesting observation was that the local people were not keen on preserving the mangrove. Though people have realized the importance of mangroves in their day to day life, there are still threats to mangroves from local inhabitants, which is reflected in their unhappiness and anger after the Bhitarkanika area was declared a national park by the government in 1998 (Badula, 2002). The mangroves of Bhitarkanika region probably survived when state protection was missing because of the presence of ferocious animals and interior location of the area. After the announcement of sanctuary and national park, government protection and strict implementation of laws have been able to protect the mangroves there.

17.4   Data

The paper analyses three types of asset damage due to the super cyclone, i.e. human lives, residential houses and livestock, which are collected from various sources (see Das, 2007 for detail). The data set for the human casualty model is at the village level and it consists of 1180 villages. The house damage analysis is based on heterogeneous units covering 451 villages and 138 *Gram Panchayats* and the analysis for livestock is based on data at a *Gram Panchayat* level analysis covering 216 *Gram Panchayats*. These differences in units and coverage area are due to the limitations of data which was only available in that format and for those specific areas.

Estimated cyclone damage models from Das (2011), which were based on Das (2007), are used in this paper to estimate the storm protection value of mangroves. Das (2007) did extensive testing for determinants of human death, three types of house damage (fully collapsed, partially collapsed, and swept away houses), and five types of livestock loss (cattle, buffaloes, sheep, goat, and poultry) suffered during the October 1999 super cyclone. Results for different sample areas were compared to infer the effectiveness of mangrove protection. Sample 1 was the entire study area excluding villages that never ever had mangroves in their coastal distance (called the mangrove non-habitat villages). Mangrove non-habitat areas were excluded as they can never be protected by mangroves or storm protection value of mangroves is meaningless for them. Secondly, by leaving them, I control for the topographic and bathymetric features of the study area[4] as my treatment villages (the ones protected by mangroves) and the control villages (the ones not having mangroves in their coastal distance during the 1999 cyclone, but which used to have mangroves that were destroyed over time) have similar bathymetry and topography. Sample 2 is sample 1 minus the areas falling under the cyclone eye. The wind direction inside the cyclone eye area being circular (anticlockwise before the eye passes and clockwise afterwards), the forest can provide little protection. Hence, the expectation is that the storm protection value per unit of mangroves is accurately captured in sample 2 and sample 1 is the entire area protected by mangroves.[5] Storm damage models based on sample 1 and 2, not others, are used in the paper. For estimating the storm

---

[4] Mangroves come up in areas with similar topographic and bathymetric features.

[5] In Das (2007), samples 3, 4, and 5 were parts of sample 2 that were within 10 km distance from coast, beyond 10 km distance from coast and suffered storm surge inundation during the cyclone, respectively. These samples 3, 4, and 5 are not discussed in the present paper.

protection value, I consider only those damage models of Das (2007) for the above two areas where the mangrove was found to have a statistically significant effect, i.e. human death, fully collapsed houses, partially collapsed houses, and losses of both cattle and buffaloes. See Das (2007, Tables 1, 2, 3, 4, 7, 8, and 10) for description of variables and regression results used.

17.5 Methods

First the physical estimates of damage avoided due to mangrove presence have been calculated, and then in step 2, this averted damage is valued to estimate the storm protection value. Averted damage is defined as the difference between the actual damage witnessed and the predicted damage in absence of mangroves. These are measured for different sample areas described above and for three different scenarios, i.e. no mangroves, if historical mangroves were present and if historical mangroves were present and mangrove habitat villages were not there. After measuring the averted damage for the three assets, i.e. human life, houses, and livestock, these damages are valued and summed to measure per unit storm protection value. In the valuation process, the differences in units and coverage of study areas are carefully taken into account to arrive at a realistic and representative value void of ambiguities and biases. Local prices prevailing in the study area and value of statistical life generated for India are used in valuation.

17.6 Results

17.6.1 Averted Damage

In total, 392 persons lost their lives during the 1999 cyclone in sample 1 area but the toll may have been 603 in the absence of the mangrove (Table 17.1). Thus, 211 deaths (54% of the lives lost in that area) were possibly averted due to the presence of the mangroves. The mangroves provided greater protection to areas of sample 2, where 217 deaths (82% of lives lost in sample 2) were estimated to have been averted by mangroves.

If the historical mangrove forest (as existed in 1944) had not been cleared by 1999, only 31 persons would have probably died instead of 392 in sample 1 area, even if the 89 forest villages would have been where they are. However, if the 89 coastal villages had not been permitted in the mangrove area, there would probably have been only 17 casualties.

In the absence of the mangroves, the number of fully collapsed houses may have been higher by 19,936; partially collapsed houses lower by 14,049 indicating that some of the partially collapsed houses would have been completely damaged (see

Table 17.1 Averted human death due to mangrove forests

	Actual deaths 1	Predicted death if mangrove = 0 (Assump-1) 2	Predicted death if mangrove = mhabitat (Assump-2) 3	Predicted death if mangrove = mhabitat and mangrove habitat villages = 0 (Assump-3) 4	Averted deaths (1–2) 5	Averted deaths (1–3) 6	Averted deaths (1–4) 7
Sample-1 (N = 840)	392	603	31	17	211 (54%)	361 (92%)	375 (96%)
Sample-2 (N = 711)	266	483	25	11	217 (82%)	241 (91%)	255 (96%)

Table 17.2 Volume of house damage and livestock[a] loss averted due to the mangrove forests (figures are numbers)

Damage type	Assumption-1		Assumption-2		Assumption-3	
	Sample-1	Sample-2	Sample-1	Sample-2	Sample-1	Sample-2
Fully collapsed houses	19,936	13,110	178,660	82,225	165,975	74,675
Partially collapsed houses	−14,049	−12,657	−125,900	−79,376	−119,702	−72,087
Buffaloes	704	683	1320	994	1399	1100
Cattle	3844	4668	17,946	12,993	17,385	12,312

[a]Swept away houses, goat, sheep, and poultry have been left out as mangrove was insignificant for them in all models

Table 17.2). Similarly, buffalo and cattle loss would have been higher by 704 and 3844, respectively, in sample-1 area. These figures would have been 13,110, −12,657, 683, and 4668 in sample 2 area. If the 1944 forest had been there, not a single house would have fully collapsed in both the sample areas.[6] We would probably have witnessed only partially collapsed houses.

17.6.2 Storm Protection Value of the Mangroves

The valuation of damage is done with the aim of understanding: (a) the saving in government compensation disbursed to victims and (b) the social benefit of

---

[6] This is inferred from the derivation that the number of averted fully collapsed houses (due to historical mangroves) is higher than the actual number of fully collapsed houses in those areas.

mangroves when valued at market price. Accordingly, the damages are valued @compensation paid, @revised compensation rates, and @prevailing market prices of damaged assets in 1999. What prices are used and how the value of statistical lives is adjusted to value human deaths are described in Das (2009).

17.6.2.1 Average Storm Protection Value

The mangrove variable was measured as kilometre width of the forest, and thus, the average storm protection value (ASPV) of every kilometre width of the existing and historical mangrove forest to a village are measured for sample 1 and sample 2 areas under the three assumptions. First these are measured for each of the damages separately and then added across the damages to measure weighted average storm protection (WASP) value to a village. These are shown in Tables 17.3 and 17.4. The ASPV to a village in sample 1 is Rs. 2239 for protecting human lives and Rs. 1157 for reducing house damage[7] (see Table 17.3, situation 1). If the 1944 forest were still there along with the villages subsequently established (Situation 2), these values would be Rs. 1207 and Rs. 2315, respectively. In situation 3, the corresponding values would be Rs. 1496 and Rs. 2488, respectively. These values are higher for sample 2 areas compared to the sample 1 area for every type of damage and situation. This suggests that the protective services of mangroves are more effective in the cyclone outer eye areas. The areas falling under the cyclone eye receive the strongest winds which are also circular and mangroves can provide little protection there. Thus, our hypothesis of using sample 2 as a more accurate valuation scheme for storm protection services by mangroves is supported by these findings. Another observation is that the average value of present mangroves is much higher than historical mangroves for every sample area but only for averting deaths (both human lives and livestock), whereas the reverse is the case for house damages. The average width of present mangrove is much smaller (approximately 1 km) compared to historical mangrove (approximately 4 km). This suggests that the relation between mangrove width and protection from different types of damages may not be linear. Having more mangroves may not help in averting more deaths but seems to avert more house damages. This allows for calibrating mangrove size depending on the social objective, and an optimum width of the forest can be defined to act as buffer during cyclones.

The WASP value provided by a kilometre of present mangrove in a village is Rs. 3928.43 when valued at market prices (see Table 17.4). However, if government compensation rates were used to determine these values (in terms reduced compensation to be paid), it varies between Rs. 46.55 (@actual amounts paid) and Rs. 183.63 (@revised house damage compensation rates). The average storm protection values of kilometre width of historical mangroves, shown in columns 3 and 4, varies between

---

[7] This is computed as value for reduced FC houses (Rs 1331)—value for increased PC houses (Rs 174).

Table 17.3 Average storm protection value per village provided by every km width of present mangrove and historical mangrove (in Rs.)

Type of damage	Value/km of present mangrove/village		Value/km of hist. mangrove/village (coastal villages remaining)		Value/km of hist. mangrove/village (coastal villages removed)	
	Sample-1	Sample-2	Sample-1	Sample-2	Sample-1	Sample-2
Human death	2239.35	2743.95	1207.87	1132.44	1495.68	1478.28
Fully collapsed houses	1331.53	1368.55	2663.52	3143.39	2873.11	3235.85
Partially collapsed houses	−174.42	−245.59	−348.88	−564.03	−385.15	−580.85
Buffaloes	5.91	8.77	2.32	3.97	2.62	4.83
Cattle	26.87	49.94	26.32	43.20	27.17	45.07

Notes Rates (market prices) used are: Value of Statistical life @ Rs. 10,918,132/; Price of FC house @ Rs. 53,800/; Price of PC house @ Rs. 10,000/; Price of Buffalo @ Rs. 6000/; Price of Cow @Rs. 5000/ and) Price of Sheep @ Rs. 1200/

Table 17.4 Weighted average storm protection value for a village by every km width of present mangroves and historical mangroves (in Rs.)

Value @ different valuation rates	Value/km of present mangrove	Value/km of hist. mangrove (coastal villages remaining)	Value/km of hist. mangrove (coastal villages removed)
Value @ government compensation paid	46.55	68.69	72.9
Value @ revised government compensation for house damage	183.63	385.42	399.06
Value @ market price and VSL with ε = 0.35	3928.43	3761.4	4185.68

Note: ε represents the income elasticity of marginal willingness to pay

Rs. 69/ and Rs. 4186/, and the values are the highest if the coastal villages established in mangrove habitat areas are relocated (situation 3).[8]

17.6.2.2   Total Storm Protection (TSP) Value

There are around 1250 villages in Kendrapada district and of which 850 villages had mangrove historically between them and the coast (sample-1) and 580 of these villages were outside the cyclone eye (sample-2). Sample 1 being the entire area that receives storm protection from mangroves, we multiply the unit values of present mangroves shown in Table 17.2 by 850 to get the TSP value (for protecting human lives, residential houses and livestock) of every kilometre width of the forest to the state exchequer and the society.

Dividing the value of total avoided damages of sample 1 area by the mangrove area (17,900 ha), total savings to the state exchequer and to the society by every hectare of the present forest were also calculated (see Table 17.5). [9]

A 1 km width of the forest saved Rs. 3,339,166 for the economy and Rs 3968 to the state government in the form of reduced compensation liability (Table 17.3).[10] In comparison, the savings by every hectare of mangroves forests are Rs.182, 080/

---

[8] The volume of damages averted due to mangrove presence being low for the mangrove habitat area villages, the unit values increase as these villages are removed from the analysis.

[9] As mentioned before, the per hectare values are the simple averages. To get the value at market price, we simply added the market values of different averted damages of sample 1 area and then divided it by the area of the present mangroves. Only sample 1 area was considered as that is the entire area benefited by mangroves. We did similarly to get values at other valuation rates.

[10] The savings to the state government by the present mangroves would have been Rs 156,083/ if the revised compensation rate was used.

Table 17.5 Total storm protection value (for Kendrapada) by every km width and by every hectare of present mangroves

	Value of damage averted per km (width) of mangrove	Value of damage averted per ha (area) of mangrove
Saving to state government in compensation paid in 1999	Rs. 39,568/(USD 943)	Rs. 2339/(USD 56)
Saving to state government if revised compensation for house damage would have been applicable in 1999	Rs. 156,083/(USD 3716)	Rs. 8550/(USD 204)
Saving to district economy (value of damages at market prices)	Rs. 3,339,166/(USD 79,504)	Rs. 182,080/(USD 4335)

Notes The exchange rate used is 1USD = 42 INR as prevalent in 1999

to the district economy for reducing human death, damage to residential houses, and loss of livestock.[11]

On the basis of these values, we try to analyse one important policy question, i.e. should the remaining mangroves be preserved to receive storm protection given high demand for land for alternate uses?

17.6.3   Is Mangrove Preservation Economically Justified?

This question is analysed by comparing the land price of agricultural land in cleared forest area (opportunity cost of preserving forest) to the storm protection value per ha of the forest. The average land price in Mahakalpada tehsil of Kendrapada, where maximum of the mangrove forests were converted to other uses, was Rs. 172, 970 per hectare during [12] 1999–2000. The partial storm protection value of a hectare of mangroves at market prices being Rs. 18,208 (Table 17.5) to the district for protecting only three assests (human lives, livestock and houses), prima facie, there is a strong case for the preservation of the forest. However, we also compare the annualized returns of these two values.

We assume the three types of averted damages discussed in this paper to constitute one-tenth of the total averted damages of mangroves by a conservative estimate.[13]

---

[11] Every hectare of mangrove saved the state exchequer Rs. 2339 (actual compensation paid) or Rs.8550 (revised rates) in the form of reduced compensation.

[12] The land price as reported by the land registration office varied between Rs. 70,000/ to Rs. 100,000/ per acre around 1999 (Personal communication with Jatindra Dash, IANS), and the land price in mangrove adjacent area being on lower side, we use the lower limit, i.e. Rs 70,000 per acre and this calculates the price per hectare as Rs. 172,970.

[13] Badola (2002) estimated the total storm protection value of Bhitarkanika Mangroves of Orissa during the same super cyclone of Oct 1999 by considering the protection of mangroves from multiple damages and found the value to be equivalent of USD 116.28 per household. As the average number

By this assumption, the storm protection value of a hectare of mangrove during super cyclone of October 1999 works out to be Rs. 1,820,800 which is much higher than the land price.

17.6.3.1 Probability of Extreme Events and Annualized Benefits

The study area is highly cyclone prone and records of the past 200 years reveal that the frequency of very severe cyclonic storms has gone up significantly in the last 3–4 decades. In between 1903 and 1999, Orissa witnessed 52 cyclones of which eight were Very Severe Cyclonic Storms and one was a Super Cyclone (Chittibabu et al, 2004). Moreover, six of the nine devastating cyclones occurred in the last 30 years so the annual probability of occurrence of a devastating cyclone is 0.2. Thus, the probability adjusted annual storm protection value of a hectare of mangrove (Rs. 364,160) is more than twice the market price of land cleared of forest. If we assume an interest rate at 8% per annum,[14] the annual opportunity cost of preserving mangrove forest at 1999 prices works out to be Rs. 13,837 or Rs. 20,756 if we assume a very high return @ 12% per annum. The annual benefit from protecting forest is therefore 18–26 times higher than the annual opportunity cost of preserving the forest. These findings support protection of mangrove forest to get storm protection benefit as a socially desirable strategy. Even if we use a lower annual probability of any cyclone (0.09 per annum), the mangrove preservation will still be justified. Under these rates and with the lower cyclone probability (0.09 per annum), the net present benefit to society or welfare gain to society from preserving mangrove forest is Rs. 143,393 and Rs. 215,089 per ha with 12 and 8% discount rates, respectively. These numbers indicate a very high benefit from preservation of the remaining mangroves.

17.6.4 Land-Use Change

Was the destruction of mangrove forest in the past economically justifiable? As mentioned earlier, 12,866 ha of mangroves were converted between 1950 and 1999 mainly for agriculture. We now estimate the net loss in protective cover that could have been averted if the mangrove of 1944 level was not destroyed. We calculate this as the difference between the market values of avoided damages ($\sum$VAD) with historical mangroves and the present mangroves ($\sum$VAD$_{1944}$ − $\sum$VAD$_{1999}$).

---

of household in her study villages is 37, this gives the total storm protection value as USD 4302 per village which is 45 times higher than the highest storm protection value per village obtained in the present study. So a 10 times escalation of benefits to estimate total benefits is still on the conservative side.

[14] In the absence of information on rate of return from agriculture in coastal Kendrapada, we calculated annual return @ 8% which is the average of the estimated range of real discount rates (7.6–9.7%) from the Indian labor market studies and also comparable to financial market rates in 1990s (Shanmugam, 2006).

Dividing the above value by the area of the lost mangrove forest (12,866 ha), the extra burden for destroying every hectare of forest comes out to be Rs 706,882 for only three damages. The benefit of forest destruction, which is captured by per hectare land price, is much lower than this. Under the assumption that these three averted damages are one-tenth of the total averted damages of the mangroves, the extra burden for destroying every hectare is Rs 7,068,820. If we multiply this value by the annual probability of devastating cyclones (0.2), the probability adjusted annual burden due to loss of storm protection cover comes out to be nearly seven times higher than the benefit from forest destruction (i.e. the land price of cleared forest land).

We may infer that the social benefit of retaining the forest cover is much higher than the current land value (Rs1, 72,970 per ha). As noted earlier, the benefits estimated are lower bound values, and therefore, actual benefits are likely to be much higher than indicated here.

17.7 Conclusion and Policy Implications

The study quantifies the storm protection services of mangroves of Odisha and the storm protection value of every km width of present mangrove to have been Rs 39,568 to the state exchequer in the form of reduced compensation and Rs 3,339,166 to society for saving human life, livestock, and preventing house damage. The per hectare benefits (for just averting the three damages) were estimated to be Rs 182,080. These three damages are a small proportion of the total damages averted by the mangroves. Making some conservative assumptions, we find the cyclone probability adjusted annual storm protection value per hectare of mangroves to be more than twice the market price of cleared mangrove forest land and 18–26 (or nearly 20) times higher than the annual return from land. All these suggest the preservation of remaining mangroves as a socially and economically viable strategy to receive storm protection services.

Mangroves save lives and properties in the vulnerable coastline areas and thus provide static resilience to society during natural disasters like storms. This is also found by Mahmud et al., (2021, Chap. 20 of this volume) in the context of Bangladesh. Climate change makes it imperative to conserve the mangroves and policy makers need to make arrangements for their protection. Usually, people living in areas around the mangrove do not realize the importance of mangroves as most of the ecosystem services are invisible and indirect. Awareness generation can go a long way in ensuring mangrove conservation, especially in vulnerable coastal areas like the state of Odisha.

Acknowledgements This paper is a follow up of the special article "Examining the Storm Protection Services of Mangroves of Orissa during the 1999 Cyclone" published in Economic and Political Weekly, Vol. XLVI No. 24, 11 June 2011 and SANDEE working papers 25-07 and 42-09 by the author. It uses some result tables and figures published in the EPW paper and some results from

SANDEE working papers to measure the averted damage by mangroves, values them and finally examines the question of mangrove conservation to build resilience. An earlier version of this paper titled "The Case for Mangrove Conservation: Valuing Damage Averted in Orissa's 1999 Super Cyclone" was presented at the 4th World Congress of Environmental and Resource Economists (WCERE) held in Montreal, Canada, during 28th June to 2nd July 2010. The paper has been enriched after addressing the comments received from reviewers, discussant, and from the floor.

References

Badula, R. (2002). *Valuation of the Bhitarkanika mangrove ecosystem for ecological security and sustainable resource use* (EERC Working Paper Series, WB 1).

Bann, C. (1997). *The economic valuation of mangroves: A manual for research*. International Development Research Centre.

Barbier, E. B. (2007). Valuing ecosystem services as productive inputs. *Economic Policy, 22*(49), 177–229.

Chan, H. T., Ong, J. E., Gong, W. K., Sasekumar, A. (1993). The socio-economic, ecological and environmental values of mangrove ecosystems in Malaysia and their present state of conservation. In B. F. Clough (ed.), *The economic and environmental values of mangrove forests and their present state of conservation in south-east Asia/Pacific region* (Vol. 1, pp. 41–81) (Okinawa, Japan: International Society for Mangrove Ecosystems, International Tropical Timber Organisation and Japan International Association For Mangroves, 1993).

Choudhaury, B. P. (1990). Bhitarkanika: Mangrove swamps. *Journal of Environment and Science, 3*(1), 1–16.

Chittibabu, P., et al. (2004). Mitigation of flooding and cyclone hazard in Orissa, India. *Natural Hazards, 31*, 455–485.

Costanza, R., et al. (2008). The value of coastal wetlands for hurricane protection. *Ambio, 37*(4), 241–248.

Das, S. (2011). Examining the storm protection services of mangroves of Orissa during the 1999 cyclone. *Economic and Political Weekly, XLVI*(24), 60–68. https://www.epw.in/journal/2011/24/special-articles/examining-storm-protection-services-mangroves-orissa-during-1999.

Das, S. (2009). *Economic valuation of a selected ecological function—Storm protection: A case study of mangrove forest of Orissa* (PhD Thesis). University of Delhi.

Das, S. (2007). *Storm protection by mangroves in Orissa: An analysis of the 1999 super cyclone* (SANDEE Working Paper No. 25-07).

Das, S., & D' Souza, N. (2019). Identifying the local factors of resilience during cyclone Hudhud and Phailin on the east coast of India. *Ambio*. https://doi.org/10.1007/s13280-019-01241-7. Available at https://rdcu.be/bSFcf

Farber, S. (1987). The value of coastal wetlands for protection of property against Hurricane wind damage. *Journal of Environmental Economics and Management, 14*, 143–151.

Field, C. B., et al. (1998). Mangrove biodiversity and ecosystem function. *Global Ecology and Biogeography Letter, 7*(1), 3–14.

Ghosh, S., & Roy, S. (2022). Resilience to climate stresses in south India: Conservation responses and exploitative reactions. In A. K. E. Haque, P. Mukhopadhyay, M. Nepal, & M. R. Shammin (Eds.), *Climate change and community resilience: Insights from South Asia*. Springer.

Kathiresan, K., & Rajendran, N. (2005). Coastal mangrove forest mitigate Tsunami. *Estuarine, Coastal and Shelf Sciences, 65*, 601–606.

Kousky, C., & Shabman, L. (2015), *A proposed design for community flood insurance*. Resources for the Future.

MA (Millennium Ecosystem Assessment). (2003). *Ecosystems and human well-being: A framework for assessment*. Island Press.

Mohanty, N. C. (1992). *Mangroves of Orissa*. Project Swarajya Publication.

Mahmud, S., Haque, A. K. E., & De Costa, K. (2021). Climate resiliency and location specific learnings from coastal Bangladesh. In A. K. E. Haque, P. Mukhopadhyay, M. Nepal, & M. R. Shammin (Eds.), *Climate change and community resilience: Insights from South Asia*. Springer.

Sathirathai, S. (1998). *Economic valuation of mangroves and the role of local communities in the conservation of natural resources: A case study of Surat Thani, South of Thailand* (EEPSEA Research Report). http://703.116-43-43-77/publications.research1/ACF9E.html.

Shanmugam, K. R. (2006). Rate of time preference and the quantity adjusted value of life in India. *Environment and Development Economics, 11*, 569–583.

Spanink, F., & van Beukering, P. (1997). *Economic valuation of mangroves eco-systems; potential and limitations* (CREED Working Paper No. 14). International Institute for Environment and Development.

Tri, N.H., Adger, N., Kelly, M., Granich, S., & Nimh, N. H. (1996). *The role of natural resource management in mitigating climate impact: Mangrove restoration in Vietnam* (Working Paper, GEC 96-06). CSERGE (Centre for Social and Economic Research on Global Environment).

Chapter 18
Using Climate Information for Building Smallholder Resilience in India

Madhavan Manjula, Raj Rengalakshmi, and Murugaiah Devaraj

Key Messages

- User-specific, relevant seasonal climate forecast (SCF) information has the potential to induce risk-reducing decisions for players across the agricultural value chain.
- The lead time and forecast requirement for SCF are diverse across the value chain and might induce competitive/complementary decisions across the value chain.
- SCF needs to be bundled with institutional support for better uptake and competitive advantage to the smallholder farmers.

Disclaimer: The presentation of material and details in maps used in this book does not imply the expression of any opinion whatsoever on the part of the Publisher or Author concerning the legal status of any country, area or territory or of its authorities, or concerning the delimitation of its borders. The depiction and use of boundaries, geographic names and related data shown on maps and included in lists, tables, documents, and databases in this book are not warranted to be error-free nor do they necessarily imply official endorsement or acceptance by the Publisher, Editor(s), or Author(s).

M. Manjula (✉)
School of Development, Azim Premji University, Bengaluru, India
e-mail: manjula.m@apu.edu.in

R. Rengalakshmi · M. Devaraj
JRD Tata Ecotechnology Centre, M. S. Swaminathan Research Foundation, Chennai, India

R. Rengalakshmi
e-mail: rengalakshmi@mssrf.res.in

M. Devaraj
e-mail: devaraj@mssrf.res.in

© The Author(s) 2022
Haque et al. (eds.), *Climate Change and Community Resilience*,
https://doi.org/10.1007/978-981-16-0680-9_18

18.1 Introduction

Shifting rainfall patterns, increasing rate of extreme events changing temperature regimes are some of the outcomes of climate variability and climate change influencing agricultural production. While all the factors pose threat to crop cultivation, climate variability during the crop season is a major source of agriculture production risk (Legler et al., 1999; Paz et al., 2006) in rainfed systems. The impact of climate variability on agricultural productivity is manifested through crop loss due to floods or seasonal droughts induced by excessive or insufficient rains during the crop season. Variability in temperature and relative humidity during the growing season triggers sporadic pest and disease outbreaks resulting in massive crop losses (Jones et al., 2017; Tanyi et al., 2018). The impact of climate variability on the farming system has a multiplier effect on the other players across the agricultural value chain ultimately posing a challenge to the local agricultural economy and food security (Lipper et al., 2014; Wheeler & Von Braun, 2013).

Knowledge of climate variability prior to crop season and its incorporation in management decisions across the value chain is touted as the key adaptation strategy for intra-seasonal climate variability. In India, in addition to the existing short and medium range weather forecasts disseminated by the Indian Meteorological Department (IMD), access to reliable intra-seasonal and seasonal climate forecasts (month to multi-month time frames) could induce a set of adaptive responses that might help to reduce production risks posed by climate variability (Meinke et al., 2006). Sivakumar (2006) reported that access to forecasts of meteorological risks and timely agro-meteorological advisories can assist farmers to take appropriate strategic and tactical decisions to cope with changing climate. Climate consortiums in South America, Africa and West Asia have in the past attempted to generate and disseminate seasonal climate forecast (SCF) (Hansen & Sivakumar, 2006; Pulwarty, 2007). But these attempts have not yielded the desired results due to various reasons (Bruno Soares & Dessai, 2016; Hansen et al., 2011; Rickards et al., 2014)

The reasons range from low predictive skill of the forecasts at finer spatial and temporal scales, lack of easy access to SCF, challenges in interpreting probabilistic SCF and inability of the end user to take up adaptive responses based on SCF, and attitude and psychology to taking risks (McCrea et al., 2005; Vogel & O'Brien, 2007). The other challenge is communicating probability-based climate forecasts to the end users. Several studies on application of SCF information indicate the risk of a deterministic interpretation of probabilistic forecast information and the use of empirical approaches in explaining the probabilistic nature of SCF (Hansen et al., 2007; Suarez & Patt, 2004). Communication tools which adopt a participatory approach in disseminating SCF have been reported to be effective in Africa and India (Hansen et al., 2007; Meinke et al., 2006). Another challenge is meeting the diverse seasonal climate information needs of different actors involved in the crop value chains. To improve the relevance of climate forecasts, it is imperative to identify the decision-relevant attributes of forecast information for specific activities and actors and incorporate that into the forecasts (Stern & Easterling, 1999).

Experiences from across the globe show that SCF information should have reasonably good predictive skill, be locale specific, be tailor-made to end user requirement and be communicated through a participatory approach using customised communication tools to have better uptake (Vogel & O'Brien, 2007). This is particularly useful to know as pointed out by Bahinipathi and Patnaik (2021, Chap. 4 of this volume) when they examined the factors that determine farmer adaptation behaviour.

This paper is drawn from a multi-country multidisciplinary project that aims to investigate the use of SCF in enhancing food security in South Asia. The project also aimed to develop a blueprint for improved seasonal climate information across case study regions in the Indian Ocean Rim Countries. The paper details the project experience in the state of Tamil Nadu, India.

18.2   Study Area and Methods

The study was conducted in Dindigul district, located in south-western Tamil Nadu and geographically spread over 6266.64 km^2. The economy is predominantly agrarian. The major crops are maize and other cereals, vegetables, pulses, cotton, oilseeds, paddy and sugarcane. Agriculture is predominantly rainfed, and the main agriculture season is from October to December. The district receives a mean annual rainfall of 845.6 mm. The region benefits more from north-east monsoon (53%), and the maximum rainfall is between October and December. January and February are the months, which receive minimum (49.6 mm) rainfall. The coefficient of variation in the inter-annual period is around 30%. The farmers in this region have a strong perception of climate variability and relate it to the frequent water-deficit years, premature end, late onset and uneven distribution of rains (Fig. 18.1).

The study adopted a value chain approach and included on-farm and off-farm players across the agricultural value chain of the major crops cultivated in that region. Mixed methods were used to elicit information on existing climate risk management, nature of climate information needed, and dissemination networks. Semi-structured interviews were conducted to collect quantitative information from farmers. Samples were drawn using stratified sampling technique. A total of 242 marginal and small farmers were surveyed. Participatory appraisal tools such as resource mapping, social mapping, seasonal calendar, trend analysis, time use studies and gender analysis were used to elicit qualitative information. The off-farm players were engaged in the study through consultative workshops and one-on-one interviews. Reflective learning methods like decision analysis were used to communicate SCF information.

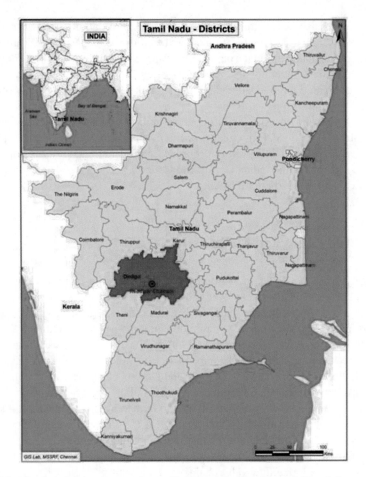

Fig. 18.1 Location of the study area. *Source* M. S. Swaminathan Research Foundation

18.3   Results and Discussion

18.3.1   Stakeholder Perception of Climate Risks

Rainfall was the major climate risk articulated by players across the value chain. However, there were striking differences in the aspect of rainfall that was considered critical for farming and business by the respective stakeholders. Extreme weather events like droughts and floods were cited as climate risks that were impacting at a regional level. Farming was sensitive to late/early onset of monsoon, unequal intra-seasonal distribution of rainfall, extended dry spell after sowing, extended dry spell/excess rainfall during flowering, excess rainfall at harvest, untimely and inadequate rain during the season (Table 18.1).

Table 18.1 Climate risks experienced by farmers in the last three agricultural seasons

Climate risk	Number of farmers articulating respective climate risks
Late/early onset of monsoon	108
Unequal distribution of rainfall in the season	51
Extended dry spell after sowing	55
Excess rainfall in peak flowering/harvesting season	122
Untimely rains	53
In adequate amount of rainfall during the season	79

Source Field Data

Note Total will not add up to 242 as there are multiple responses from each respondent

18.3.2   Key Decisions and Climate Information of Relevance

The key decisions and the climate variable that would influence these decisions for farmers and extension agents who provide crop management advisories to farmers are detailed in Table 18.2.

Table 18.2 Key on-farm decisions and influence of climate information

Key decisions	Key climate variable that informs the decisions
Time of sowing	Onset of monsoon
Choosing of crops/crop variety	Total rainfall and its intra-seasonal distribution
Irrigation management—timing of irrigation and quantity of water to be applied	Total rainfall and its intra-seasonal distribution
Resource use allocation—labour and finance	Total rainfall and its intra-seasonal distribution
Fertiliser application—quantity, type and stage of application	Distribution of rainfall across the crop growth stages
Timing of pesticide application	Wind direction and speed; rainfall distribution across crop growth stages
Time of harvest	Distribution of rainfall during the crop maturation stages

Source Field Data

Aspects like total amount of rainfall during the season, onset of monsoon and intra-seasonal distribution of rainfall are said to influence the business decisions of input dealers. The major business decisions affected by climate/weather parameters for an input dealer are as given below:

- Stocking of inputs (seeds/fertilisers/pesticides)—quantity, type and time
- Transport of inputs (seeds/fertilisers/pesticides)—time
- Supply of inputs (seeds/fertilisers/pesticides)—quantity, type and time.

18.3.3 Current Source of Climate Information Across Stakeholders and Its Utility

The stakeholders have access to nowcast (very short term, up to 2 h), and short and medium range forecasts given by the Indian Meteorological Department (IMD). The forecasts are received through television, radio, newspapers and mobile phones. In addition to these public sources, farmers in the study area receive medium range weather forecast information through the Village Knowledge Centre (VKC) and Farmer Producer Company (FPC) network. Majority of farmers also rely on traditional knowledge for climate information. This is evident from the use of proverbs and folk songs that refers to climate parameters like rainfall, wind direction and wind speed in this region. The major sources of traditional knowledge are the older farmers, local astrologers and the almanac. However, traditional knowledge was said to be losing its relevance in a changing climate. The medium-range forecasts given by IMD for Dindigul district are based on the readings from Agro-Meteorological Field Unit (AMFU) located within Reddiarchatram block within the district. Hence, the forecasts and the advisories based on these forecasts come with 70–80% accuracy and are said to be useful and reliable for decision-making. Interestingly, the VKC-FPC network is also a major source of non-climatic information related to agriculture and allied sectors in the region. The FPO also plays the role of input supplier and produce aggregator to its shareholders. For shareholding members, the FPO is the nodal agency for accessing locale-specific information, technology and inputs and output markets. Thus, the VKC-FPC network supports the farmers to act on the climate information-based agro-advisories.

18.3.4 Strengthening Reach of Existing Climate Information

At present, weather/climate information is being communicated through mass media like television, radio and newspaper. Among these channels, television has the highest reach. But the weather bulletins disseminated through television are usually given at the end of the news bulletin and mostly go unnoticed. Measures like telecasting the weather report at the beginning of the news bulletin, having a dedicated weather channel and running scrolling text on district-specific climate information at frequent intervals throughout the day, were articulated as strategies to increase the reach of weather/climate information through television.

Communicating weather/climate information through mobiles was felt to be more effective since mobile technology had better reach in the region. It was suggested that the government should make it mandatory for all service providers to disseminate weather information through mobiles and emphasise this as a criterion for granting licence for operation.

18.3.5 Seasonal Forecast Requirement (Parameters and Lead Times) Across Stakeholders

Different stakeholders articulated different requirements for the seasonal forecast information (Table 18.3). The climatic parameters demanded by farmers are total rainfall, onset and distribution of total rainfall across the season. Of this, distribution of rainfall across seasons was articulated as important climate information by about 66% of respondents. Climate forecasts with a maximum of one-month lead time were thought to help in strategic on-farm decisions with more than 50% respondents echoing this. Seasonal climate forecasts with longer (2–6 months) lead times were said to have no relevance for farmers in the region.

On the other hand, off-farm players like input dealers, insurance agents and credit institutions required information with 3–6-months lead time to make strategic decisions. Input companies and district-level wholesale dealers of inputs saw a lot of potential for strategic business planning and risk reduction if forecasts are given 6 months before the start of the season. The sub-dealers at the block level demanded forecast information with a lead time of 1–2 months. The local village-level traders required information on total rainfall and its distribution with a one-month lead time.

Table 18.3 Forecast and lead time requirements expressed by farmers

Forecast requirement	No. of farmers articulating
Total rain for the season	68
Onset of rain for the season	78
Distribution of rain (intra-seasonal)	160
Lead time in months/days before sowing	No. of farmers articulating
One	135
Two	40
Three	3
3–4 days before sowing	81

Source Field Data

Note Total may not add up to 242 as there are multiple responses

Seasonal forecasts specific to a region were found to be of less significance for players beyond farmgate such as district-level traders in agricultural commodities. Their scale of operation was much beyond the region, and hence, climate variability in one region would not affect their business.

18.3.6 Communicating Seasonal Climate Forecasts: Role of Decision Analysis

Communicating SCF is a challenge since the forecast is probabilistic in nature. In SCF when one says there will be 40% chance of a normal rainfall season, one needs to understand that there is 60% chance of this not happening. Conventional methods of forecast communication like weather bulletins through mass media are largely in deterministic mode (provides quantitative value or range of weather parameters expected for a given time and an area). A different mode of communication is needed to communicate SCF given as tercile probabilities, which is an estimate of the likelihood of the rainfall that may occur within the given lead time. A decision analysis framework is more useful to communicate probabilistic SCF, as it is developed based on the principle of reflective learning. The decision analysis framework serves as a decision support tool in assessing the value of seasonal climate forecasts against multiple criteria. It is useful in working out trade-offs between competing objectives and helps to compare relative profitability of the probable decisions/choices that the respondents make based on the forecasts (Carberry et al., 2000). The framework was applied to on-farm decisions with farmers and off-farm decisions with members of the FPC.

Decision support tools like decision tree, decision graph and wonder bean were used to communicate SCF to farmers (Harrison & Williams, 2008; Liguton & Hayman, 2010). These tools help the end user to visualise the outcomes of each decision vis-a-vis the associated resource cost and the economic implication of decisions. The decision tree also allows for adding complexities to the decision process. Climatically risky decisions at on-farm and off-farm level were integrated and analysed using the tool. The tool helps visualise the possible scenarios and the outcomes and relate it to their own farming situations. The participants were asked to articulate their assumption on the potential yield and economic returns given the SCF scenarios of normal, above normal and below normal rainfall.

Capacity of the participants on deciphering probabilistic seasonal climate forecasts was built through representing different probabilities of rainfall for the season. The process started from a climatology scenario of equal chances of normal, dry and wet season and varying the probabilities to a great degree among the different options. In order to define the normal season, farmers were asked to share their perceptions based on practical experiences. According to them, normal rainfall amounted to 15 plough rainfall which is 375 mm on the metric scale (one plough rainfall is 25 mm). This corroborated with the 30-year historic average monthly rainfall data for the region during September/October to December.

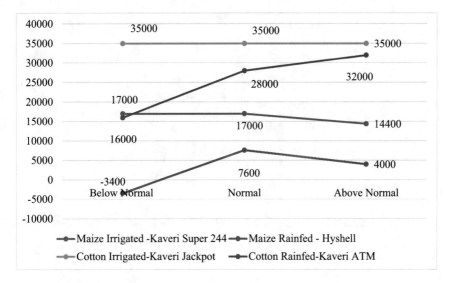

Fig. 18.2 Gross margins of crops and crop varieties across different forecast scenarios. *Source* Field Data

Decision tree framework which was used to compare the economics of different decisions was explained using excel sheets. They were given a probability forecast of 60% chance of a normal season, 20% of a dry season and 20% of a wet season. These forecast values were provided by the climate scientists. Given this scenario, the farmers were asked to articulate their crop choices as a first step and the varietal choices for the crops chosen as an added complexity in the second step. The outcome of the exercise is shown in Fig. 18.2. Irrigated cotton variety Kaveri jackpot was the best choice given the relative profitability of the crop across the SCF scenarios.

18.3.7 Decision Tree Analysis: Off-Farm Level

The off-farm actor considered here is the Reddiarchatram Sustainable Agriculture Producer Company Limited (RESAPCOL)—a farmer producer company based in Kannivadi, Reddiarchatram block. Input supply (seeds, fertilisers and pesticides) is one of the key essential services offered by the company to its members at the off-farm level. Here, managing the necessary seed stock by networking with private seed dealers is crucial. The demand side decision of critical concern is advance booking of seeds for the forthcoming cropping season with the private dealers by RESAPCOL. Amount of seeds of the different crops and the varieties to be stocked, phasing out distribution and the nature of transaction (cash/credit) to be extended to sub-dealers are the key supply side decisions. Table 18.4 gives the outcomes of the off-farm decision analysis.

Table 18.4 Outcome of decision analysis exercise with farmer producer company

Rainfall scenario	Probability (%)	Advance booking: seed quantity (ton)	Contingency arrangement			Quantity purchased 2 months before the season	Quantity distributed to sub-dealers	Mode and time of payment
Below normal	20	3.5	50% maize	45% cotton	5% ladies finger	0.9	25% of demand	Cash on delivery
Normal	60	5	60% maize	30% cotton	10% ladies finger	3	60% of demand	Credit
Above normal	20	7	55% maize	35% cotton	10% ladies finger	5	70% of demand	Credit

Source Field Data

18.3.8   Relevance of SCF and the Communication Tool for Strategic Planning Across Stakeholders

The importance of non-climatic factors on strategic decisions was emphasised by most of the players. In the case of farmers, it was the resourcefulness of the individual farmer and their coping capacity that were key drivers of crop choice or varietal choice decisions. The farmers articulated that the production decisions and outcomes are determined by climate variables as well as agronomic and economic factors such as labour, animal traction and credit. The communication cum planning tool used does not account for the non-climatic factors. Hence, the decisions/outcomes of these exercises may not be pertinent to and often are disconnected from the real-life agricultural decisions.

While facilitating the session, it was observed that farmers felt difficulty in articulating their strategic decisions with respect to committed and non-committed costs. This was because farmers normally based their committed and non-committed cost decisions on observation and assessment of actual climatic occurrences in the field. The climate information given to them (probability of a normal/below normal/above normal rainfall year) did not tell them anything about the intra-seasonal distribution and variations. Several of the non-committed cost decisions like intercultural operations, plant protection and top dressing are taken based on the intra-seasonal distribution and variation of rainfall.

Further, it emerges that the majority of key farm decisions are taken in very short lead times. Irrespective of any forecast (scientific or traditional), farmers prepare the land in anticipation of the season. Given the edaphic conditions, choice of the crop as well as variety is based on the onset of the rainfall prior to sowing. With respect to crop choices under rainfed situations, they take up either cotton or maize and the determining factors articulated are a) market price of the previous year/season and b) availability of groundwater to give one or two critical rounds of irrigation.

With respect to sowing decisions, if the onset is normal—i.e. receiving at least 50 mm of rainfall within 3–5 days in late September or early October—farmers start purchasing seeds, mostly hybrid seeds. These seeds are available in the market for immediate cash payment as well as credit. Planning for intercultural operations, top dressing and harvesting are purely based on how the season progresses. Such decisions do not require prior preparation and are taken based on actual occurrences.

For the off-farm player—the Farmer Producer Company and multinational input companies—strategic planning was done on the basis of their historic sales data, the standard acreage under each crop in the region, the ground-level data supplied by their field assistants and data on the existing market share of their competitors. The wholesale dealers will not incur heavy losses even if the forecast information goes wrong because bookings and commitments are initiated before the beginning of the season. Actual transaction of commodities happens after the commencement of the season or just about when it starts.

Table 18.5 Probable decisions based on SCF for different players

Off-farm player	Probable decision based on SCF
Input wholesalers	Stocking rate (managing inventory)
Producer (farmer)	Land allocation and crop choice
Produce aggregator (cotton)	Plan/influence produce turnover volumes
Produce aggregator (vegetables)	Forecast volume of incoming stock
Processor (cotton)	Help predict cotton quality
Financial institution (banks)	Decide on lending ceilings for different crops
Wholesale marketing	Not useful

Source Field Data

18.3.9 Probable Decisions Based on SCF Across the Value Chain

Choice of crops and crop variety are the decisions that farmers would base on SCF information (Table 18.5). For the wholesale input dealer at the district level, SCF was more useful in determining the stock of seeds to be maintained before the beginning of the season. It helps them decide on the stock of crops, varieties within these crops and the proportion of different crops. It also helps them to decide the nature and volume of transactions with the sub-dealers. If SCF forecasts a good season, they would extend inputs on credit to the sub-dealers, and if a bad season is predicted, they would prefer a cash-and-carry method. The credit institutions would use the information to decide the target for crop loans for the region as well as loan ceiling for each crop. They will also use the information for withholding or pushing crop loans. Similar experiences have been reported from Brazil (Lemos et al., 2002) and Zimbabwe (Phillips et al., 2002). Likewise, the insurance agents will also use the information to plan their targets and compensation payouts. Several of these decisions by players at the higher end of the agricultural value chain are competitive and often end up being counterproductive to the interests of the resource poor small farmer.

18.3.10 Suggestions for Making SCF Relevant for Stakeholders

Improving the capacities of the end user to interpret and use SCF for decision-making is a primary requirement for the climate forecast to be relevant. Further, SCF with 6-month lead time needs to be followed up with shorter lead times ranging from 1 to 3 months. This will cater to the forecast requirement of the different stakeholders and increase its utility in planning and management decisions. Downscaling to block level would increase the confidence of the stakeholders on the forecast. SCF needs to be packaged as a seamless forecast that combines with the existing short, medium and extended range weather forecast. SCF models should incorporate dynamics of the microclimate of the region and changes in climate variables post-extreme climate events like flood and drought. SCF should combine traditional knowledge of the region with climate science for improving its communication. SCF should be given along with advisories that build mitigation aspects directed at climate risk reduction. This has been envisaged and integrated in the National Action Plan on Climate Change under the implementation section as part of institutional arrangements for managing climate change agenda (Rattani, 2018).

Resource capacity of the end user plays a major role in determining the utility and uptake of the extended range forecast information. SCF information needs to be backed up with necessary input, and infrastructural and logistic support for the farmers/users to translate to on-field risk reducing action. A strong institutional mechanism to disseminate and implement SCF will help achieve this. The FPO which serves as a nodal agency for shareholding farmers can provide institutional support in terms of input, credit, infrastructural and logistic support to utilise SCF for reducing risk in farming.

18.4   Conclusions

Seasonal climate forecasts help in inducing risk reducing decisions across the agricultural value chain. However, SCF is only one suite of information on climate and weather that players across the agricultural value chain might use to make decisions. Hence, SCF generated and disseminated for a region needs to be flexible in terms of lead times and complement the existing short and medium range weather forecasts. But incorporation of SCF in decision-making depends a lot on the capacity of the end users in accessing and understanding the forecasts. Further, field-level adaptation action based on these decisions is dependent on the resource capacity of the end users and the presence of an enabling environment (Hansen, 2002). To enhance the usefulness of climate forecast information and advisories, it is essential to identify the decision-relevant attributes of forecast information for specific activities and players in the value chain, and provide forecasts incorporating those attributes. Communication is another crucial element that decides the utility of SCF; hence, appropriate participatory methods need to be adopted to communicate SCF. Capacity building of the end user in understanding, interpreting and using the forecast information for decision-making needs to be taken up for realising better utility of SCF.

The flip side of SCF is that it might sometimes undermine food security in the region by adding to the vulnerability of the primary producers. The forecast information is being put to use for different purposes by players across the value chain. Some of these decisions may be complementary to the primary producer, while some can be competitive and counterproductive to the primary producer. For example, the credit agencies' decision to limit crop lending in anticipation of a forecasted bad season can be very detrimental to the primary producers. A case in point is the experience in Brazil and Zimbabwe of financial institutions engaged in extending agricultural credit tightening their credit lending in response to a forecast for increased probability of drought. Similarly, the decision not to promote crop insurance in anticipation of a forecasted bad season by an insurance company can be counterproductive to the primary producers. Hence, efforts for strengthening climate resilience at a regional level need to factor in all these complementary and competitive engagements among the different stakeholders to ensure a win-win situation for all the players across the agricultural value chain.

Acknowledgement The authors acknowledge research funding from Australian Agency for International Development (Grant No. 59553).

References

Bahinipati, C. S., & Patnaik, U. (2021). What motivates farm level adaptation in India? A systematic review. In A. K. E. Haque, P. Mukhopadhyay, M. Nepal, & M. R. Shammin (Eds.), *Climate change and community resilience: Insights from South Asia*. Springer.

Bruno Soares, M., & Dessai, S. (2016). Barriers and enablers to the use of seasonal climate forecasts amongst organisations in Europe. *Climatic Change, 137*, 89–103.

Carberry, P., Hammer, G., Meinke, H., & Bange, M. (2000). The potential value of seasonal climate forecasting in managing cropping systems. In G. L. Hammer, N. Nicholls, & C. Mitchell (Eds.), *Applications of seasonal climate forecasting in agricultural and natural ecosystems* (pp. 167–181). Atmospheric and Oceanographic Sciences Library (Vol. 21). Springer.

Hansen, J. W., & Sivakumar, M. V. K. (2006). Advances in applying climate prediction to agriculture. *Climate Research, 33*(1), 1–2.

Hansen, J. W., Baethgen, W., Osgood, D., Ceccato, P., & Ngugi, R. K. (2007). Innovations in climate risk management: Protecting and building rural livelihoods in a variable and changing climate. *Journal of Semi-Arid Tropical Agricultural Research, 4*(1), 1–38.

Hansen, J., Mason, S., Sun, L., & Tall, A. (2011). Review of seasonal climate forecasting for agriculture in sub-Saharan Africa. *Experimental Agriculture, 47*(2), 205–240.

Harrison, M., & Williams. J. B. (2008). Communicating seasonal forecasts. In A. Troccoli, M. Harrison, D. L. T. Anderson, & S. J. Mason (Eds.), *Seasonal climate: Forecasting and managing risk* (pp. 293–314). NATO Science Series. Springer.

Jones, L. M., Koehler, A. K., Trnka, M., Balek, J., Challinor, A. J., Atkinson, H. J., & Urwin, P. E. (2017). Climate change is predicted to alter the current pest status of *Globodera pallida* and *G. rostochiensis* in the United Kingdom. *Global Change Biology, 23*(11), 4497–4507.

Legler, D. M., Bryant, K. J., & O'Brien, J. J. (1999). Impact of ENSO-related climate anomalies on crop yields in the US. *Climatic Change, 42*, 351–375.

Lemos, M. C., Finan, T. J., Fox, R. W., Nelson, D. R., & Tucker, J. (2002). The use of seasonal climate forecasting in policymaking: Lessons from Northeast Brazil. *Climatic Change, 55*(4), 479–507.

Liguton, J. P., & Hayman, P. (2010). Communicating and using seasonal climate forecasts: A challenge crossing national, organizational, and disciplinary boundaries. *Philippine Journal of Development*.

Lipper, L., Thornton, P., Campbell, B. M., Baedeker, T., Braimoh, A., Bwalya, M., Caron, P., Cattaneo, A., Garrity, D., Henry, K., Hottle, R., Jackson, L., Jarvis, A., Kossam, F., Mann, W., McCarthy, N., Meybeck, A., Neufeldt, H., Remington, T., Sen, P. T., Sessa, R., Shula, R., Tibu, A.,

& Torquebiau, E. F. (2014). Climate-smart agriculture for food security. *Nature Climate Change,* *4*, 1068–1072.

McCrea, R., Dalgleish, L., & Coventry, W. (2005). Encouraging use of seasonal climate forecasts by farmers. *International Journal of Climatology, 25*, 1127–1137.

Meinke, H. B., Nelson, R., Kokic, P., Stone, R., Selvaraju, R., & Baethgen, W. (2006). Actionable climate knowledge: From analysis to synthesis. *Climate Research, 33*(1), 101–110.

Paz, J. O., Fraisse, C. W., Hoogenboom, G., Hatch, L. U., Garcia y Garcia, A., Guerra, L. C., & Jones, J. W. (2006). Peanut irrigation management using climate-based information. In *Computers in Agriculture and Natural Resources—Proceedings of the 4th World Congress* (pp. 660–665).

Phillips, J. G., Deane, D., Unganai, L., & Chimeli, A. (2002). Implications of farm-level response to seasonal climate forecasts for aggregate grain production in Zimbabwe. *Agricultural Systems, 74*, 351–369.

Pulwarty, R. S. (2007). Communicating agro climatological information, including forecasts for agricultural decision. In World Meteorological Organization (Ed.), *Guide to agricultural meteorological practices* (pp. 17.1–17.13).

Rickards, L., Howden, S., Crimp, S., Fuhrer, J., & Gregory, P. (2014). Channelling the future? The use of seasonal climate forecasts in climate adaptation. In J. Fuhrer, & P. Gregory (Eds.), *Climate change impact and adaptation in agricultural systems: Soil ecosystem management in sustainable agriculture* (pp. 233–252).

Sivakumar, M. V. K. (2006). Dissemination and communication of agrometeorological information. *Global Perspectives Meteorology Applications, 21*–30.

Suarez, P., & Patt, A. G. (2004). Cognition, caution and credibility: The risks of climate forecast application. *Risk Decision Policy, 9*(1), 75–89.

Stern, P. C., & Easterling, W. E. (1999). *Making climate forecasts matter.* National Academy Press.

Tanyi, C. B., Ngosong, C., & Ntonifor, N. N. (2018). Effects of climate variability on insects pests of Cabbage: Adapting alternate planting dates and cropping patterns as control measures. *Chemical and Biological Technologies in Agriculture, 5*, 25.

Rattani, V. (2018). *Coping with climate change: An analysis of India's national action plan on climate change.* Centre for Science and Environment.

Vogel, C., & O'Brien, K. (2007). Who can eat information? Examining the effectiveness of seasonal climate forecasts and regional climate-risk management strategies. *Climate Research, 33*, 111–122.

Wheeler, T. & von Braun, J. (2013). Climate change impacts on global food security. *Science (New York, N.Y.), 341*(6145), 508–513.

Chapter 19
Farmer Adaptation to Climate Variability and Soil Erosion in Samanalawewa Catchment in Sri Lanka

E. P. N. Udayakumara

Key Messages

- Mean annual rainfall has decreased while mean annual temperature has increased from 1922 to 2008.
- It is also observed that the general perception of the farmers is also similar in terms of the overall trend.

19.1 Introduction

At present, poverty is a growing concern in most of the developing and least developed countries. More than one billion people around the world live in extreme poverty. Nearly 77.4% of Sri Lanka's population live in rural areas and depend upon their local environments for their daily survival (Central Bank of Sri Lanka, 2020). Environmental degradation is a major cause of poverty among rural communities around the world. Thus, the natural resources can be considered as part and parcel of these communities and sustainable management of natural resources is very crucial for

E. P. N. Udayakumara (✉)
Department of Natural Resources, Faculty of Applied Sciences, Sabaragamuwa University of Sri Lanka, Belihiloya, Sri Lanka
e-mail: udayaepn@appsc.sab.ac.lk; udayaepn@gmail.com

© The Author(s) 2022
Haque et al. (eds.), *Climate Change and Community Resilience*,
https://doi.org/10.1007/978-981-16-0680-9_19

both enhancing rural livelihoods and conserving the environment for the future (Das, 2021, Chap. 17 of this volume; Vidanage et al., 2022, Chap. 15 of this volume).

Sri Lanka is an agricultural country and nearly 25.3% of people still depend on land-centered activities for their survival (Central Bank of Sri Lanka, 2020). At present, most of the country's natural resources, viz. land, water, forest, coastal, biodiversity, air, etc., are significantly threatened. Among them, land is one of the crucial and greatly threatened natural resources in the country (Ministry of Forestry & Environment, 2001).

From 1881 to 1900, forest cover in Sri Lanka decreased from 85 to 70% of the total land area because central hills were cleared for plantations crops and dry zone forests were logged for timber by the British rulers. By the year 1948, when Sri Lanka got its independence, forest cover had declined to 44% (De Zoysa, 2001), by 1981 it was 25% (Bandarathilleke, 1991) and it further declined to only 19.9% of the country's total land area by the year 2006 (Central Bank of Sri Lanka, 2006).

Several studies (Chandrapala, 1996; Domroes, 1996) have reported that the amount of rainfall in Sri Lanka has been declining gradually. For instance, during the period 1931–1960, the average annual rainfall was estimated to be 2005 mm. This has since declined to 1861 mm from 1961 to 1990 (Domroes & Schaefer, 2000). A study of the upper catchments in the central hill country around Nuwara Eliya also indicates a declining trend in annual rainfall from 1900 to 2002 (Madduma Bandara & Wickramagamage, 2004).

Although there are no studies on land degradation or land-use cover changes (LUCC) or on climate change impact in the Samanalawewa catchment, these phenomena are evident in the area. These changes can cause many environmental and socioeconomic problems like habitat destruction and loss of biodiversity, reduced or uneven water flows, water stresses, deterioration of water quality, the decline in agricultural production and its wealth, downstream economic damage due to sedimentation, change in the size and spatial distribution of the human population, and changes in livelihood (Technology Evaluation and Management Services, 1992a). Presently numerous drivers and forces, such as human population growth, existing land-use policies, growth of demand in agriculture, unsecured land tenure systems, infrastructure development, urbanization, destruction of forest, chena cultivation also known as shifting cultivation or swidden cultivation, soil erosion, and natural calamities pose a threat to the Samanalawewa catchment area as well as the country's natural resources.

Most developing countries have constructed dams to create small or large reservoirs to meet the demand for hydropower. However, most recent studies have argued that these artificially constructed reservoirs have provided only short-term benefits while creating other socioeconomic hardships for the local population that is dependent on agriculture (Kim, 2007).

The construction of the Samanalawewa reservoir was started in 1988 and completed in 1992 to divert the Walawe River to generate electricity. No environmental impact assessment (EIA) was undertaken while constructing the Samanalawewa reservoir, even though this was a requirement of the National Environmental (Amended) Act No. 56 of 1988, and the socioeconomic and environmental impacts of its early stages are not well documented. However, it is important to note that while the project had unexpected negative impacts on the traditional livelihood of people in the catchment; it also had positive effects on the region and on the national economy (Technology Evaluation and Management Services, 1992a).

Studies revealed that the Samanalawewa reservoir had impacted the livelihood of the farmers in the downstream regions while it had a relatively low impact on the upstream people. In particular, there was a deterioration of traditional paddy farming, decline of agricultural income, increasing scarcity of agricultural labor, rural–urban migration, and erosion of community cohesion due to inadequate resettlement plans for evacuees in the Samanalawewa reservoir area (Ceylon Electricity Board, 2006).

It is known that large-scale reservoirs constructed for irrigation and hydropower developments not only have socioeconomic impacts but also create lasting environmental damages including changes in the weather patterns (Rosenberg et al., 1995). People in developing countries are the most vulnerable to climate variability impacts because they are more exposed to weather extremes without the ability to safeguard themselves and because of low adaptive capacity to climate variability (Tol et al., 2004).

The Samanalawewa reservoir diverted the waterways, removed hundreds of hectares of land from the landscape, caused loss of natural high forest cover ($\leq$ 430 m), led to extensive agricultural expansion on eroded lands, caused land salinity, and loss of habitat (Central Engineering Consultancy Bureau, 1991). In addition, as the reservoir was being filled in 1992 a major leak (7.5 m^3 s^{-1}) caused a landslide approximately 300 m downstream of the dam. To control that leak, the "Wet Blanketing" method was used. There is still a leak which is about 1.8 m^3 s^{-1} (Laksiri et al., 2005).

Several reports and studies suggest that climate variability, LUCC and land degradation are taking place in Samanalawewa catchment as well as other catchments in Sri Lanka at present (Udayakumara et al., 2010). However, there is a dearth of adequate and reliable information and understanding related to climate variability and land degradation due to soil erosion which limits our ability to monitor, mitigate and also develop appropriate policies for the long-term sustainability of the catchment area. Therefore, this study tries to assess the status of climate variability, land degradation and farmers' adaptation of soil and water conservation (SWC) measures in the study area so that options for better management of catchment resources can be developed in the future.

19.2 Study Area and Methods

19.2.1 Study Area

The Samanalawewa catchment is one of the most important catchments in the country because of its diverse land uses. The study area is situated in the Ratnapura District

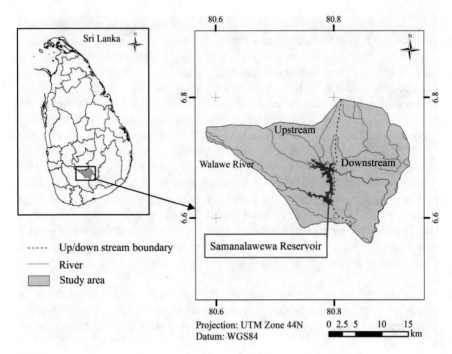

Fig. 19.1 Location map of the study area. *Source* Udayakumara et al. (2012)

of Sri Lanka, stretching between 80.58° and 80.92° east longitude and 6.56° to 6.80°
north latitude covering an area of about 536 km² (Fig. 19.1). The upper part of
the Samanalawewa catchment is situated on the southern face of the rise to Horton
Plains and Peak Wilderness sanctuary in the Nuwara-Eliya Divisional Secretariat
Division (DSD) of the Nuwara-Eliya District, while the lower part inclusive of the
Samanalawewa reservoir (897 ha) lies within the DSD of Imbulpe and Balangoda.

The region is mainly drained by the Walawe River and its tributaries, such as
Belihul Oya (Tributary), Denagan Oya, Diyawini Oya, Kuda Oya, and Hulanda
Oya. During 1987–1992, Walawe River was dammed at the confluence of the Belihul
Oya and the Walawe River leading to the formation of the Samanalawewa reservoir
extending over 897 ha surface area. The Samanalawewa reservoir generates 60–
120 MW of hydropower and is the main man-made water body in the catchment
(Technology Evaluation and Management Services, 1992b). This single-purpose
(only hydropower) project (Technology Evaluation and Management Services,
1992a) has caused changes to the ecosystem though water diversion and is one of the
important factors behind the socioeconomic and associated environmental impacts
in the catchment through. The population of 114,743 in the catchment area (Department of Census & Statistics Sri Lanka, 2008) is not evenly distributed because of
terrain and variation in land conditions.

Climatically the catchment belongs to the three zones, viz. the wet (rainfall over
2500 mm), intermediate (1750–2500 mm), and the dry (below 1750 mm). The upper

catchment (upstream or upper area of the catchment from the dam) is in the wet zone while the lower catchment area is in the intermediate zone. Geographically, the region consists of rocks belonging to the Highland group, comprising quartzite, marbles, and undifferentiated metasediment such as garnet, granulite, charnockite, and biotic gneisses (Technology Evaluation and Management Services, 1992a).

19.2.2 Types and Sources of Data

This section briefly explains the types and sources of data used in this research study. Data was collected from a variety of sources in the form of tabular data, reports, and field surveys. The basic data used in the study is presented in Table 19.1.

The study area consists of 23,304 households (HHs) in 67 upstream villages and 4638 HHs in 9 downstream villages (Department of Census & Statistics Sri Lanka,

Table 19.1 Data types and sources

Data	Data type	Source
Soil	– Minimum and the maximum soil erosion rates (mm yr^{-1}) at 30° slopes from January to December 2008 and – Bulk density	Field experiments
RS	– ALOS image (sensor: AVNIR-2, scene ID: ALAV2A111173460, resolution: 12 m spatial, acquisition date: February 25, 2008)	Geoinformatics Centre, AIT, Thailand
	– ALOS image (sensor: AVNIR-2, scene ID: ALAV2A111173470, resolution: 12 m spatial, acquisition date: February 25, 2008)	Geoinformatics Centre, AIT, Thailand
GIS	– Digital elevation data (scale 1:10,000)	Department of Survey, Sri Lanka
Meteorological	– Yearly rainfall from 1922 to 2008 and Temperature from 1973 to 2008	Department of Meteorology, Sri Lanka and Weather Stations of the study area
Socioeconomic	– HH and livelihood-related information from 1986 to 2008	Field survey
Other published/unpublished	–	Various publications, documents, reports, etc.

Source Udayakumara (2011)

2007; Udayakumara et al., 2012). For this study, 15 villages were selected randomly from upstream and 6 villages from downstream. The total number of HHs in the 21 selected villages was 7269. For this study, 201 households were sampled from the watershed.

The sampled 201 HHs were interviewed from February to June 2009. A pretested structured questionnaire was used to ascertain information on socioeconomic conditions, farmer's perceptions of the climate variability of the area. The interviews were conducted with the help of five trained field enumerators.

19.2.3 Research Methods

19.2.3.1 Assessing Climate Variability

Average annual rainfall and temperature data were collected from nine locations established in the study area (Fig. 19.2) namely Agarsland Estate, Balangoda Estate, Non-Pareil Estate, Rye Estate, Sabaragamuwa University of Sri Lankan, Samanalawewa Dam Site Office, Irrigation Office-Kaltota, Agriculture In-service Training Institute-Rajawaka and Balangoda Post Office. Farmer's perceptions on climate variability data such as rainfall (amount, intensity, and duration/storm event), temperature (daytime temperature, nighttime temperature, number of daytime hot

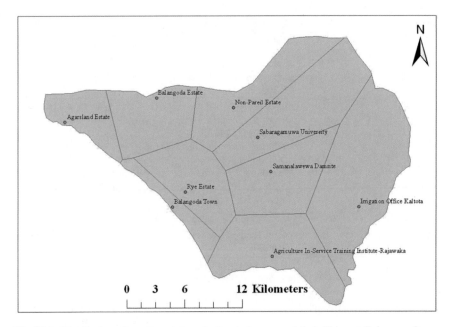

Fig. 19.2 Distribution of weather stations in the study area and their Thiessen Polygons. *Source* Udayakumara (2011)

days, and number of nighttime hot days) were collected through a survey. The data was analyzed using SPSS™ 16 and MINITAB™ 14 statistical software.

To find out the best-fitted rainfall and temperature models, trend analysis was carried out using three types of time series trend models such as linear, exponential growth, and quadratic. Then considering each model's accuracy measures (mean absolute percentage error or MAPE, mean absolute deviation or MAD, and mean squared deviation or MSD), the best-fitted rainfall model and temperature models (minimum, maximum, and average annual) were selected. Finally, based on questionnaire survey data, farmer's perception on climate variability from 1997 to 2008 was analyzed.

19.2.3.2   Assessing Soil Erosion

Due to lack of availability of soil erosion assessment-related data, it was limited to use the data-intensive soil erosion models for the study area. Thus, the empirical soil erosion model developed by Honda (1993) was used to assess the annual rate of soil erosion in the study area (Udayakumara et al., 2010). The model has given satisfactory results in other areas with similar climatic and/or geomorphologic conditions to this study area (Udayakumara, 2011), like the Ashio region of Japan (Honda, 1993), the Siwaliks in Nepal (Honda et al., 1996), and the mountainous area of Northern Thailand (Hazarika & Honda, 2001). The strength of the model is that it is not data-intensive and can be used in an area like the Samanalawewa catchment, where data is scarce (Udayakumara et al., 2010).). In terms of model performance, Honda et al. (1996) reported that this model has produced results similar in quality to that produced by extensive fieldwork. The model is mainly governed by slope gradient, vegetation index, and maximum and minimum rates of soil erosion at 30° slopes (Udayakumara, 2011).

19.2.3.3   Assessing Farmers' Adaptation of SWC Measures

As farmers experience soil erosion in the catchment area, they might resort to several different types of soil and water conservation (SWC) measures (Udayakumara et al., 2012). In this study, the adoption of SWC measures was a dichotomous-dependent variable (Y) that signifies whether or not a household is willing to adopt SWC measures. $Y = 1$, if a HH is willing to adopt SWC measures, and $Y = 0$ otherwise. The selected 15 explanatory variables that are hypothesized to have an association with the willingness to adopt SWC measures were used in the study to explain their decisions. The findings of the past studies and questionnaire surveys and existing theories were used to select the 15 explanatory variables and structure the working hypotheses (Udayakumara et al., 2012). The potential explanatory variables, which are hypothesized to influence farmers' willingness to adopt SWC measures in the study area, are farmers' perception of soil erosion problem, household head's age, education, and gender; security of land tenure, past experiences, land size, off-farm income, training on SWC, and advice from extension officers. A logistic regression model was used to the set of explanatory variables mentioned above where the dependent variable (adoption of SWC measure) is binary.

19.3 Results and Discussion

This section presents the results and discussion of climate variability, land degradation due to soil erosion and farmers' adaptation of SWC measures.

19.3.1 Rainfall Variability

The respective values of mean, minimum, maximum, and range for average annual rainfall during the period of 1922–2008 were found to be 2247.7 (SD = 565.8), 612.3, 3500.0, and 2887.7 in mm. The maximum and minimum rainfall amounts were in the years 1922 and 1981, respectively. The obtained standard deviation value (SD) for the average annual rainfall is quite high, which implies that rainfall variation is fairly high within the considered period.

The trend analysis yielded three rainfall models, viz. linear, exponential growth, and quadratic for the average annual rainfall. Considering these three models accuracy measures (MAPE, MAD and MSD), a model with the least MAPE, MAD, and MSD values was selected as the best rainfall model for the study area. According to the above three models, the linear trend model has the least values of MAPE, MAD, and MSD, compared to the other two models. Results suggest that the average annual rainfall of the area has decreased by 5.5 mm per year from 1922 to 2008 (Fig. 19.3). Deforestation, agricultural expansion, and development of infrastructure in the study area are mainly attributed to the reduction in rainfall (Technology Evaluation and Management Services, 1992a). According to the previous literature, Chandrapala (1996) has disclosed that the average annual rainfall over Sri Lanka has also decreased 4.8 mm per year from 1961 to 1990. Based on the estimates, the forecasted average annual rainfall value in the year 2030 would be 1893.5 mm.

According to Table 19.2 for the period from 1997 to 2008, 77% of respondents said the amount of rainfall had decreased while 13% felt there was no change, and 10% said it had increased. 65% of respondents said the intensity of rainfall had decreased followed by 23% who said there was no change, while 12% claimed it had increased. Duration of rainfall is also considered to have decreased by 67% of respondents, followed by 18% who reported no change and 15% reporting an increase (Table 19.2).

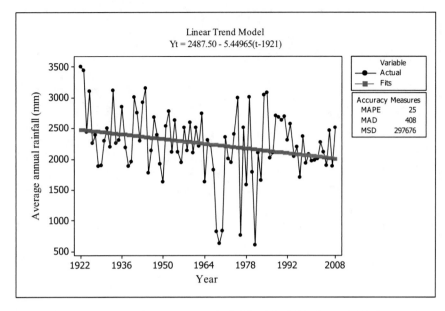

Fig. 19.3 Trend analysis plot for average annual rainfall (1922–2008). *Source* Udayakumara (2011)

Table 19.2 Perceived rainfall changes from people (1997–2008)

Perception on rainfall	Decreased	Increased	No change
	% Respondents		
Amount	77	10	13
Intensity	65	12	23
Duration (storm event)	67	15	18

Source Udayakumara (2011)

19.3.2 Temperature Variability

The respective values of mean, minimum, maximum, and range for average annual minimum temperature were found to be 19.6 (SD = 0.7), 18.1, 21.0, and 2.9 in °C. The trend analysis yielded three average annual minimum temperature models, viz. linear, exponential growth, and quadratic. According to these three models, the exponential growth model has the least values of MAPE, MAD, and MSD, compared to the other two models. Thus, the best-fitted average annual minimum temperature model for the study area is an exponential growth model shown in Fig. 19.4. The model shows that the average annual minimum temperature of the study area has increased from 1973 to 2008. Thus, the forecasted average annual minimum temperature value in 2030 would be 20.8 °C.

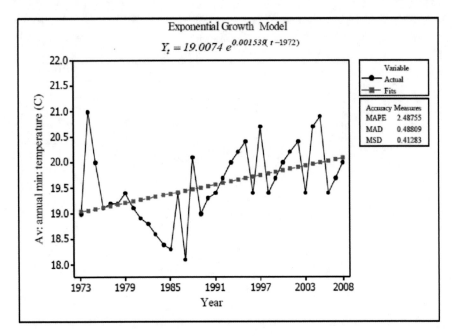

Fig. 19.4 Trend analysis plot for average annual minimum temperature (1973–2008). *Source* Udayakumara (2011)

For the average annual maximum temperature, mean, minimum, maximum, and range were found to be 27.67 (SD = 0.40), 26.8, 28.6, and 1.8 in °C, respectively. The trend analysis yielded three average annual maximum temperature models, viz. linear, exponential growth, and quadratic. According to these three models, the linear trend model has the least values of MAPE, MAD, and MSD, compared to the other two models. Thus, the best-fitted average annual maximum temperature model for the study area is a linear trend model (Udayakumara, 2011) as shown in Fig. 19.5.

The model shows that the average annual maximum temperature of the study area has increased from 1973 to 2008. As the rate of temperature increment is about + 0.02 °C yr-[1], the forecasted average annual maximum temperature value in 2030 would be 28.3 °C (Udayakumara, 2011).

While the respective values of mean, minimum, maximum, and range for average annual temperature were found to be 23.6 (SD = 0.4), 22.8, 24.5, and 1.7 in °C, the trend analysis yielded three average annual temperature models: linear, exponential growth, and quadratic. According to these three models, the linear trend model has the least values of MAPE, MAD, and MSD, compared to the other two models. Thus, the best-fitted average annual temperature model for the study area is a linear trend model (Udayakumara, 2011) as shown in Fig. 19.6 as the rate of temperature increment is about 0.02 °C yr-[1]. Thus, the forecasted average annual minimum temperature value for the year 2030 would be 24.5 °C. In general, the global average temperature is projected to rise in the range of 1.4–5.8 °C by the year 2100 under the

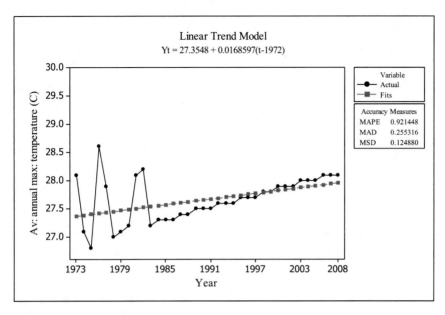

Fig. 19.5 Trend analysis plot for average annual maximum temperature (1973–2008). *Source* Udayakumara (2011)

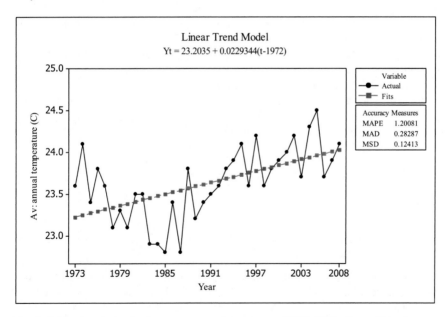

Fig. 19.6 Trend analysis plot for average annual temperature (1973–2008). *Source* Udayakumara (2011)

Table 19.3 Perceived temperature changes from people (1997–2008)

Perception on temperature	Decreased	Increased	No change
	% Respondents		
Daytime temperature	5	72	23
Nighttime temperature	7	66	27
Number of daytime hot days	2	81	17
Number of nighttime hot days	23	63	14

Source Udayakumara (2011)

different emission scenarios (Intergovernmental Panel on Climate Change, 1995). However, according to this study, in this area, this value would also be within the IPCC range, i.e., about 2.1 °C (Intergovernmental Panel on Climate Change, 1995). This is a sign of global warming.

According to Table 19.3 and as shown by 72% of respondents, the daytime temperature increased during the period from 1997 to 2008, followed by 23% who said there was no change and 5% reporting a decrease. About 66% of respondents, nighttime temperature was received to have increased followed by 27% saying there was no change and 7% reporting a decrease. As per the perception of the respondents, the number of daytime hot days increased by 81%, which is followed by 17% of no change and 2% perceived it to have decreased. According to 63% of respondents, the number of nighttime hot days has also increased whereas 23% thought it has decreased and 14% said there was no change.

According to the weather stations' temperature analysis results, the temperature has actually increased which is similar to the perception of the majority of people. Clearly, they have noticed the changes in the number of daytime and nighttime hot days in the area.

19.3.3 Soil Erosion

Maximum, minimum, and mean annual soil erosion rates were 22.1 mm yr^{-1}, 0.0 mm yr^{-1}, and 0.33 mm yr^{-1} (SD = 0.31), respectively. Based on soil erosion calculated at each pixel, six preliminary categories of soil erosion ranging from very low to extremely high were identified (Table 19.4). The study area's gross rate of soil erosion ranged from 0.0 to 289.0 t ha^{-1} yr^{-1} with an average rate of 4.3 t ha^{-1} ya^{-1} (SD = 4.1) (Fig. 19.7).

Furthermore, 65% of the surveyed people perceived soil erosion that had decreased followed by 30% who said it had increased and only 5% reported no change. Most

Table 19.4 Status of soil erosion in the study area, 2008

Soil erosion status	Range (t ha^{-1} yr^{-1})	% area
Very low	0–4	55.3
Low	4–8	30.2
Moderately high	8–12	10.1
High	12–16	2.9
Very high	16–20	0.9
Extremely high	20–289	0.6

Source Udayakumara (2011)
Note Total area (536 km^2)

Fig. 19.7 Soil erosion map of the study area. *Source* Udayakumara (2011)

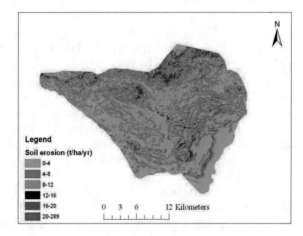

believe that the main reason for this decline in erosion and landslides is due to the decline of rainfall.

19.3.4   Farmers' Adoption of SWC Measures

Out of the 201 sample respondents, 121 (60.2%) were found to be willing to adopt the SWC measures, whereas the remaining 80 (39.8%) respondents were not willing to adopt those conservation measures. The adopted SWC measures include agronomics (e.g., the use of mulch and organic manure, soil surface/subsurface treatments), vegetative (e.g., increasing trees, shrub, and grass cover), structural (e.g., constructing terraces, bunds, and ditches), and management (e.g., changing species composition of crops, controlling cropping intensity, and observing fallow period).

Moreover, the results also indicated that 19.8% of the farmers have been adopting SWC measures for more than 20 years, with 12.4% of farmers adopting the measures

Table 19.5 Impacts of adopted SWC measures

Adoption of SWC measures (years)	% (n)	Average SWC cost USD ha^{-1} yr^{-1}	Average rate of soil erosion t ha^{-1} yr^{-1} (SD)	Soil erosion versus soil generation
< 5	24.8 (30)	85.5	10.5 (5.7)	48
5–10	18.2 (22)	75.5	7.8 (5.2)	35
10–15	24.8 (30)	66.5	3.5 (0.8)	16
15–20	12.4 (15)	45.7	2.5 (0.7)	11
> 20	19.8 (24)	30.8	1.8 (0.5)	8

Source Udayakumara (2011)
Note n Number of HHs according to years of SWC measures adopted, *UD$* US dollar, *SD* Standard deviation

for 15–20 years, 24.8% for 10–15 years, 18.2% for 5–10 years, and another 24.8% for less than 5 years (Table 19.5).

Since the farmers in the study area have been adopting the SWC measures, the average SWC costs, the average rate of soil erosion, and the ratios of soil erosion vs. natural soil generation have declined drastically. Table 19.5 shows that the farmers who have adopted the SWC measures for more than 20 years incurred the least conservation cost (30.8 USD ha yr^{-1}) as well as the least soil erosion rate (1.8 t ha yr^{-1}) compared with the farmers who have adopted SWC measures only recently (less than 5 years).

Finally, out of the thirteen correlated variables in the binary logistic model, the following six variables are included in the binary logistic model: four variables, viz. farmer's perception of soil erosion problem, gender of the HH head, training on SWC, and advice from the extension officers are significant at 0.01%; the other two variables such as past awareness about technology and off-farm income are significant at 0.05%.

19.4 Conclusions

Average annual rainfall anomalies have shown a significant decreasing trend from 1922 to 2008 in the study area. The rate of decrease of average annual rainfall for the above period was in the order of 5.5 mm per year. It is also observed that the general perception of the farmers is also similar in terms of the overall trend. Average annual minimum, average annual maximum, and average temperatures have shown significant increasing trends from 1973 to 2008. This is also similar in terms of the farmer's perception of temperature. The results indicate that the general perception is in line with the trends in the scientific data. The facts and the experiences coincide, and as a consequence of this situation, at present dwellers of the catchment encounter many negative impacts such as scarcity of water for agriculture, loss of biodiversity, and unexpected droughts and rains.

According to the Kyoto Protocol, the climate change issue is a global problem, and therefore, it needs the cooperation of all countries. Thus, to arrest the decrease of average annual rainfall and increment of average temperature in Sri Lanka, proper programs/policies have to be worked out. Among them, implementation of reforestation and afforestation programs for carbon sequestration, prevention of deforestation, formulation of policies to curtail unnecessary gas emissions, reduction of the use of fossil fuel by improving the efficiency of energy usage and by substituting renewable energy resources are very important.

Soil erosion is also a serious concern in most parts of Sri Lanka. In the Samanalawewa catchment, the rate of soil erosion ranges from 0 to 289 t ha^{-1} yr^{-1} with a catchment level average of 4.3 t ha^{-1} yr^{-1}. The household survey reveals that inappropriate soil and crop management practices and increased economic activities due to population growth, poverty, and inadequate labor are the major indirect causes of soil erosion.

As soil erosion is one of the prominent types of land degradation in the study area at present, adoption of new and improved soil conservation measures by farmers has significant effects on soil productivity on-site as well as downstream. According to the survey, 60.2% were found to be willing to adopt the improved SWC measures. The study found that farmers' perception of soil erosion as a problem, gender of the head of the household, level of awareness about soil conservation technology, off-farm income, training on SWC measures, and regular advice from extension officers are major determinants of adoption of improved SWC measures by the farmers. Finally, the farmers who have adopted the SWC measures incurred the least conservation cost as well as the least soil erosion rate in their lands compared with the farmers who have adopted SWC measures only recently.

References

Bandarathilleke, H. M. (1991). *National forest policy strategies for conservation of forest resources of Sri Lanka.* Paper presented at the Proceedings of the Second Regional Workshop on Multiple Trees, Kandy, Sri Lanka.

Central Bank of Sri Lanka. (2006). *Annual report, 2006.* Central Bank of Sri Lanka.

Central Bank of Sri Lanka. (2020). *Sri Lanka socio-economic data 2020.* Statistics Department, Central Bank of Sri Lanka.

Central Engineering Consultancy Bureau. (1991). *Samanalawewa hydro electric project environment study* (Final Report) (p. 159).

Ceylon Electricity Board. (2006). *Long term generation expansion planning studies 2006–2015.* Transmission and Generation Planning Branch, Ceylon Electricity Board.

Chandrapala, L. (1996). Long term trends of rainfall and temperature in Sri Lanka. *Climate Variability and Agriculture,* 153–162.

Das, S. (2021). Valuing the role of mangroves in storm damage reduction in coastal areas of Odisha. In A. K. E. Haque, P. Mukhopadhyay, M. Nepal, & M. R. Shammin (Eds.), *Climate change and community resilience: Insights from South Asia.* Springer.

De Zoysa, M. (2001). A review of forest policy trends in Sri Lanka. *Policy Trend Report, 2001,* 57–68.

Department of Census and Statistics Sri Lanka. (2007). *Handbook of census and statics-2007*. Department of Census and Statistics.

Department of Census and Statistics Sri Lanka. (2008). *Handbook of census and statics-2008*. Department of Census and Statistics.

Domroes, M. (1996). Rainfall variability over Sri Lanka. *Climate Variability and Agriculture, 163–179*.

Domroes, M., & Schaefer, D. (2000). *Trends of recent temperature and rainfall changes in Sri Lanka*. Paper presented at the Proceedings of the International Conference on Climate Change and Variability.

Hazarika, M., & Honda, K. (2001). *Estimation of soil erosion using remote sensing and GIS, its valuation and economic implications on agricultural production economic implications on agricultural production*. Paper presented at the Sustaining the Global Farm 10th International Soil Conservation Organization Meeting Purdue University and the USDA-ARS National Soil Erosion Research Laboratory, USA.

Honda, K. (1993). *Evaluation of vegetation change in the Ashio Copper Mine using remote sensing and its application to forest conservation works* (PhD thesis). University of Tokyo.

Honda, K., Samarakoon, L., Ishibashi, A., Mabuchi, Y., & Miyajima, S. (1996). *Remote sensing and GIS technologies for denudation estimation in a Siwalik watershed of Nepal*.

Intergovernmental Panel on Climate Change. (1995). *The regional impacts of climate change: An assessment of vulnerability*: Cambridge University Press.

Kim, S. (2007). Evaluation of negative environmental impacts of electricity generation: Neoclassical and institutional approaches. *Energy Policy, 35*(1), 413–423.

Laksiri, K., Gunathilake, J., & Iwao, Y. (2005). *A case study of the Samanalawewa Reservoir on the Walawe river in an area of Karst in Sri Lanka*.

Madduma Bandara, C. M., & Wickramagamage, P. (2004). *Climate change and its impact on upper watershed of the hill country of Sri Lanka. Climate Change Secretariat Working Paper*. Environmental Economics and Global Affairs Division, Ministry of Environment and Natural Resources.

Ministry of Forestry and Environment. (2001). *Sri Lanka: State of the Environment in 2001*. Department of Government Printing.

Rosenberg, D. M., Bodaly, R. A., & Usher, P. J. (1995). Environmental and social impacts of large scale hydroelectric development: Who is listening? *Global Environmental Change, 5*(2), 127–148.

Technology Evaluation and Management Services. (1992a). *Samanalawewa hydro electric project*. Environmental Post Evaluation Study (Interim Report) (p. 15).

Technology Evaluation and Management Services. (1992b). *Samanalawewa hydro electric project. Environmental post evaluation study* (Draft Final Report) (p. 250).

Tol, R. S. J., Downing, T. E., Kuik, O. J., & Smith, J. B. (2004). Distributional aspects of climate change impacts. *Global Environmental Change Part A, 14*(3), 259–272.

Udayakumara, E. P. N. (2011). *Land degradation and rural livelihood in the Samanalawewa catchment, Sri Lanka* (PhD). Asian Institute of Technology, School of Environment, Resources and Development, Thailand, Asian Institute of Technology, School of Environment, Resources and Development.

Udayakumara, E. P. N., Shrestha, R. P., Samarakoon, L., & Schmidt-Vogt, D. (2010). People's perception and socioeconomic determinants of soil erosion: A case study of Samanalawewa watershed, Sri Lanka. *International Journal of Sediment Research, 25*(4), 323–339.

Udayakumara, E. P. N., Shrestha, R., Samarakoon, L., & Schmidt-Vogt, D. (2012). Mitigating soil erosion through farm-level adoption of soil and water conservation measures in Samanalawewa Watershed, Sri Lanka. *Acta Agriculturae Scandinavica, Section B-Soil & Plant Science, 62*(3), 273–285.

Vidanage, S.P., Kotagama, H.b., & Dunusinghe, P. (2022). Sri Lanka's small tank cascade systems: building agricultural resilience in the dry zone. In A. K. E. Haque, P. Mukhopadhyay, M. Nepal, & M. R. Shammin (Eds.), *Climate change and community resilience: Insights from South Asia*. Springer.

Chapter 20
Climate Resiliency and Location-Specific Learnings from Coastal Bangladesh

Sakib Mahmud, A. K. Enamul Haque, and Kolpona De Costa

Key Messages

- From the perspectives of coastal areas of Bangladesh, *location-specific learning effects* have a strong connection with socially heterogeneous coastal population's climate resiliency efforts in terms of their gradual investments toward storm-resistant homes.
- *Location-specific learning effects* are more evident among the socially heterogeneous coastal households that are located close to government-sponsored embankments and cyclone shelters, nearest vehicular road, primary school, and a natural capital, such as the mangrove forest.

Electronic supplementary material The online version of this chapter (https://doi.org/10.1007/978-981-16-0680-9_20) contains supplementary material, which is available to authorized users.

S. Mahmud (✉)
School of Business and Economics, University of Wisconsin-Superior, Superior, USA
e-mail: smahmud@uwsuper.edu

A. K. E. Haque · K. De Costa
Department of Economics, East West University, Dhaka, Bangladesh
e-mail: akehaque@gmail.com

K. De Costa
e-mail: kona_bd@yahoo.com

© The Author(s) 2022
Haque et al. (eds.), *Climate Change and Community Resilience*,
https://doi.org/10.1007/978-981-16-0680-9_20

- Targeted policies on post-disaster relief and rehabilitation efforts for most vulnerable coastal households living in high storm-risk zones should be supported with policies to ensure access to external finance for home improvements, access to natural capital through extensive mangrove forest coverage along the coastlines, and dissemination of information on best practices to build cost-efficient storm-resistant homes.

20.1 Introduction

Climate adaptation and climate risk literature reveals that social heterogeneity plays a big role in determining the measures taken by households against natural disasters (Cutter & Finch, 2008; Cutter et al., 2013; Haer et al., 2017; Koks et al., 2015; Smith et al., 2006). Households' capacity to adapt and respond to natural hazards is largely a function of their socio-demographic status that is related to their social vulnerability (Cutter et al., 2013; Koks et al., 2015; Smith et al., 2006). Hence, social heterogeneity in terms of differences in social characteristics of households living in high climate risk areas can be considered an important factor in determining the feasibility of climate mitigation and adaptation policies. Given this observation, this chapter explores how social heterogeneity and location-specific learning shapes the adaptation strategies of coastal households in Bangladesh. Findings from our study not only reveal the factors influencing the capacity of coastal households to adapt and respond to major cyclone events but also develop a better understanding of the policy implications on individual and community risk mitigation, and evacuation plans. We hope our findings might lead to adoption of insurance coverage against natural disaster risks, currently non-existent in developing economies and help government shape other policies to support community-based mitigation of climate change.

Coastal areas around the world are experiencing increasing frequency and severity of tropical storm events (Knutson et al., 2013; IPCC, 2014; Vitousek et al., 2017; Walsh et al., 2016). It is expected that the resiliency of vulnerable coastal communities will be severely tested in future (IPCC, 2014, 2019; World Bank, 2011). Governments which act as *insurers of last resort* against natural disasters in developing countries with weak catastrophe insurance markets also do not have sufficient funds to support climate-resilient programs to protect coastal communities (IPCC, 2014, 2019; World Bank, 2011). Given such constraints, coastal communities are now investing their time and resources on private defensive strategies (IPCC, Mahmud & Barbier, 2016; World Bank, 2011). However, effectiveness of private defensive strategies, a form of individual and community-based adaptation strategies, might be influenced by a host of factors. They are: (i) *publicly sponsored climate-resilience programs*, such as embankments and cyclone/storm shelters (Kunreuther et al., 2016; Lewis & Nickerson, 1989; Mahmud & Barbier, 2016); (ii) *access to a natural capital*, such as a mangrove forest (Das, 2021; Mahmud & Barbier, 2016; Chap. 17 of this volume) and asset ownership (Prowse & Scott, 2008; Vatsa, 2004); (iii) *access to financial and social capital* (Adger et al., 2003; Mahmud & Barbier, 2016); and

(iv) *access to domestic and foreign remittances* (Clarke & Wallsten, 2003; Yang & Choi, 2007). Our literature review reveals that there are few comprehensive studies that identify the most important factors influencing private adaptation strategies of vulnerable poor coastal communities in developing countries. To fill this research gap, we performed a more inclusive empirical analysis using cross-sectional household survey data collected from twelve villages in three south-western coastal districts of Bangladesh that were exposed to two severe cyclonic storm events: *2007, Cyclone Sidr* and *2016, Cyclone Roanu.*.[1] Besides information on household demographics, consumption, and socio-economic information, our household survey data include information on damages inflicted by each storm event, households' perception of their homes being exposed to future storm event, domestic and foreign remittances, access to external finance and non-governmental organizations (NGOs), migration, household loans, and land ownership. In addition, the survey data includes information on household home structure before each storm event (i.e., before 2007, Cyclone Sidr and before 2016, Cyclone Roanu) and during the survey period of October–November 2016. Our data allowed us to construct an *asset index* and separate *home index before each storm event* for every household.

We selected the Bangladesh coastal area for our analysis due to its extreme vulnerability to tropical storm (cyclones) events which are expected to increase in frequency and severity (IPCC, 2014; Rawlani & Sovacool, 2011; World Bank, 2011), and, also, the authors' familiarity with the terrain. In addition, there is well-documented evidence of indigenous private defensive strategies adopted over generations by the coastal communities of Bangladesh. (Garai, 2017; Hasan et al., 2017; Mahmud & Barbier, 2016). Examples of individual and community-based adaptation behavior include converting a mud house to a brick house (to resist storm and tide-related erosion of walls and floors), raising both the plinth-height, increasing the number of floors, refurbishing the walls, installation of tube well for water, modernization of toilets, improvement of animal sheds, ponds, the boundary of the house, raising the plinths, etc.[2]

---

[1] Cyclone Sidr, with maximum wind speed of 260 km/h, is a category 5 cyclone under Saffir-Simpson scale (greater than 252 km/h) and a severe cyclonic storm with hurricane intensity (greater than 118 km/h) under the Bangladesh Meteorological Department (BMD) scale. It made landfall in southern coastal districts of Bangladesh in November 15, 2007. The storm caused large-scale evacuations with 3447 deaths and an approximate US$1.7 billion damages (GOB, 2008). On the other hand, Cyclone Roanu, with a maximum speed of 110 km/hour, made landfall in in May 21, 2016, in the same area. Through its early warning and evacuation system, Bangladesh was able to successfully reallocate around two million people. Consequently, the death toll was kept at 26 with one lakh houses were damaged, and about 150,000 families were affected (GOB, 2017).

[2] All these private storm-protection actions are identified and listed based on our field and household surveys. Prior to asking households their private storm-protection actions for home improvement against major storms, our survey questions include whether households perceive their homes to be at risk from flooding and tidal surge from a major storm event. These questions are to confirm that the sole purpose for adopting private storm protection actions by the coastal households is to reduce their risks from storm-inflicted damages to homes and other properties. Compared to households that are located inland and living further away from coast, some of these private actions for home

Using household investments for home improvements, which is an important adaptation mechanism to cope against severe cyclones in coastal areas in developing countries and considering the importance of building climate-resilient communities along the Bangladesh coast, we address the following research questions in our study.

(1) Do government-sponsored climate resilience programs, such as embankments, access to vehicular roads, primary schools in the event of cyclones, and cyclone shelters, have significant influence on household decisions to invest more on private adaptation to reduce damages from a major cyclone (storm) event?

(2) Does asset ownership significantly influence household investment on private defensive strategies against storm-inflicted damages?

(3) Does access to external financing through private and public donations (charities), and non-government organizations (NGOs) play a major role in household defensive expenditures to reduce storm-inflicted damages?

(4) Does total remittance from home and abroad lead to higher private defensive expenditures for households?

(5) Do households living close to a natural forest, such as mangroves, invest less on private defensive strategies against major storm or cyclone events?

These are very timely questions since climate risk vulnerabilities are expected to rise for Bangladesh and other coastal communities around the world (GOB, 2008; IPCC, 2019; World Bank, 2011). By finding patterns of private defensive behavior and its relationship with key identifiers, this study would help establish effective coastal adaptation policies for Bangladesh and other developing nations.

The rest of the paper is organized as follows. Section 20.2 discusses the scope of our research. Section 20.3 reports the results based on our empirical analysis. Section 20.4 outlines conclusions and policy recommendations.

20.2 Scope of the Study and Social Heterogeneous Nature of the Study Area

For our study area, we considered Bhola, Barguna, and Patuakhali districts of the Barisal division since the Disaster Management Bureau (DMB) of Bangladesh identified these districts as the zones most affected by frequent cyclones.[3] From each district, we selected an upazila (sub-district). Our selected upazilas are Monpura from Bhola, Amtoli from Barguna, and Kapara from Patuakhali. After selection of the upazilas, we selected two affected unions (the lowest tier of administration in rural

---

improvement might look similar, but they were merely displaying the non-risk factor of improved social standing from the perspectives of the non-coastal households located in these low-risk areas.

[3] Administratively, Bangladesh has 6 divisions, 64 districts or zilas, 508 upazilas, and 4466 unions (Source: Statistical Pocketbook of Bangladesh, 2019). The term 'union' refers to the lowest administrative unit in the rural areas of Bangladesh. Under the Village Chaukidari Act of 1870, villages were grouped into unions to provide for a system of watches and wards in each village.

Fig. 20.1 Study area along with the tracks of Cyclone Sidr and Cyclone Roanu

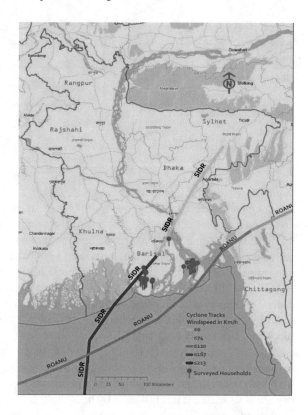

communities) from each upazila based on the available data from the DMB on number of households by exposure to the most recent severe storm events, Cyclone Sidr and Cyclone Roanu, which made landfall on November 15, 2007, and May 23, 2016, respectively. Applying the *two-stage sampling methods,* we selected 600 households for our survey from twelve villages covering three coastal districts. Figure 20.1 shows our study area along with the tracks of Cyclone Sidr and Cyclone Roanu.

We employed structured questionnaires to conduct interviews as part of our household survey. Prior to the main household survey, a pilot survey was conducted to improve the final version of the questionnaire. Table 20.1 summarizes the distribution of the sample households. For details on demographic and socio-economic characteristics of the respondent households in the survey, please see the supplementary table (Table S20.1).

For adaptation against major storm events, the majority of the households increased the number of floors (25%), sank new tube wells for water (24%) and improved their pond areas (12%) after Cyclone Sidr. The same adaptation strategies were applied post-Cyclone Roanu. Survey results show that income, savings, and donations were the major sources of funds for adaptation after these two major storm events. On perceptions of public (government-sponsored) adaptation measures

Table 20.1 Distribution of ample households

Districts	Upazila	No. of selected unions	No. of selected villages	Total number of households
Bhola	Monpura	2	4	200
Barguna	Amtoli	2	4	200
Patuakhali	Kapara	2	4	200
Sum total		6	12	600

Source Field survey

implemented by the government, around 22% of the households think building embankments with stone and cement blocks is the most effective strategy.

The next most effective strategy is raising the height of the embankment (16.46%) followed by building new cyclone shelters or expanding the existing ones (14%), building raising floors or heights of the house (13.44%), and raising plinths (11.55%). Interestingly, other public adaptation measures, such as building clay embankment (9.15%) and afforestation (7.86%), did not get much approval from the survey respondents. Our survey data tells that 93.44% of the households experienced flooding or water logging affecting their houses. For more detailed information, please see the supplementary table (Table S20.2).

Table 20.2 summarizes the sources of funds to support private adaptation measures for home improvements post-Cyclone Sidr and post-Cyclone Roanu. In both cases, we find external finance through public and private donations to be the second most important source of funds followed by household savings.

To understand household investment on different adaptation initiatives, we collected information on adaptation costs for the households based on their location. Table 20.3 shows that houses that are built outside the embankment (31.15% of the total households surveyed), spent Bangladesh Taka (BDT) 6169 more on average on adaptation costs compared to houses that are built outside the embankment. Interestingly, houses that are protected by the natural forest (11.94% of the total households surveyed) spent on average BDT 5272 less on adaptation costs than the houses that

Table 20.2 Sources of funds for adaptation

For adaptation after Cyclone Sidr (2007)	Percentage (%)	For adaptation after Cyclone Roanu (2016)	Percentage (%)
Savings	35.15	Savings	46.70
Loan	16.01	Loan	12.83
Donation	29.02	Donation	21.21
Help from friends/ relatives	6.51	Help from friends/relatives	0.71
Sold land / asset	13.31	Sold land/ asset	18.54
Total frequencies ($N = 1334$)	100	Total frequencies ($N = 561$)	100

Source Field survey

Table 20.3 Differences in adaptation costs based on household location

Location of the household	Average adaptation costs (in BDT)	Differences in adaptation costs (in BDT)	Comments
Protection from major storms due to surrounded by natural forest			
House surrounded by forest	44,915	5272***	Houses surrounded by forest spent BDT 5227 less
House not surround by forest	50,187		
Protection from major storms due to located inside polder			
House located inside the polder	47,497	6169***	Houses inside the polder spent BDT 6169 less
House located outside the polder	53,665		
Protection from major storms due to located on highland			
House located on the lowland	48,814	3227*	Houses located on mid-and- highland spent BDT 3227 less
House located on the high land	45,587		

Source Field survey
Note Independent samples t test with $N = 610$ observations. Statistically significance levels: ***1%, **5%, *10%

are not protected by the natural forest. Among the households that are located in lowland areas (33.45% of the total surveyed), average spending on adaptation costs is BDT 3227 more compared to households located in mid- and highland areas. We also report household perception on their homes being affected by tidal surge from a major cyclone event.

Table 20.4 reveals household perceptions on their homes being affected by flooding and tidal surge. More than 95% of the total households surveyed felt their homes are at risk of being exposed to flooding and tidal surge from major cyclones.

Table 20.4 Households perception on flooding and tidal surge from major cyclones

	Flooding/water logging Total 'yes' responses (% in brackets out of 607 valid responses)	Tidal surge Total 'yes' responses (% in brackets out of 607 valid responses)
Study area	579 (95.38%)	583 (96.05%)
Patuakhali	194 (97.48%)	199 (100%)
Borguna	211 (100%)	198 (94%)
Bhola	174 (88.32%)	186 (4.41%)

Source Field survey

20.3   Survey Results and Discussion

For empirical analysis, our econometric estimations are based on households'
response to private storm-protection actions on home improvement after *Cyclone
Sidr* and *Cyclone Roanu*. All our results are reported as supplementary materials
under Tables S20.3, S20.4 and S20.5. Table 20.5 presents the results of ordinary
least squares (OLS) regression analyses using the full sample of our survey. Our

Table 20.5 Adaptation costs using OLS estimates

Dependent variable	Adaptation costs for home improvement after Cyclone Sidr	Adaptation costs for home improvement after Cyclone Roanu
Independent variables	(3)	(6)
Income (monthly, in log)	59,684.74 $(2.99)^{***}$	5080.83 $(3.86)^{***}$
Foreign and domestic remittance received (in Tk)	− 16,499.09 $(− 0.72)$	− 2725.72 $(− 1.81)^{**}$
Home improvement loan (in Tk.)	− 0.0763 $(− 0.29)$	-0.0314 $(− 1.83)^{**}$
Children in household	− 3710.07 $(− 0.26)$	− 2809.27 $(− 2.96)^{***}$
Female in household	7477.65 (0.88)	1648.53 $(2.96)^{***}$
Household earning members	20,812.88 $(1.79)^{**}$	− 325.51 $(− 0.43)$
Member of an NGO group (=1, 0 otherwise)	39,578.7 $(1.46)^{*}$	863.85 (0.63)
Access to finance access for home improvements (=1, 0 otherwise)	28,595.51 (1.02)	7790.72 $(4.19)^{***}$
Home index before Cyclone Roanu		− 1642.87 $(− 2.05)^{**}$
If household located outside the polder/embankment (=1, 0 otherwise)	− 5198.07 $(− 0.17)$	6475.37 $(3.14)^{***}$
If household located at low elevation (=1, 0 otherwise)	− 67,774.5 $(− 1.86)^{**}$	4761.47 $(1.93)^{**}$
If household located close to a natural forest (=1, 0 otherwise)	8043.37 (0.34)	− 5875.38 $(− 3.80)^{***}$
Number of obs	610	610
R-squared	0.0925	0.2376
Prob > *F*	0.0001	0.000

t tests are shown in parentheses beneath coefficient estimates. Significance levels: ***1%, **5%,
*10%. For more regression estimates, see the supplementary materials

exact OLS regression specifications are included in the supplementary materials. The household adaptation costs on storm-resistant home improvements are as follows:

The dependent variables are the adaptation costs for home improvements after Cyclone Sidr, and the adaptation costs for improvements for Cyclone Roanu, in Bangladesh Taka (Tk.). To capture the influence of social networks, we included household access to external finance for home improvement and if any household member has connections with at least one non-governmental organization (NGO) that is actively operating within the community. To capture the influence of household psychology behind their adaptation behavior on home improvements, we incorporated household perception of being exposed to damages from future major storm events (cyclones bearing Category 8 or 9 signals), and whether the households experienced loss of assets from 2007, Cyclone Sidr and 2016, Cyclone Roanu events. Lastly, to capture the influence of asset holdings, we created an *asset index* based on household ownership of farmland, livestock, pond, orchard, poultry, mechanized vehicle and boat, rickshaw or van, shop, rental home, etc. We also created a separate *home index before Cyclone Sidr* and a similar *home index before Cyclone Roanu*. These home indices comprised the quality of home structures for wall, roof, and floor materials. A greater value of the home indices indicates that the home structure has improved in terms of its ability to withstand the wrath of a cyclone. We added controls to deal with *exogenous sources of variations* on households' location preferences, which is the distance of the household from the nearest primary school, vehicular road, embankment, and the cyclone shelter, respectively. For robustness checks of our results, we applied full information maximum likelihood (FIML) of the sample selection model using Heckman (1979).

Results from our empirical analysis using different regression specifications reveal that under a social heterogeneous framework, a household's relative physical location and distance from government-sponsored programs, such as embankments, primary school, and vehicular road are important determinants of household adaptation expenditure for home improvements. In addition, a household's location close to a forest, a natural form of storm-protection, is also an important determinant as it leads to lower adaptation costs for home improvements, and similarly, better-built houses (or higher home index value) had lower impacts. Estimation results from our analysis show that access to external finance through public and private donations positively influences household defensive behavior on converting existing houses to become more storm-resistant. Asset ownership, access to credit, and NGO membership are also influential, but they are not statistically significant in most of the regression specifications using different estimators. Same applies to the remittance variable regarding its insignificant influence on households incurring higher adaptation costs for storm-resistant homes.

Due to data limitations, we acknowledge the fact that there could be some missing variables that might be relevant for future analysis. For example, data on how much money allocated for private adaptation or defensive strategies against major cyclone (storm) events out of total remittance received as well as from different sources of external financing would further improve the analysis. There is also room for improvements by further working on the regression specifications that are considered for empirical analysis. Despite the limitations, we think our empirical findings have the potential to contribute to disaster-related literature on household-coping mechanisms and resiliency against climate change-induced developments. The results are consistent across the Cyclone Sidr and the Cyclone Roanu experiences; however, we see a clear recall problem related to Cyclone Sidr data because it had happened about 10 years ago.

20.4 Conclusion

Given that the influence of social heterogeneity on coastal households' capacity to adapt and respond to climate change-induced natural disaster risks, our paper examines the major factors that might be influential on private defensive behavior of poor coastal households against major storm events. Our estimates show that households save nearly ৳6475 taka[4] or $75.2 dollar in terms of housing improvement costs due to the protection that the forests provide to them. It is ৳5875 taka or 68.3 USD per household living outside the protection of the embankment. Similarly, higher the home index value, lower is the cost of home improvement after the cyclone.

On policy implications, our empirical results support the influence of government-sponsored climate resilience programs on vulnerable and poor coastal households' adaptation choices against major cyclone events. Having embankments and ensuring access to nearest vehicular road, primary school, and cyclone shelters, etc., reduce cyclone-inflicted damages to home and other properties. Although these government-sponsored initiatives allow the coastal households that are living near the public infrastructure to allocate resources in areas other than home improvements, these results also indicate that local and national governments should *target* their post-disaster relief and rehabilitation efforts for those households that could be identified as most vulnerable to disasters due to their locations away from the infrastructure. Considering the presence of *location-specific learning effects*, the government can coordinate with the local communities, non-governmental organizations, and donor agencies to impart information and knowledge on best practices in building storm-resistant homes at the least possible cost. By encouraging the coastal communities to pursue private defensive strategies or by developing mangroves along the coast-line, the government can make its climate-resilient programs including post-disaster relief and rehabilitation programs more effective. Therefore, communities should be pursued to maintain the forests along the coastline to protect them against cyclones. This will reduce their private defensive costs on home improvement.

Since our findings also reveal the importance of access to credit, governments can streamline and simplify regulations for financial institutions and NGOs, so that they can lend credits to households willing to invest in storm-resistant homes and other adaptation strategies against major cyclone events. Although the influence of remittances on household adaptation costs for home improvements turned out to

---

[4] 1$ = ৳86 taka (2020).

be somewhat weak in our estimation results, one cannot rule out its importance in shaping household private defensive behavior in future. Furthermore, our findings indicate that governments should reserve development funds by forming public–private partnerships with key stakeholders of the communities. Such development funds could be simultaneously coordinated with policies that encourage long-term adaptive capacities of the socially heterogeneous coastal households to develop more efficient and equitable climate-resilient programs along the Bangladesh coast.

References

Adger, W. N., Huq, S., Brown, K., Conway, D., & Hulme, M. (2003). Adaptation to climate change in the developing world. *Progress in Development Studies, 3*(3), 179–195.

Adida, C. L., & Girod, D. M. (2011). Do migrants improve their hometowns? Remittances and access to public services in Mexico, 1995–2000. *Comparative Political Studies, 44*(1), 3–27.

Clarke, G. R., & Wallsten, S. (2003, January). Do remittances act like insurance? Evidence from a natural disaster in Jamaica. *Development Research Group, the World Bank.*

Cutter, S. L., Emrich, C. T., Morath, D. P., & Dunning, C. M. (2013). Integrating social vulnerability into federal flood risk management planning. *Journal of Flood Risk Management, 6*(4), 332–344.

Cutter, S. L., & Finch, C. (2008). Temporal and spatial changes in social vulnerability to natural hazards. *Proceedings of the National Academy of Sciences, 105*(7), 2301–2306.

Das, S. (2021). Valuing the role of mangroves in storm damage reduction in coastal areas of Odisha. In A. K. E. Haque, P. Mukhopadhyay, M. Nepal, & M. R. Shammin (Eds.), *Climate change and community resilience: Insights from South Asia.* Springer.

Garai, J. (2017). Qualitative analysis of coping strategies of cyclone disaster in coastal area of Bangladesh. *Natural Hazards, 85*(1), 425–435.

Government of the People's Republic of Bangladesh. (2008), Super Cyclone Sidr 2007: Impacts and strategies for interventions, a report prepared by the Ministry of Food and Disaster Management, Bangladesh Secretariat.

Government of the People's Republic of Bangladesh. (2017). Report on Cyclone 'Roanu,' a report prepared by the Bangladesh Meteorological Department (BMD).

Haer, T., Botzen, W. W., de Moel, H., & Aerts, J. C. (2017). Integrating household risk mitigation behavior in flood risk analysis: An agent-based model approach. *Risk Analysis, 37*(10), 1977–1992.

Hasan, I., Majumder, M. S. I., Islam, M. K., Rahman, M. M., Hawlader, N. H., & Sultana, I. (2017). Assessment of community capacities against cyclone hazard to ensure resilience in South central coastal belt of Bangladesh. *International Journal of Ecological Science and Environmental Engineering, 4*(1), 1–14.

Heckman, J. J. (1979). Sample selection bias as a specification error. *Econometrica: Journal of the Econometric Society,* 153–161.

Intergovernmental Panel on Climate Change (IPCC). (2014). *Climate change 2014: Impacts, adaptation, and vulnerability, fifth assessment report (AR5)* (p. 1132). Cambridge University Press.

Intergovernmental Panel on Climate Change (IPCC). (2019). In V. Masson-Delmotte, P. Zhai, H.-O. Pörtner, D. Roberts, J. Skea, P. R. Shukla, A. Pirani, W. Moufouma-Okia, C. Péan, R. Pidcock, S. Connors, J. B. R. Matthews, Y. Chen, X. Zhou, M. I. Gomis, E. Lonnoy, T. Maycock, M. Tignor, T. Waterfield (Eds.), *Global warming of 1.5 °C.* An IPCC special report.

Koks, E. E., Jongman, B., Husby, T. G., & Botzen, W. (2015). Combining hazard, exposure and social vulnerability to provide lessons for flood risk management. *Environmental Science & Policy, 47*, 42–52.

Knutson, T. R., Sirutis, J. J., Vecchi, G. A., Garner, S., Zhao, M., Kim, H. S., Bender, M., Tuleya, R. E., Held, I. M., & Villarini, G. (2013). Dynamical downscaling projections of twenty-first-century Atlantic hurricane activity: CMIP3 and CMIP5 model-based scenarios. *Journal of Climate, 26*(17), 6591–6617.

Kunreuther, H., & Pauly, M. (2004). Neglecting disaster: Why don't people insure against large losses? *Journal of Risk and Uncertainty, 28*(1), 5–21.

Kunreuther, H., Michel-Kerjan, E., & Tonn, G. (2016, December). Insurance, economic incentives and other policy tools for strengthening critical infrastructure resilience: 20 proposals for action. A study conducted through the *Center for Risk Management and Decision Processes, The Wharton School, University of Pennsylvania.*

Lewis, T., & Nickerson, D. (1989). Self-insurance against natural disasters. *Journal of Environmental Economics and Management, 16*(3), 209–223.

Mahmud, S., & Barbier, E. B. (2016). Are private defensive expenditures against storm damages affected by public programs and natural barriers? Evidence from the coastal areas of Bangladesh. *Environment and Development Economics, 21*(6), 767–788.

National Oceanic and Atmospheric Administration. (NOAA, 2019). *What is a hurricane, typhoon, or tropical cyclone?* Hurricane Research Division. http://www.aoml.noaa.gov/hrd/

Paul, S. K., & Routray, J. K. (2011). Household response to cyclone and induced surge in coastal Bangladesh: Coping strategies and explanatory variables. *Natural Hazards, 57*(2), 477–499.

Prowse, M., & Scott, L. (2008). Assets and adaptation: An emerging debate. *Institute of Development Studies (IDS) Bulletin, 39*(4), 42–52.

Rawlani, A. K., & Sovacool, B. K. (2011). Building responsiveness to climate change through community-based adaptation in Bangladesh. *Mitigation and Adaptation Strategies for Global Change, 16*(8), 845–863.

Roy, A. K. D., & Gow, J. (2015). Attitudes towards current and alternative management of the Sundarbans Mangrove Forest, Bangladesh to achieve sustainability. *Journal of Environmental Planning and Management, 58*(2), 213–228.

Smith, V. K., Carbone, J. C., Pope, J. C., Hallstrom, D. G., & Darden, M. E. (2006). Adjusting to natural disasters. *Journal of Risk and Uncertainty, 33*(1–2), 37–54.

Statistical Pocketbook of Bangladesh. (2019).

Stock, J. H., Wright, J. H., & Yogo, M. (2002). A survey of weak instruments and weak identification in generalized method of moments. *Journal of Business and Economic Statistics, 20*(4), 518–529.

Vatsa, K. S. (2004). Risk, vulnerability, and asset-based approach to disaster risk management. *International Journal of Sociology and Social Policy, 24*(10/11), 1–48.

Vitousek, S., Barnard, P. L., Fletcher, C. H., Frazer, N., Erikson, L., & Storlazzi, C. D. (2017). Doubling of coastal flooding frequency within decades due to sea-level rise. *Scientific Reports, 7*(1), 1399.

Walsh, K. J., McBride, J. L., Klotzbach, P. J., Balachandran, S., Camargo, S. J., Holland, G., Knutson, T. R., Kossin, J. P., Lee, T. C., Sobel, A., & Sugi, M. (2016). Tropical cyclones and climate change. *Wiley Interdisciplinary Reviews: Climate Change, 7*(1), 65–89.

World Bank (2011, April). Vulnerability, risk reduction, and adaptation to climate change: Bangladesh. *Climate Risk and Adaptation Country Profile.*

World Bank. (2018, April). Migration and remittances: Recent developments and outlook, special topic: Transit migration. *Migration and Development Brief, 29.*

Yang, D., & Choi, H. (2007). Are remittances insurance? Evidence from rainfall shocks in the Philippines. *The World Bank Economic Review, 21*(2), 219–248.

Part V
Urban Sustainability

Chapter 21
Making Urban Waste Management and Drainage Sustainable in Nepal

Mani Nepal, Bishal Bharadwaj, Apsara Karki Nepal, Madan S. Khadayat, Ismat Ara Pervin, Rajesh K. Rai, and E. Somanathan

---

Disclaimer: The presentation of material and details in maps used in this book does not imply the expression of any opinion whatsoever on the part of the Publisher or Author concerning the legal status of any country, area or territory or of its authorities, or concerning the delimitation of its borders. The depiction and use of boundaries, geographic names and related data shown on maps and included in lists, tables, documents, and databases in this book are not warranted to be error-free nor do they necessarily imply official endorsement or acceptance by the Publisher, Editor(s), or Author(s).

The original version of this chapter was revised. The correct chapter author names are now listed in order. The correction to this chapter is available at
https://doi.org/10.1007/978-981-16-0680-9_30

M. Nepal (✉)
South Asian Network for Development and Environmental Economics (SANDEE), International Centre for Integrated Mountain Development (ICIMOD), Kathmandu, Nepal
e-mail: mani.nepal@icimod.org

B. Bharadwaj
University of Queensland, St. Lucia, 4072 Queensland, Australia
e-mail: s4441329@student.uq.edu.au

A. Karki Nepal
International Centre for Integrated Mountain Development (ICIMOD), Kathmandu, Nepal
e-mail: apsara.nepal@icimod.org

M. S. Khadayat
Kathmandu, Nepal
e-mail: madankdyt@yahoo.com

I. A. Pervin
Institute of Water Modelling, Dhaka, Bangladesh
e-mail: iap@iwmbd.org

R. K. Rai
School of Forestry and Natural Resource Management, Institute of Forestry, Tribhuvan University, Kathmandu, Nepal

© The Author(s) 2022, corrected publication 2022
Haque et al. (eds.), *Climate Change and Community Resilience*,
https://doi.org/10.1007/978-981-16-0680-9_21

Key Messages

- Cities in Nepal are facing urban flooding and waterlogging risks which can be addressed by more investment on constructing drainage systems.
- Without proper solid waste management, the returns from investment on urban infrastructure would be low after few years.
- Additional user fees, revenue from recycling waste and additional import duty on importing plastic raw materials provide enough resources for managing solid waste which helps appreciating urban property prices.

21.1 Introduction

Cities in developing countries are facing increasing threats from waterlogging, flooding and water contamination aggravated by improperly managed solid waste. Several factors are responsible: unplanned urban growth has led to expansion of city areas into low-lying flood plains; climate change is leading to intense rain-fall events which risk overwhelming city drainage systems; informal dumping sites may result in toxic waste entering rivers and groundwater following heavy rainfall; and indiscriminate dumping of untreated solid waste can clog the drainage system.

Waterlogging and water pollution have a disproportionate impact on the poorest sections who usually live near flood plains or in the slums in the less developed sections of the city (Ahmed, 2016; Sharma et al., 2022, Chap. 22 of this volume). Climate change is likely to exacerbate these problems.

The suffering caused by frequent waterlogging has created public pressure to develop strategies. The most popular strategy is to invest in urban drainage infrastructure in order to eliminate drainage congestion. However, an analysis of the causes of waterlogging illustrates that investment in physical infrastructure alone is not enough (Pervin et al., 2020). Cities also need to work with inhabitants to improve the 'urban culture' of solid waste disposal. This requires a change in behaviour in terms of reduction and disposal of municipal solid waste. Stakeholders' survey from South Asia suggests that South Asians prefer waste-to-energy strategies in order to address municipal solid waste management issue (Haque et al., 2019).

There are several known solutions to this: (a) development or creation of a market for recycling; (b) creating conditions for reuse of waste; (c) developing strategies for reducing the volume of waste; and (d) proper disposal of solid waste. In established cities, waste from businesses and households is collected for a fee while city councils take responsibility for dumpsites, while such a system is absent in many emerging cities in developing countries.

e-mail: rjerung@gmail.com

E. Somanathan
Indian Statistical Institute, New Delhi, India
e-mail: som@isid.ac.in

After the promulgation of a new Constitution in 2015, Nepal designated more than 220 new municipalities. Several of these new municipalities have no formal waste management system. Even if they have some form of waste collection system (formal or informal), proper management of the collected waste has been another challenge as these cities lack sanitary-landfill sites. Though most have informal and unorganized markets for recycling metal, paper and plastic, they rarely have segregation at source. As a result, informal agents like rag-pickers are engaged in sorting and segregating the tradable products from other wastes at the dumpsite or at the secondary transfer stations. The rag-pickers are often women and children from poor families (Bharadwaj, Baland, & Nepal, 2020) who are exploited as they are unorganized and not protected by rules or regulations. In addition, they are also exposed to health risks during the process of separating the tradable products.

Against this background, protecting cities require reducing the volume of waste, improving collection systems, investment in drainage, replacing riverside and wetland dumping with engineered landfills and identifying sustainable mechanisms to finance these improvements in waste management for enhancing urban resilience.

In Bharatpur, the informal collection system involves collection from houses and businesses and dumping at designated sites. Rag-pickers use dumpsites to collect recyclable or reusable materials (Rai et al., 2019). The city has no sanitary-landfill site, and hence, there is a further risk of health hazards through water contamination (both surface and underground water aquifers) and foul smell.

This chapter summarizes the key learning from a large-scale research project coordinated by the South Asian Network for Development and Environmental Economics (SANDEE) Secretariat at the International Centre for Integrated Mountain Development (ICIMOD) under the Cities and Climate Change research,[1] where we: (a) examine the existing situation of waterlogging and flooding in Bharatpur under different climatic scenarios and the contribution of proper solid waste management in avoiding flooding and waterlogging risks, (b) examine the effectiveness of information and waste bins on the streets on cleanliness in Bharatpur, (c) estimate the value of cleaner neighbourhoods in Bharatpur and other urban centres across Nepal, (d) describe the extent of citizens' willingness to pay for improvements in waste management in the city and (e) explore avenues for sustainable financing of municipal solid waste management in Nepal for enhancing urban resilience and adapting to the changing climate. For detailed analysis of these issues, readers are advised to refer to Rai et al. (2019); Bharadwaj, Baland, and Nepal (2020); Nepal et al. (2020); Nepal et al. (2021); and Pervin et al. (2020).

21.2 Study Area and Data

This case study is based on the Bharatpur Metropolitan City (BMC), located in the southern plains of Nepal. Bharatpur is a fast growing medium-sized city with

---

[1] https://www.icimod.org/sandee-idrc/.

a population of over 300,000 in its core area. The city is on the bank of a major river, Narayani, which also facilitates the drainage system of the city. There are two supporting natural drainage systems in the city: Pungi (6.6 km) and Kerunga (28.4 km) canals. Since the city does not have adequate man-made drainage systems, the encroachment of the natural canals has impacted drainage so that even moderate rainfall results in waterlogging in the city (Pervin et al., 2020).

For the study, we collected different types of data from Bharatpur during 2017–2018 and complemented it with secondary data. For an understanding of waterlogging and the flooding potential, primary data on rainfall, water level, water discharge, existing drainage networks and their cross sections was collected from the field during the summer of 2017. This information was supplemented by historical hydrological and climate data for developing a mathematical drainage network model (Pervin et al., 2020).

For understanding existing social and behavioural issues and household preferences on solid waste management, we conducted a primary survey of 150 Tole Lane Organizations (TLOs), (small community-level organizations for development activities consisting of 100 households on average, which elect a small executive committee of 9–11 members for addressing community issues including municipal solid waste management.). The 150 TLOs were selected randomly from over 350 TLOs. These TLOs were divided into two groups—control and intervention—for understanding the effectiveness of the interventions. As baseline, we surveyed seven households using systematic random sampling from each TLO (Rai et al., 2019). Altogether 1050 households were surveyed from 14 wards of the city's core area before the intervention. After the intervention, we collected two additional rounds of data from the same households for examining the impact of information campaigns and low-cost waste bins on the streets for keeping the city neighbourhood cleaner. The same baseline survey was used for estimating the value of clean neighbourhoods for Bharatpur. For urban centres across the country, we used a nationally representative household survey (Nepal Living Standards Survey 2010/11). The detailed information is available in Nepal et al. (2020). For sustainable financing of municipal solid waste management, we used information collected by the Asian Development Bank and supplemented it with field data (Bharadwaj, Rai, & Nepal, 2020).

21.3 Adapting to Urban Flooding

Cities in developing countries are particularly susceptible to urban flooding and waterlogging since they have limited drainage infrastructure that is often mismanaged and congested for different reasons including improper disposal of solid wastes in the canals and drains (Haque, 2013). Excessive rainfall exacerbates waterlogging, especially in low-lying areas where slums are generally located in fast growing unplanned cities. Consequently, the groundwater gets contaminated, increasing the likelihood of public health problems (ten Veldhuis et al., 2010).

Our analysis included rainfall data collected by the Department of Hydrology and Meteorology (DHM) for Bharatpur city at Bharatpur station (from 2000 to 2016), water-level data for the Narayani river at Devghat station and for the Rapti river at Rajayia station collected by the research team. The bankfull levels of the Narayani and Rapti rivers were considered as the design water level for Bharatpur. The surveyed cross sections of the Narayani and Rapti rivers were interpolated to generate bankfull water-level data at the outfalls of the Pungi and Kerunga canals. We used MIKE11 modelling tools from the Danish Hydraulic Institute (DHI) to develop the drainage model. The model was calibrated against the water level and discharge data of the canals for 2017.

We developed some climate change scenarios for Bharatpur based on changing rainfall patterns as the climate change models predict that the volume of rainfall would increase during the monsoon (May–October) and reduce during the dry season (December–March) in the Ganges basin, whereas Bharatpur is one of the sub-basins (Pervin et al., 2020). In the study area, high-intensity and short-duration rainfall is projected to increase, overwhelming the existing drainage infrastructure and resulting in more waterlogging and flooding in future. We also consider the rapid urbanization and land-use change in Bharatpur while developing the model scenarios.

Based on several focus group discussions, we developed the adaptation scenarios for Bharatpur. Under the existing conditions, about 13% (Fig. 21.1a: Current scenario) of the land area is under the risk of flooding and waterlogging. With appropriate structural interventions, the flooding risks could go down to 5% (Fig. 21.1b: Improved scenario). These results are shown in Fig. 21.1a and b.

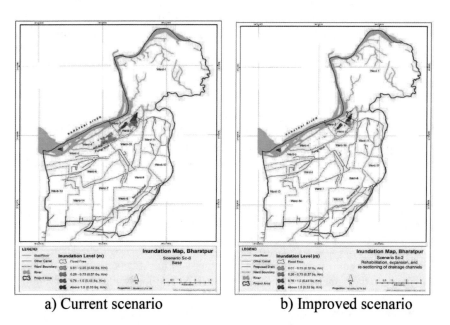

a) Current scenario   b) Improved scenario

Fig. 21.1 Waterlogging risks with current and improved scenarios. *Source* Pervin et al. (2020)

The scenario analysis suggests that without proper solid waste management, the city area at risk of flooding returns to 8% in five years time, diminishing the returns from investment in physical infrastructure.

21.4   Effectiveness of Information and Street Bins in Keeping the Neighbourhoods Cleaner

Since proper solid waste management is the key to reducing flooding and waterlogging risks in the city, we looked at some of the low-cost interventions that would help cities keep neighbourhoods clean. Our intervention involved providing information to the households on how to manage household waste better rather than dumping it or burning it in the intervention TLOs. We also installed street waste bins as part of the intervention. These waste bins are 20 L bins intended for the people who could put their leftover food and other unwanted items while travelling around the city, but not intended for solid waste generated by the households. For this purpose, we randomly selected two groups of 75 Tole Lane Organizations (TLOs) each, with one designated as a control group for comparison.

After a baseline survey, households and executive committee members of each intervention TLO participated in a workshop which provided information on managing household waste, including segregation, recycling and composting of the waste and why burning or throwing waste is detrimental. We organized one workshop for each intervention TLO. At the same time, 20-litre street waste bins were installed in the intervention TLOs. We conducted a post-intervention survey in the first half of 2018 and a final round of post-intervention survey in the second-half of 2018.

Since we have three rounds of data from each household, we used (a) a simple linear regression model with indicator variables for 'intervention' and 'round' for last two rounds of data (without baseline) and (b) a difference-in-difference approach using all three rounds of data in order to see the impact of intervention on the cleanliness of the neighbourhoods. More specifically, the outcome variables are: cleanliness of the neighbourhood, waste given to the collectors instead of burning or throwing and segregation of the waste at source. Since we use an experimental design to randomly assign control and treatment neighbourhoods, our results have a causal interpretation.

The key findings of the research are summarized below (Nepal, Karki Nepal et al., 2021):

- *Neighbourhoods and streets are cleaner due to the intervention*: There was a 9% point increase in the perceived cleanliness of the streets, and 14–22% point increase in the perceived cleanliness of the neighbourhoods after the intervention.
- *More households started giving their wastes to the waste collectors*: Due to the intervention, there was an 8–10% point increase in households that give household waste to the collectors instead of burning or throwing it out.

- *No impact on the at-source segregation*: However, our results suggest that there is no impact of the intervention on the at-source segregation of the waste.

Since the intervention was intended to inform households on how to segregate at source and why not to throw or burn the waste, the first two results are expected, but the third result is surprising given that the households had been sensitized how to segregate the waste at source. However, this is possibly due to the fact that there is no provision for collecting degradable and non-degradable wastes separately even though households segregate it at source. In the absence of such separate collection service, households tend to unlearn the learning from the sensitization workshops. Perhaps, if households do not see a likely consequence of their action, they may ignore the instruction or information, suggesting that any programmes or policies should be consequential to the households for them to change their behaviour (Nepal et al., 2009).

21.5 Households' Preferences for Municipal Solid Waste Management

The city has been managing solid waste collection through two private companies. They charge monthly fees (ranging from NPR 30 to NPR 100 for the households, and NPR 200 to NPR 4000 for commercial outlets, where the service is based on the volume of waste and frequency of collection). About 95% of the households in the sampled wards subscribe to the collection service. The city provides additional resources to the private companies since the collected fees do not cover the operational costs. The city also refunds 30% of the collected fees to the concerned TLOs so that they can invest in cleaning up their communities.

Household survey results suggest that the majority (53%) of the city residents were not satisfied with the waste collection services that they were getting. This is mainly because they were not getting the services on the given day at a pre-specified time as there was no fixed time or date for picking up the wastes from the households. The private service providers have estimated that only 60% households pay waste collection tariff regularly. Our choice experiment analysis suggests that households were willing to pay 10–28% additional fee on top of what they were paying if the service provider collects the waste on a given day and time, and if the city installs waste bins on the streets that pedestrians could use for disposing unwanted litters. This could result in an additional income of NPR 5 million additional for the city. A progressive tariff increase based on the number of floors of the individual houses can generate more revenue, since willingness-to-pay increases by 8.2% on average for every additional floor of a house (Rai et al., 2019).

Our analysis suggests that the city needs to spend around 17% additional resources in addition to the existing expenditure in order to provide the expected services (on time collection on a given day and placing bins on the streets) to the city residents. A further 10% is required for separate collection trips for degradable and non-degradable waste. Our analysis suggests that with some additional efforts, the city could meet the expectation of the residents and could collect additional revenues for managing the municipal solid waste better. If all households pay the collection fee as a result of improved services, then the additional collected tariff would outweigh the additional expenditure required for improved services (Table 21.1). In addition, a separate collection service generates additional income from the recycled materials, particularly plastic. This can also lead to increased longevity of landfill sites. Overall, there is some revenue gap under the current system of municipal solid waste management, where composting of organic waste is not considered as an option. When segregation at source promotes composting at home, then waste transportation cost will be reduced from the estimated 10% since the volume of waste will also decline under this scenario.

In short, city residents are concerned about the municipal solid waste management and willing to pay additional service fee if the current situation improves. For this, the municipality has to improve waste collection service to increase the participation of city residents, for which they are ready to pay additional fees. It is also true that the increased revenue due to the improved services may cover the additional cost, but it is less likely to cover the total cost of municipal solid waste collection services at the existing tariff rate. However, reduced waste due to segregation and longevity of landfill sites increases the social and environmental benefits. We discuss how the revenue shortfall could be overcome in the following section.

Table 21.1 Annual costs and revenue of Bharatpur municipal solid waste management

S.No.	Cost		Revenue	
	Particular	Cost (NPR in millions)	Particular	Revenue (NPR in millions)
1	Current operational cost	39.19	Current service fee collection	18.00–21.60
2	Improved service (collection on time)	6.54	If all residents pay, additional revenue	14.40
3	Improved service (collection of degradable and non-degradable separately)	4.09	Additional payment for waste bins and on time service	0.60–5.00
4	Total	49.82		33.00–4100

Source Field survey 2017–18[2]

---

[2] These estimates are based on the field survey 2017–2018, choice experiment analysis, focus group discussion and key-informant interviews with Bharatpur Metro Officials and private companies engaged in waste collection in Bharatpur.

21.6 Sustainable Financing of Municipal Solid Waste Management

To examine the revenue potential of materials recovery and recycling for the municipalities in Nepal, we use data collected by the Asian Development Bank in 2013 from 58 municipalities across the country.

Based on existing solid waste management across the municipalities, we examine the materials recovery and recycling rates of plastics, papers and other wastes. We focus mainly on plastic wastes as discarded plastic remains in the environment for a long time and harms the whole ecosystem—land, rivers and oceans. Although plastic bags ban policy with sufficient penalty and strict enforcement can reduce the use of plastic bags, it cannot address the increasing use of plastic for other purposes (Bharadwaj, Baland, & Nepal, 2020). Plastic-related waste constitutes about 10% of household solid waste. Under different scenarios, we find that recovered plastic ranges from 0.36 to 1.62% of total solid waste produced in Nepal. Under the optimistic solid waste management scenario—where the plastic material recovery rate is 15% and the collection efficiency is 66.7%, the revenue generated from recycling plastic waste—where price of recovered plastic is NPR 30 per kg and solid waste management costs is NPR 2347 per ton—covers 138% of the management costs of the plastic-related waste and prevents 4220 t of plastic entering the environment annually in the form of solid waste. Figure 21.2 exhibits the mechanism of the plastic recovery and recycling market. However, under the more practical scenario (average

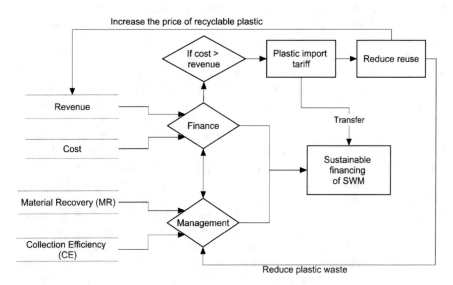

Fig. 21.2 A framework for sustainable financing of municipal solid waste management. *Source* Bharadwaj, Rai, and Nepal (2020)

of best and worst cases), the revenue generated from recycling plastic covers only 82.85% of the plastic-related waste management cost.

Since there would be a resource gap under this scheme for financing plastic waste management, we also estimated an appropriate tariff rate on importing plastic items (including plastic granules and other raw materials) that could help finance the plastic waste management across the country. Our estimate suggests that an additional tariff of 1% on the imported plastic materials would generate enough revenue for financing the cost of managing plastic wastes (Bharadwaj, Rai, & Nepal, 2020).

The additional tariff would not only generate required revenue for managing plastic waste, but would also reduce the use of plastic items by making them costlier, thus spurring the demand for recycled plastic in one hand also promoting the alternatives to plastic on the other. The incentive for recycling will reduce the plastic material in the waste stream, which would otherwise end up being released into the environment given the weak dumping site management which results in waste being dumped in the forest or on the riverbanks.

21.7   Value of Cleaner Neighbourhood

Improper management of municipal solid waste not only contaminates groundwater and impacts health but it also diminishes the aesthetic value of neighbourhoods and lowers the value of housing property. In general, the price of a housing unit depends on several factors called attributes. Three main categories of the attributes are considered valuable while buying a housing unit. They are structural (land areas, built-up areas, types of roof, wall and floor materials, number of rooms and bathrooms and floors, etc.), neighbourhood (crime rate, school quality and distance, distance to the markets and hospital) and environmental (presence of open space and parks, distance from the forest and rivers, distances from the waste dumping sites and presence of waste collection services) characteristics. Even though these characteristics are important, the buyers may not pay separate prices for each of these attributes as a housing unit is a bundle of all these attributes. We have used the Hedonic price model to disentangle the value of each attribute and examine the effects of municipal solid waste collection services at the neighbourhood level on housing property values (Nepal et al., 2020).

For this purpose, we used two sets of data—one is collected from Bharatpur and the other is collected by the Nepal Central Bureau of Statistics (the third wave of the Nepal Living Standards Survey 2010/11). Hedonic studies in developed countries use market transaction data of housing units. However, market transaction data in Nepal generally understates the housing property price as buyers and sellers have all kinds of incentives to underreport property prices for avoiding taxes unless the buyer gets credit from a bank while buying a house. Therefore, we use self-reported housing values from both sets of data. The study sites are indicated in Fig. 21.3, where red colour indicates Bharatpur Metro, and blue colour indicates other municipalites (2011 Nepal Census).

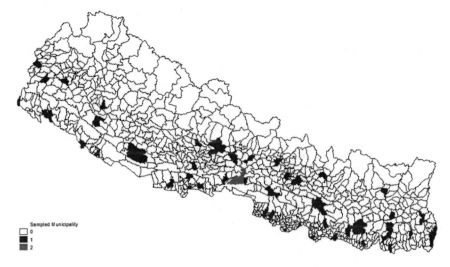

Fig. 21.3 Sampled municipalities across the country. *Source* Nepal et al. (2020)

We estimated the implicit price of presence of solid waste management service, a proxy for cleaner neighbourhood, in three market segments—urban hills, urban Terai and Bharatpur. We found that in these market segments, the housing price could be 25–57% higher in communities where solid waste management services are in place while the housing price could be 11% lower in a community where there are open drains nearby (Nepal et al., 2020). That is, city residents place a high price premium on cleaner neighbourhoods and a price penalty when the drainage system is open.

Figure 21.4 indicates that the housing price could be greatly enhanced with the provision of municipal solid waste management in the community. Hence, making solid waste management a top priority is rewarding not only to the home owners but also to the cities where the tax base includes the assessed value of property, and a part of the tax revenue could be used to finance the revenue shortfall for managing municipal solid waste better.

21.8   Discussion and Conclusion

This chapter highlights some of the key findings from a large-scale research project, and the detailed methods and results are published in several papers (Rai et al., 2019; Bharadwaj, Rai, & Nepal, 2020; Nepal et al., 2020; Nepal et al., 2021; Pervin et al., 2020; Nepal, Karki Nepal et al., 2021). Flooding is a major hazard for cities like Bharatpur, and this problem will become more serious with climate change, requiring improved drainage systems. This involves not only an improvement in the physical infrastructure of canals and drains but also in solid waste management to

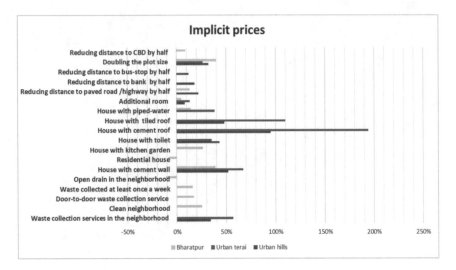

Fig. 21.4 Implicit prices of the housing attributes. *Source* Authors' calculation

prevent clogging of the drainage network. The latter requires the involvement of the communities.

In the case of Bharatpur, while physical investment in drainage systems is estimated to reduce the flood-prone area by 38% (comes down to 5% from 13%), improved waste management is estimated to reduce it by an additional 37% (prevents going back to 8% flood-prone area from 5%). Women tend to manage household solid waste better in terms of segregation and composting that reduces the volume of waste going to landfill sites, suggesting that targeted sensitization programmes would help making the communities cleaner and less prone to waterlogging (Nepal et al., 2021; Rakib et al., 2022, Chap. 24 of this volume).

The randomized intervention consisting of provision of information and sensitization to waste management as well as provision of street bins resulted in cleaner streets and neighbourhoods and more waste being given to collectors rather than being dumped nearby or burnt. However, segregation at source did not improve probably because residents did not perceive it as being effective in the absence of a system for collecting segregated waste separately by the collection service.

At present, as in most cities in low-income countries, the revenue from municipal waste collection fees in Bharatpur is insufficient to cover the cost of collection and disposal. This makes it difficult to improve solid waste management. However, our analysis of household survey data shows that people are willing to pay more for better service. This can partially close the revenue gap. Additional revenue could be raised by recycling plastic materials and increasing the tariff on imported plastic in Nepal by a modest 1%. This would raise sufficient revenue for managing plastic wastes.

A further avenue for raising revenue to cover waste management costs that promotes adapting to climate change in the cities for enhancing urban resilience arises from the effect that cleaner neighbourhoods seem to have on housing prices.

If some of this increase in pricing of houses is captured by the municipalities through an increase in revenue from property taxes, then this would easily bring in enough revenue to finance the necessary improvements in municipal solid waste management.

Acknowledgements The authors gratefully acknowledge the International Development Research Center (IDRC), Ottawa, Canada for providing financial support (Grant #08283-001) under the Cities and Climate Change research (2017–2020) to conduct this study. The International Centre for Integrated Mountain Development (ICIMOD), where the first author is affiliated, acknowledges with gratitude the support of the Governments of Afghanistan, Australia, Austria, Bangladesh, Bhutan, China, India, Myanmar, Nepal, Norway, Pakistan, Sweden, and Switzerland. However, the views as well as interpretations of the results presented in this research are those of the authors and should not be attributed to their affiliated organizations, their supporters or the funding agency.

References

Ahmed, I. (2016). Building resilience of urban slums in Dhaka, Bangladesh. *Procedia—Social and Behavioral Sciences, 218*, 202–213.

Bharadwaj, B., Baland, J. M., & Nepal, M. (2020). What makes a ban on plastic bags effective? The case of Nepal. *Environment and Development Economics, 25*(2), 95–114.

Bharadwaj, B., Rai, R. K., & Nepal, M. (2020). Sustainable financing for municipal solid waste management in Nepal. *PLOS One, 15*(8), e0231933.

Haque, A. K. E. (2013). *Reducing adaptation costs to climate change through stakeholder-focused project design: The case of Khulna city in Bangladesh* (No. G03520). International Institution for Environment and Development. http://pubs.iied.org/G03520/?k=enamul.

Haque, A. E., Lohano, H. D., Mukhopadhyay, P., Nepal, M., Shafeeqa, F., & Vidanage, S. P. (2019). NDC pledges of South Asia: Are the stakeholders onboard? *Climatic Change, 155*(2), 237–244.

Nepal, M., Berrens, R. P., & Bohara, A. K. (2009). Assessing perceived consequentiality: Evidence from a contingent valuation survey on global climate change. *International Journal of Ecological Economics and Statistics, 14*(P09), 14–29.

Nepal, M., Rai, R. K., Khadayat, M. S., & Somanathan, E. (2020). Value of cleaner neighbourhoods: Application of hedonic price model in low income context. *World Development, 131*, 104965.

Nepal, M., Cauchy, M., Karki Nepal, A., & Gurung Goodrich, C. (2021). Household waste management and the role of gender in Nepal. In A. Achaarya (ed.). *Environmental Economics in Developing Countries: Issues and Challenges* (forthcoming), Routledge.

Nepal, M., Karki Nepal, A., Khadayat, M. S., Pervin, I. A., Rai, R. K., & Somanathan, E. (2021). *Low-cost strategies to improve municipal solid waste management in developing countries – Learning from an experiment in Nepal*. Working Paper, SANDEE, Kathmandu, Nepal.

Pervin, I. A., Rahman, S. M. M., Nepal, M., Haque, A. K. E., Karim, H., & Dhakal, G. (2020). Adapting to urban flooding: A case of two cities in South Asia. *Water Policy, 22*(S1), 162–188.

Rai, R. K., Nepal, M., Khadayat, M. S., & Bhardwaj, B. (2019). Improving municipal solid waste collection services in developing countries: A case of Bharatpur Metropolitan City, Nepal. *Sustainability, 11*(11), 3010.

Rakib, M., Hye, N., & Haque, A. K. E. (2022). Waste segregation at source: A strategy to reduce waterlogging in Sylhet. In A. K. E. Haque, P. Mukhopadhyay, M. Nepal, & M. R. Shammin (Eds.), *Climate change and community resilience: Insights from South Asia*. Springer.

Sharma, U., Brahmbahtt, B., & Panchal, H. N. (2022). Do community-based institutions spur climate adaptation in urban informal settlements in India? In A. K. E. Haque, P. Mukhopadhyay,

M. Nepal, & M. R. Shammin (Eds.), *Climate change and community resilience: Insights from South Asia.* Springer.

ten Veldhuis, J. A. E., Clemens, F. H. L. R., Sterk, G., & Berends, B. R. (2010). Microbial risks associated with exposure to pathogens in contaminated urban flood water. *Water Research, 44*(9), 2910–2918.

Chapter 22
Do Community-Based Institutions Spur Climate Adaptation in Urban Informal Settlements in India?

Upasna Sharma, Bijal Brahmbhatt, and Harshkumar Nareshkumar Panchal

Key Messages

- Empirical evidence in this paper for women led CBOs to lead the way in increasing awareness about climate change and spurring climate adaptation action in the informal settlements in developing countries.
- While CBOs by definition are inclusive institutions, yet they may need to be more sensitive towards participation of marginalized in the slum setting.
- CBOs may have the potential to be the bridges between slum dwellers and local governments to access development and adaptation benefits.

22.1 Introduction

Half of the world's urban population resides in Asia (UN-Habitat, 2015; United Nations Human Settlements Programme, 2011) and Asian cities are among the densest and most populous cities in the world. The resource strapped city governments in most Asian cities face an overwhelming challenge of creating the necessary

Electronic Supplementary Material The online version of this chapter (https://doi.org/10.1007/978-981-16-0680-9_22) contains supplementary material, which is available to authorized users.

U. Sharma (✉) · H. N. Panchal
School of Public Policy, IIT Delhi, New Delhi, India
e-mail: upasna@iitd.ac.in

H. N. Panchal
e-mail: ppz198390@sopp.iitd.ac.in

B. Brahmbhatt
Mahila Housing Sewa Trust, Ahmedabad, Gujarat, India
e-mail: bijalb@mahilahsg.org

infrastructure for the ever-expanding areas and are typically not able to keep up with the demand. This has resulted in one in seven people living in poor quality and congested habitats with limited or no access to necessary infrastructure and services and lack of legal land tenure (Mitlin & Satterthwaite, 2013; Satterthwaite, 2007; United Nations Human Settlements Programme, 2011). In India, every sixth urban Indian resides in slums (MoHUPA, 2013). Many informal settlements are located in geographically disadvantaged, high-risk areas, more prone to climate-related hazards (like floods or landslides). In addition, informal settlements are resource scare, often in dilapidated conditions and do not have essential infrastructure that can withstand climate shocks, making them more vulnerable to climate risk. Adaptation to climate risk is an important strategy for reducing impacts of climate hazards, which requires development of capacity to adapt.

National plans and strategies for reducing vulnerability to climate risk, typically in the form of planned adaptations to climate change, are often formulated and designed with no involvement of the urban residents—many of whom live in informal settlements (Satterthwaite, 2007). Additionally, the people living in informal settlements often lack capacities to implement adaptation options. It is hypothesized that if the slum residents are provided with a supporting system to act, including access to knowledge, they may develop learning capacities which may lead to enhancement of capacities to achieve desired outcomes. All these point to the need for appropriate institutions to facilitate the learning process and hence the process of enhancement of adaptive capacity of slums.

The residents of slums are socially marginalized from access to critical services or resources, and this is closely linked to the institutional exclusion in urban areas. Informal settlement residents usually have difficulties engaging with local governments (Mitlin & Satterthwaite, 2013). In India, despite the implementation of the 74th amendment act, which pertains to the urban governance and the role of citizens in collective decision-making, the voice of urban citizens has been scarcely included (De & Nag, 2016). A community-based organization at the level of a slum can help to voice the concerns enabling behaviour of collective action and collective decision-making strengthening capacity to adapt.

One of the difficulties that community-based adaptation faces in cities is that much of the population lives in informal settlements on land that is occupied without formal authorization. Local governments may refuse to provide risk-reducing infrastructure because of this apparent 'illegality' (Satterthwaite, 2007). The interaction of community-based organizations with the formal government institutions would help strengthen bottom-up decision -making and collaboration between the governmental service providers and slum residents.

The objective of the study is to understand whether community-based organizations (CBOs) like women's community action groups enhance the ability of people living in urban informal settlements to take action to reduce vulnerability to climate risk. This has some similarities with the type of intervention attempted in Bangladesh as documented by Rakib et al. (2022, Chap. 24 of this volume). We draw from the

'Women-led Resilience Building of Urban Poor in South Asia' project which was developed by Mahila Housing SEWA Trust (MHT) and its partners as a part of the Global Resilience Partnership (GRP) challenge (MHT, 2018). The research question that we raise is: Are CBOs effective in (a) raising *awareness* of climate change and adaptation options, and (b) help slum households to reduce climate impacts due to heat stress, flooding and vector-borne diseases? Another objective of this paper is to assess some preliminary evidence on their ability to engage meaningfully with local government.

22.2 Mahila Housing Trust (MHT) and the Global Resilience Project

The 'Women led Resilience Building of Urban Poor in South Asia' project aimed to build the resilience capacities of twenty five thousand low-income families living in slums/informal settlements in seven cities of South Asia, to take the lead in action against four climate risks. These four climate stressors are (a) heat waves; (b) flooding and inundation; (c) water scarcity; and (d) increased climate change-related incidence of water- and vector-borne diseases (MHT, 2018).

The project worked to create an integrated model wherein women take a lead through collective action and technology incubation, to devise locally relevant pro-poor and gender-sensitive climate resilient solutions and promote a culture of sustainable development and resilience among the urban poor in South Asia (MHT, 2018). To achieve this required, among other things, the formation of Community Action Groups (CAG) where women were trained on different aspects of climate change and urban governance.

22.3 Methods

22.3.1 Framework for the Study

There are three main elements in the framework for this study—one is the vulnerability of the system to external climate stimuli such as temperature and rainfall, two is the adaptive capacity of the system and the third is the adaptations made or actions taken by community members or households to address the climate risk (see Fig. 22.1).

This chapter uses the notion of 'vulnerable situation' conceptualized by Füssel (2005), where 'vulnerability' can only be used meaningfully with reference to a particular vulnerable situation (i.e. assessment context). Four fundamental dimensions describe the vulnerable situation—(i) the system (or region and/or population group and/or sector, e.g. slum communities in this study), (ii) the hazards (or threats

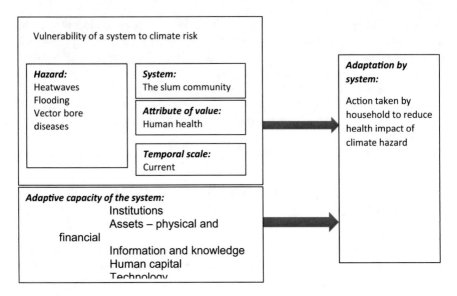

Fig. 22.1 Framework for the study

or stressors) considered (e.g. heatwaves, flooding and vector-borne diseases such as malaria and dengue), (iii) the consequences (or effects or valued attributes or variables of concern, e.g. impact on health and economic activity) and (iv) a temporal reference (e.g. current time period instead of future time period in this study).

Adaptive capacity of a system is defined as the ability of the system or its units to perceive the risk (need for change in behaviour), formulate a response (in terms of required change in characteristics or behaviour) and then implement the response (i.e. bring about a change in the characteristics or behaviour) with the view to reducing climate risk Patwardhan et al. (2003). The adaptive capacity inherent in a system is determined by (Brooks & Adger, 2005)—the resources available for adaptation such as natural resources, financial capital, human capital, knowledge of risk, appropriate social institutions for managing risks and appropriate technology, and, by the ability and willingness of those who need to adapt to deploy these resources effectively.

Adaptation involves an alteration in something (the system of interest, activity, sector, community or region) in response to something (the climate-related stress or stimulus) (Smith et al. 2001). The forms that adaptation takes can be classified in many ways. One way of classifying adaptation options can be based on Morgan's (1981) suggested four conceptual approaches to reducing technology-induced risk through modifying the effects and the exposure. They are:

- The natural and human environment can be modified—for example by raising plinth level to avoid flooding of house.
- The exposure processes can be modified or avoided—for example bed nets and netting screens on windows and doors can be installed to avoid mosquito bites.

Table 22.1 Four cities included in this paper

City	State	Climate	Population (lakhs)	Slum population (%)
Ahmedabad	Gujarat	Hot semi-arid	55.7	4.49
Bhopal	Madhya Pradesh	Humid subtropical	17.9	26.68
Jaipur	Rajasthan	Hot semi-arid	30.4	10.62
Ranchi	Jharkhand	Humid subtropical	10.7	20.8

Source Compiled from Census of India (2011)

- The effects processes can be modified or avoided—for example taking oral rehydrating salts (ORS) in case of dehydration in heat waves, taking medicines for malaria and dengue if affected by it.
- The effects, once they occur, can be mitigated or compensated for—for example compensation given by local governments for loss of property during extreme climate events.

22.3.2   Study Area

The study was conducted in four cities in India—Ahmedabad, Bhopal, Jaipur and Ranchi. In Ahmedabad, some slums already had well-established networks of women leaders in both the informal settlements and citywide, emerging out of MHT's long history of intervention. In the three other cities, MHT was in the process of establishing networks of women leaders at both the informal settlement and city levels, emergent out of MHT's shorter history of working in these cities. Table 22.1 presents a brief description of the four cities included in this paper. They represent different geographic destinations in India with varied climatic conditions. Each of them is the capital city of their respective states. The highest slum population is present in Bhopal followed by Ranchi, Jaipur and Ahmedabad.

22.3.3   Intervention

MHT made many interventions to achieve the objectives of its project on resilience building. However, the most pertinent one for this paper is the creation of Community Action Groups (CAGs). A total of 100 slums was selected by MHT for intervention based on a dearth of basic amenities within the slums and to provide diversity in the level of pre-existing social capital. This was verified based on a slum profile developed interactively with slum residents, focused on level of individual water supply connections, sanitation, housing pattern and existence of a Community Action Group (CAG). In the treatment slums, the project formed a CAG at slum levels by training, on an average, 10–12 women and youth leaders in each slum to act as local community advocates and climate specialists on climate risk, surveillance and vulnerability assessment, collective response action and technical solutions (MHT, 2018). In the established and emergent cities, MHT worked with CAGs to form Vikasinis (city-level women led federation of CAGs) which would represent the voice of people in their slum communities in discussion with local government and technical groups (MHT, 2018).

22.3.4 Sampling

Selection of Treatment and Control Slums

Out of 100 slums selected for intervention by MHT, 20 project intervention (treatment) slums in Ahmedabad and 15 treatment slums in the other 3 cities (5 in each) were selected randomly. Since Ahmedabad contained both established and emergent slums, the sample in Ahmedabad was stratified based on the existence or not of a Community Action Group (CAG) in the slum prior to the project. Equal number of slums with existing CAGs and ones lacking a CAG were selected randomly. The other three cities only contained emergent slums because MHT has begun working in these cities only recently and no intervention slum contained an existing CAG. Control slums were selected based on their similarity to treatment slums. Slums were matched one to one on the basis of proximity, infrastructure status and governance and vulnerability to climate risks. Data used to match control slums to treatment slums was verified through slum profiles developed with the participation of local residents (MHT, 2017).

Selection of Households

In all the treatment slums, 10 households with CAG members were surveyed mainly based on the data supplied by the MHT programme staff. 15 non-CAG households were surveyed in the treatment slums. In the case of control slums, geo-spatial coordinates within the boundaries of the slum were randomly generated to select the respondents. A list of 15 households was surveyed based on their availability and acceptance to participate in the survey during the first home visit. In all households, only adult women respondents were interviewed. Table 22.2 presents the sampling plan for the study (MHT, 2017).

22.3.5 Data Collection

Data was collected through survey using a structured questionnaire (details available in MHT, 2017). The baseline survey was conducted from June to December 2016 with a sample of 1241 households and the endline survey was conducted from October to December 2017.We included 852 observations in our sample for the purpose of

Table 22.2 Sampling plan for the study

City	Slum type	Treatment slums				Control slums	
		No. of slums	No. of CAG household per slum	No. of non-CAG household per slum	Total no. of households	No. of slums	No. of households
Ahmedabad	Established	10	10	15	250	10	150
	Emergent	10	10	15	250		
Jaipur	Emergent	5	10	15	125	5	75
Bhopal	Emergent	5	10	15	125	5	75
Ranchi	Emergent	5	10	15	127	5	75
					875		375

Source MHT (2017)

this paper. The reason for attrition of about 30% of households in the endline survey was mainly related to relocation of the family in another slum or city or because the women head of the household declined to participate in the survey, or because the same household or same respondent could not be located.

22.3.6   Description of Sample

Table 22.3 presents a brief description of key aspects of the sample in this study. The mean age of the sample is 40 years. The sample consists of 29% control, 38% non-CAG households and 33% of CAG households. Almost 31% of the respondents cannot read and write. 50% of the respondents' income is below INR 10,000 per month. About 52% of the sample lives in pucca houses and 20% in kutcha houses.

22.3.7   Techniques for Data Analysis

The present study allows us to use the difference-in-difference (DID) method to evaluate the outcomes of the interventions. DID is used to estimate the treatment effects by comparing the pre- and post-treatment differences in both groups: treatment and control. The difference-in-difference equation is as follows:

$$Y_i = \beta_0 + \beta_1 \text{ Treatment} * \text{Year} + \beta_2 \text{ Treatment} + \beta_3 \text{ Year} + \beta_4 \text{ Covariates} + \varepsilon$$

(22.1)

Here, Y_i is an outcome of interest; and 'Treatment' refers to whether the respondent is part of the control group or treatment group. Year is an indicator of endline year. Treatment * Year is an interaction of these treatment and year variables and

Table 22.3 Description of the sample

	Responses	Total	Control	Non-CAG	CAG	Control & Non-CAG	Control & CAG	Non-CAG & CAG
		(n = 852)	(n = 248)	(n = 323)	(n = 281)	p-value	p-value	p-value
City	Ahmedabad	56%	46%	60%	60%	0.005***	0.001***	1
	Bhopal	20%	28%	17%	17%	0.001***	0.002***	1
	Jaipur	12%	9%	13%	12%	0.25	0.37	0.92
	Ranchi	13%	17%	10%	11%	0.88	0.05	0.85
Slum type based on MHT operations established or emergent	Established	56%	46%	60%	60%	0.000***	0.001***	1
	Emergent	44%	54%	40%	40%	0.001***	0.001***	1
Literacy	Can't read and write	31%	35%	39%	18%	0.4	0.000***	0.000***
	Can read or write	69%	65%	61%	82%	0.4	0.000***	0.000***
Income (in INR)	Below 10 k	50%	50%	44%	56%	0.15	0.21	0.0034***
	Between 10 to 20 k	33%	29%	40%	29%	0.01	0.95	0.0069***
	Above 20 k	17%	20%	16%	15%	0.3	0.14	0.7
House type	Kutcha	20%	27%	21%	11%	0.08	0.000***	0.002***
	Semi-Pucca	29%	30%	26%	30%	0.31	1	0.29
	Pucca	52%	42%	53%	58%	0.012	0.000***	0.23
Age (in years)	Mean age	40	41	39	38	0.05**	0.003***	0.2324

Source Based on primary survey
** significant at alpha = 0.05, * significant at alpha = 0.1

ε is the stochastic error term. The coefficient of the interaction term (Treatment * Year) is the DID estimator used to assess the impact of the interventions in the slums. Covariates are the control variables. This design allows us to account for observable and time-invariant unobservable factors and thus attribute changes in outcomes to the Community Action Group intervention of the Mahila Housing Trust under the Global Resilience Project. Since the dependent variables are binary in nature (as shown in Sect. 22.4.1), we use linear probability regression for the data analysis in the DID framework.

22.4 Data Analysis and Results

22.4.1 Main Variables of Interest in the Study

Treatment (main independent variable): The treatment variable had three levels, i.e. control group, non-CAG group and CAG group. Control slums had no CAG programme. Within the treatment slum a respondent household could either be a member of Community Action Group (labelled as 'CAG') or not be a member of CAG (labelled as 'non-CAG').

Slum household's awareness of climate change (dependent variables): Three binary (yes/no) indicators were considered for assessing awareness of respondents about climate change. The first indicator was whether the respondent had *heard of climate change or not*. The second indicator on respondents' *awareness of climate change impacts* was captured through an open-ended question which was converted to a binary form based on assessment of qualitative responses. From among the respondents who gave responses different from 'don't know', majority of the respondents said that heat stress was increasing, followed by increase in disease incidence, flooding, pollution and water stress. The third indicator on the respondents' *general awareness of options to reduce impact of climate change* was captured through another open-ended question. The responses were coded 'yes' if the respondent mentioned a reasonable option to reduce the impacts of climate change, and 'no' if the respondent said, don't know or the response was not very coherent. The most common responses were afforestation or tree plantation, followed by. 'over-population must be controlled', 'vehicle use must be reduced', avoid the use of air conditioning, etc. Figure 22.2a presents the distribution of responses for awareness of climate change across control, non-CAG and CAG groups.

Action on climate change by slum households (dependent variables): Two indicators were considered for assessing the action of slum households in terms of adopting specific options to reduce climate risk. The first was whether the respondent had *heard of specific options to reduce climate risks* faced by them in the context of their slums. As part of MHT's intervention, the Community Action Groups attempted to build capacity building of agents through communication exercises and workshops on the understanding of climate change, participatory vulnerability assessment and an exposure to set of risk reducing or adaptation options for reducing climate risks faced by the community. The sample of respondents was asked, whether they had heard of certain options to reduce the heat stress in summers due to higher temperatures, reduce the mosquito menace and reduce impact of flooding. The response options consisted of 'heard' of the option, 'not heard' of the option and 'invested' in the option. For some of the options, the question was asked only in the endline year, as these were technologies which had not been introduced through the intervention in the baseline year. Figures S1 and S2 (in supplementary material) present the array of such risk reducing or adaptation options in terms of the proportion of

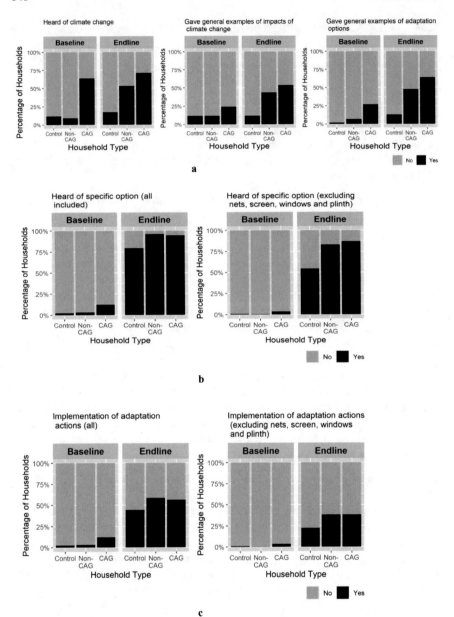

Fig. 22.2 **a** Slum household's awareness of climate change. **b** Slum household's awareness about specific adaptation options to reduce climate risk in their slums. **c** Slum household's implementation of specific adaptation options to reduce climate risk in their slums

respondents who had chosen, 'heard' and 'not heard' (invested option was converted into 'heard' category for this piece of analysis) responses across control, Non-CAG and CAG households. The second indicator was the *implementation of adaptation option(s)* by slum households which was captured by looking at the actual options in which people invested their money in baseline and endline years. The endline year had more number of options in which households invested as the Community Action Groups had sensitized their members to additional options available to reduce risk (see Fig. 22.2c).

Control variables: The awareness of climate change and actual implementation of adaptation option will depend on the vulnerable situation of households in slums and adaptive capacity of households in slums as shown in the conceptual framework for the study (Fig. 22.1). The attribute of value, i.e. human health issues due to heat stress and vector-borne disease and economic effect of flooding in terms of loss of property, loss of work and loss of schooldays for child, is included in the equation as independent variable. The greater the loss in these dimensions, the greater the expectation of a household being aware of climate risk and investing in adaptation options.

Three indicators for adaptive capacity are included here. One is related to the institutions at the local level, i.e. whether a respondent is a member of a CAG or not which is also the treatment and the main independent variable in the study. The second is the income of the households which is expected to affect the ability of the slum household to implement a particular adaptation option and is also a proxy indicator of many other dimensions of adaptive capacity related to wealth. The third indicator is literacy, with the expectation that greater the literacy, the greater the awareness of climate change and agency to implement adaptation options. Also, Table 22.3 shows that the control groups and treatment groups are unbalanced with respect to these variables. Hence, including these in the regression equation as controls is reasonable.

Another covariate included in the regression equation is whether the slum belongs to cities where MHT has established operations and is experienced in implementation of community action groups for resilience building in slums (like Ahmedabad) and cities where their operations are new and emergent (like Bhopal, Jaipur and Ranchi) to account for city fixed effects.

22.4.2 *Effect of Community Action Groups on Awareness Climate Change and Implementation of Adaptation Options*

For each outcome (dependent variable), Fig. 22.3 displays (i) the DID estimate and (ii) robust clustered standard errors of the DID estimate. The complete results of DID regression with all its covariates are available in the supplementary material as Table S2.

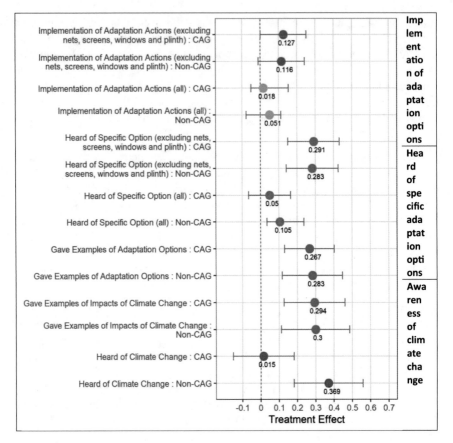

Fig. 22.3 Impact of Intervention on household awareness of climate change and implementation of adaptation actions to climate change

The DID estimates suggest that membership of Community Action Groups had a significant effect on two out of three indicators of awareness of climate change. The effect of treatment was not significant for whether the respondent had heard of climate change in the CAG group. This is because as Fig. 22.2a shows that awareness of climate change was already high among the CAG members in the intervention slums in the baseline year (probably, because MHT already had a presence in the established city (Ahmedabad) slums even before the intervention). The MHT's intervention and facilitation in these slums had included discussions on climate change even previous to the start of the Global Resilience Project. This is validated by some qualitative comments of respondents in the baseline year, when they were asked, 'Where have you heard or learned about climate change?' Majority of the respondents said that they learnt about climate change from MHT meetings. The interesting result is that even among respondents who are not members of CAG, but part of treatment slums, there is a significantly higher awareness of climate change. There seem to be spillover effects of knowledge from CAG to non-CAGs in the same treatment slum communities whereas control groups are in slum different communities with no treatment at all.

The DID estimates for awareness about specific adaptation options and implementation of adaptation actions are not significant if we include all thirteen specific adaptation options in the dependent variable. This seems counterintuitive as Figs. S1 and S2 (in supplementary material) show that for many adaptation options there seems to be a significant difference between awareness of specific adaptation options in baseline period and endline period. Delving deeper into the data, we found that the number of respondents who had adopted 'mosquito nets' and 'mosquito screens', and implemented more 'windows for ventilation' were quite large in the endline year and across not just non-CAG and CAG groups but also control group (see Tables S3 and S4 in supplementary material). The relatively large number in control group for the above three options was masking the effect of intervention on awareness and implementation of other specific adaptation options. Also, the option of construction of plinth for reducing flood risk is greater among CAG households in baseline year compared to intervention year (see Table S3 in supplementary material). Hence, we excluded the mosquito nets and screens and more windows for ventilation and construction of plinth in computing the dependent variables—'awareness of specific adaptation options' and 'implementation of specific adaptation options'.

After this, we found the effect of the intervention to be significant not only for CAG group but also for the non-CAG group, again implying positive spillover or externalities of the intervention in the treatment slums. Moreover, the effect on non-CAG group seems to be bigger in case of some dependent variables as the change in non-CAG group from baseline to endline year is a bigger change compared to the change in CAG group from baseline to endline years (the bar plots show, in many instances, in the baseline situation, the non-CAG is at a much lower level than CAG). A probable cause of more people in the control group becoming more aware of and implementing these three specific options could be the sensitization during baseline year through the questionnaire interview. Since these are commonly understood options and relatively easy to implement, we see a greater awareness and implementation of these across all three groups, irrespective of the treatment.

Though the effectiveness of CAGs to enhance the awareness of climate change, and implementation of adaptation options by slum households to reduce climate impacts and risks is established by the empirical evidence in this paper, however, the efficacy of CAGs to do the above is not perfect as is discussed in the next section.

22.4.3 Impact of Heterogeneities Within Treatment Group on the Effect of Treatment

From Fig. 22.2a and Figs. S1 and S2 (in supplementary material), it is evident that in none of the plots of climate change awareness and awareness of specific adaptation options, the proportion of CAG members who are aware of climate change or of the adaptation options is greater than 75%. This means, that at least 25% of members of CAG have either not heard of climate change and specific adaptation options or could not recall it when asked about it. There could be many reasons for this—one, that not all members necessarily attend all CAG meetings and its training programmes. The members who miss the meeting and/or training programmes naturally have lost the opportunity to learn about these issues. A second reason could be that some of the members may not comprehend fully the notions of climate change and different types of adaptation options due to lack of capacities such as education and income.

Hence, for the sample of CAG respondents only ($n = 281$), Table 22.4 presents an OLS regression with dependent variable being the number of specific adaptation options that a household had 'heard' of. The independent variables are three socio-economic indicators of capacity, i.e. being above or below BPL, literacy and house type. Literacy is positively correlated with whether a person heard of or had not heard of adaptation options to reduce heat stress and vector-borne diseases. Hence, lack of education could be a barrier in comprehending and understanding fully the various adaptation options. Similarly, people living in kuccha houses are less likely to have heard of adaptation options.

The implication of this result is that even though CAGs by definition seem to be inclusive and participatory, but even at such a decentralized level, heterogeneities in demographic and socio-economic conditions of members may mean exclusion of certain members from full and complete participation.

Table 22.4 Determinants of a CAG group respondent having 'heard' of a specific adaptation option

	Independent variables	Heard about climate actions
BPL Base: Yes	No	2.080** (0.424)
House type (Base: Kuccha house)	*Semi-pucca*	3.939*** (0.633)
	Pucca	1.9763** (0.567)
Literacy Base: can't read and write	Can read and/or write	1.197* (0.554)
Constant		2.261*** (0.651)

Observations: 281
$R^2 = 0.231$, Adjusted $R^2 = 0.219$
Residual std. error (df $= 276$) 3.349, F statistic (df $= 4$; 276) 20.685***

Source Based on data collected
*** significant at alpha $= 0.01$, ** significant at alpha $= 0.05$, * significant at alpha $= 0.1$

Fig. 22.4 Visit to local health department by respondents across control, non-CAG and CAG groups

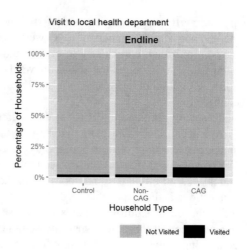

22.4.4   Community Action Group Membership and Engagement with Local Government

One of the objectives of the CAGs facilitated by MHT is to improve the ability of the women to collectively to engage with local governments in co-creating and implementing solutions to the problems that slum dwellers face. The qualitative interviews with CAG members do point to this. However, the quantitative data, which has been the focus of analysis in this paper, did not include many variables which could tap into this dimension (ability to engage with local governments) of capacity enhancement of slum households and communities. We could find only one variable in the data set that reflected this dimension—which was whether the respondent household had approached the health department to address the mosquito problem in the endline survey (see Fig. 22.4 for results).

We find that a very small proportion of respondents answered in the affirmative to this question across all the respondent categories, i.e. CAG, Non-CAG and control group. However, the odds of CAG respondents approaching the health department to address the mosquito problem was at least four times the odds of a control group respondent doing so (see Table S5 in supplementary material).

22.5   Observations and Conclusions

This paper provides empirical evidence on the effectiveness of CAGs to enhance the awareness of climate change, awareness about specific adaptation options to reduce climate impacts and to facilitate the implementation of adaptation options by slum households. We expected the household members of community action group to lead in taking climate actions because of access to information about adaptation options and various training activities conducted by the CAGs. This did hold true as we found that the household members of CAG were more likely to take climate actions than the control group, particularly in case of those options which are less commonly known such as green roofing and airlite ventilation. We even saw spillover effects to non-CAG household who also took significantly higher climate action compared to control groups.

This study further contributes to the literature on the role of institutions, particularly decentralized and participatory institutions in enhancing adaptive capacity and response actions of slum households and communities to climate change. While CBOs cannot build the trunk infrastructure (elevated roads, water, sewer and drainage mains) that all city areas and zones need, they can help install and make smaller connections and/or pipes that feed into the trunk infrastructure, thereby improving the inclusion of slum households in the use of trunk infrastructure. Example of this was seen in this study when the community together elevated an inner street connecting to the main road, so that flooding could be reduced.

We also found evidence that CBOs like the CAGs in this study can be effective in improving the capacity of slum households to engage meaningfully with city-level authorities in climate risk reduction options for their communities. This resonates with other studies like Adger et al (2003) who find that actions based on collective action can benefit and can be more effective in strengthening the voice of the people living in slums ultimately enhancing their adaptive capacity. This also adds to the literature from the global South as reviewed by Shammin and et al. (2022, Chap. 3 of this volume).

However, the efficacy of CAGs to enhance the awareness of climate change and adaptation options to reduce climate impacts and risks are not perfect as the results in Sect. 22.4.3 illustrate. The implication of this finding for functioning of CAGs is to introspect on how to increase members' participation in CAG meetings and trainings and how to make participation of members more inclusive. The role of training has been highlighted in the literature as noted in the systematic review done by Bahinipati and Patnaik (2021, Chap. 4 of this volume). Community-based models need to have higher social capital and will require collective action and continuous social learning where the vulnerable population come together for a common cause (Adger et al., 2003).

One lacuna is, that adaptive capacity building and resilience building is a sustained, long-term process, and assessments like the one made by MHT and Global Resilience Projects are one-time snapshots. Moreover, after an intervention has been made by a CBO, e.g. training related to climate change and options to address climate risk, it takes a while for the effect of these interventions to convert into desirable outcomes. Data from one year immediately after the intervention may not always capture the extent of change in capacities that has been brought about. It could be that with sustained effort, these capacities would strengthen or they could deteriorate over a period of time after the initial enthusiasm, if mechanisms to sustain the capacity building are not built into the institution.

Continuous engagement of the communities is a mounting challenge in the whole process. It should also be noted that very less empirical studies exist in measuring the resilience of the slum communities through a participatory approach. More studies in future would help to illuminate insights in improving the framework. More research is also needed into what are the variations in institutional forms of CBOs and what institutional features make some CBOs more successful than others in achieving their objectives.

References

Adger, W. N., Huq, S., Brown, K., Conway, D., & Hulme, M. (2003). Adaptation to climate change in the developing world. *Progress in Development Studies, 3*, 179–195.

Bahinipati, C. S., & Patnaik, U. (2021). What motivates farm level adaptation in India? A systematic review. In A. K. E. Haque, P. Mukhopadhyay, M. Nepal, & M. R. Shammin (Eds.), *Climate change and community resilience: Insights from South Asia* (pp. 49–68). Springer Nature.

Brooks, N., & Adger, W. N. (2005). Assessing and enhancing adaptive capacity. In *Adaptation policy frameworks for climate change: Developing strategies, policies and measures* (pp. 165–181)

Census of India. (2011). India Population Census.

De, I., & Nag, T. (2016). Dangers of decentralisation in urban slums: A comparative study of water supply and drainage service delivery in Kolkata, India. *Development Policy Review, 34*(2), 253–276. https://doi.org/10.1111/dpr.12149

Füssel, H. (2005). *Vulnerability in climate change research: A comprehensive conceptual framework*. University of California International and Area Studies. Breslauer Symposium Paper 6. Paper.

MHT. (2017). *Women's action towards climate resilience for urban poor in South Asia: Baseline report*.

MHT. (2018). *Women's action towards climate resilience of urban poor in South Asia: Evaluation, research methodology and results*.

Mitlin, D., & Satterthwaite, D. (2013). *Urban poverty in the global south: Scale and nature*. Routledge.

MoHUPA. (2013). *State of slums in India: A statistical compendium*.

Morgan, M. G. (1981). Choosing and managing technology-induced risk. *IEEE Spectrum, 18*, 53–60.

Patwardhan, A., Narayanan, K., Parthasarathy, D., Sharma, U. (2003). *Impacts of climate change on coastal zones* (pp. 326–359) Climate Change and India.

Rakib, M., Hye, N., & Haque, A. K. E. (2022). Waste segregation at source: A strategy to reduce waterlogging in Sylhet. In A. K. E. Haque, P. Mukhopadhyay, M. Nepal, & M. R. Shammin (Eds.), *Climate change and community resilience: Insights from South Asia* (pp. 369–383). Springer Nature.

Satterthwaite, D. (2007). *Adapting to climate change in urban areas: The possibilities and constraints in low-and middle-income nations*. IIED.

Shammin, M., Wang, A., & Sosland, M. (2022). A survey of community-based adaptation in developing countries. In A. K. E. Haque, P. Mukhopadhyay, M. Nepal, & M. R. Shammin (Eds). *Climate change and community resilience: Insights from South Asia* (pp. 31–47). Springer Nature.

Smith, J., Lavender, B., Smit, B., & Burton, I. (2001). Climate change and adaptation policy. *Canadian Journal of Policy Research, 2*.

UN-Habitat. (2015). *Issue paper on informal settlements*.

United Nations Human Settlements Programme. (2011). *Cities and climate change: Global report on human settlements—2011*. Routledge.

Chapter 23
A Tale of Three Himalayan Towns: Would Payment for Ecosystem Services Make Drinking Water Supply Sustainable?

Rajesh K. Rai and Mani Nepal

Key Messages

- Water scarcity has been increasing in Himalayan towns as natural springs are drying up.
- Investing in physical infrastructures alone does not help protect water sources that are mainly located upstream, or other watersheds.
- Incentive payment to the water source communities for protecting watersheds helps reducing conflict between water users and upstream communities. It requires an institution that plays a role of an intermediary.

23.1 Introduction

The Himalayas are known as the water towers of Asia and the source of freshwater to about a quarter (1.9 million) of the world's population (Bharti et al., 2020). Astonishingly, the Himalayan region in general, and its cities in particular, faces an acute drinking water shortage (Bhatta et al., 2018; Ojha et al., 2020; Rai et al., 2015; Tamang et al., 2020). Though the Himalayas are the source for most of the rivers in the region, in the hills and mountains the water from the rivers is largely used for generation of hydropower. As the fields and settlements at this altitude are higher than the rivers, usage for irrigation, household needs and other activities would

R. K. Rai
School of Forestry and Natural Resource Management, Institute of Forestry, Tribhuvan University, Kathmandu, Nepal
e-mail: rjerung@gmail.com

M. Nepal (✉)
South Asian Network for Development and Environmental Economics (SANDEE), International Centre for Integrated Mountain Development (ICIMOD), Kathmandu, Nepal
e-mail: mani.nepal@icimod.org

require heavy investment in water infrastructure. These needs are met from drawing water from natural rainfed springs. Rapid urbanization along with climate change is causing springs to dry up resulting in water shortage in the Himalayan towns (Kattel & Nepal, 2021, Chap. 11 in this volume; Rai et al., 2019a, b; Rai and Rai, 2019; Singh & Pandey, 2020).

In Nepal, settlements in the hilly regions are searching for alternative water sources, while in the more urbanized areas of lower elevation, groundwater is being extracted for domestic and agricultural uses. Most municipalities are supplied water by the Nepal Water Supply Corporation (NWSC). However, in recent years, this system has been decentralized and since 2001 the Asian Development Bank (ADB) has supported small town drinking water supply projects in Nepal (ADB 2011). Altogether, 69 towns will have their own drinking water supply project in the three-phased ADB-supported plan.

However, all these projects, whether completed or ongoing, focus mainly on supplying water with scant attention paid to protection of the water source which is a key element of sustainability. The design of the projects indicates that the supply from existing water sources is taken as a permanent supply (Bhatta et al., 2018; Rai et al., 2017, 2018). This approach ignores future uncertainties, climate change and the role of upstream communities, all of which may affect water supply.

The communities living close to the water sources (upstream community) have a significant role in the maintenance of the water sources, and the watershed and dwindling supply may lead them to draw more heavily on the sources which downstream communities depend on currently (Rai et al., 2015, 2019a).

Existing drinking water supply projects have overlooked the role of local communities in the vicinity of the watersheds given the space for conflict between the upstream and downstream communities. Studies, however, suggest that with appropriate institutional mechanisms, upstream communities can be engaged in the drinking water supply projects through a subsidiary scheme for minimizing the conflict, while sharing water resources and also maintaining the quality and quantity of the drinking water supplied to the downstream communities (Bhatta et al., 2018).

For maintaining the quantity and quality of drinking water supply to the downstream, the upstream communities are required to change their behaviour or livelihood activities. Such changes include avoiding certain upstream activities (e.g., reducing the use of chemical fertilizer in agriculture, avoiding grazing animals near the water source) that have opportunity costs. For instance, they may have to switch from conventional farming methods to organic farming practices and regulate grazing. Therefore, the upstream community should be compensated for their efforts and the income-generating activities they would have to forgo. Without such compensation, water source communities are less likely to change their activities or behaviour resulting in the degradation of quality of water or reduction in the quantity (Kosoy et al., 2007).

In principle, as the beneficiaries of this behaviour change, water users should be required to pay compensation to the water source community (Alston et al., 2013). Such compensation, termed payment for ecosystem services (PES) is a mechanism, where beneficiaries pay for the positive externalities to the managers of the ecosystem or natural resources (Engel et al., 2008; Wunder, 2005). The payment is expected to incentivize ecosystem managers to change their behaviour and protect the water sources.

The PES scheme provides incentives to ecosystem managers most of whom are farmers to undertake conservation practices and adopt new technologies which could conserve water sources while also generating income for the farmers as co-benefits (Bulte et al., 2008). Usually, households living closer to water sources are resource-dependent with low-income, thus making their opportunity cost for participating in water source protection comparatively low. More often, these households do not even need full compensation for the costs they incur while modifying their activities and behaviour for protecting water sources. A small nudge or incentive would be enough to motivate them. Therefore, PES, which does not intend to pay the upstream communities the full costs of protecting the water sources, is a cheaper strategy for providing and improving the quality of drinking water in comparison with other available options such as purchasing water from a market or spending a considerable time to collect water from alternative sources (Rai et al., 2017).

PES can therefore improve social and environmental outcomes of the project by increasing both consumer and producer surplus and improving the management of natural resources (Choi et al., 2017; Engel et al., 2008). A meta-analysis of PES schemes indicates that watershed management was the dominant programme and was widely practiced in 62 countries in 2015 (Salzman et al., 2018).

This chapter draws lessons from three cases studies of drinking water supply projects in, Dharan sub-metropolitan City; and Dhankuta and Dasharath Chand Municipalities where the possibilities of introducing a payment for ecosystem services were examined for systemizing the water source protection (Bhatta et al., 2018; Rai et al., 2017, 2018). All three towns have an ADB-supported water supply project from new sources since the existing ones are insufficient to fulfil the increasing demand of these growing cities.

23.2 Study Sites

The study was carried out in three urban centres, where local water user committees were implementing drinking water projects since existing projects were unable to fulfil the demand of urban dwellers. Out of the three urban centres, Dharan and Dhankuta are located in eastern Nepal, while Dasharath Chand is in the far west. Ecologically, Dharan is in the foothills of the Chure hills, while Dhankuta and Dasharath Chand are in the mid-hills. In terms of population, Dharan is the largest of the three towns (27,750 households) followed by Dhankuta (3130 households) and Dasharath Chand (1,473 households). Dharan is in the downstream of the Sardukhola watershed, where water comes from the upstream; while in the case of two other municipalities, the water sources and the user communities are in different watersheds where the upstream–downstream relationship does not exist.

Due to lower elevation and larger size, Dharan has been extracting groundwater to supplement the stream water for household uses, while the other two towns are using stream water for the same purpose. Due to the activities of the upstream/water source communities, the stream water that these towns are getting is not of good quality and cannot be used for drinking without further treatment. The proposed PES schemes are to incentivise the water source communities for protecting the watersheds so that the water quality gets better and water supply does not get interrupted due to anthropogenic activities.

23.3 Water Users' Preferences

Neo-classical economic theory suggests that consumers are fully aware of their preferences and select the alternative which gives them maximum utility (Ben-Akiva & Lerman, 1985). In the context of the PES scheme, success hinges on the service users paying for the conservation activities to the extent that would incentivize the service provider communities to carry out those activities. Therefore, the design of PES scheme for water supply project should consider the provisioning of the most preferred services or services with the most preferred attributes to attract water users, while ensuring that water users contribute their maximum willingness to pay (WTP). Table 23.1 provides a summary of the most preferred attributes selected by the water users of the three schemes, during the focus groups discussion and verified by local water resource experts.

The selected attributes suggest that water users in these cities are concerned with the quality as well as the quantity of available water, the distribution system, protection of water sources, and how the payment for water sources protection will be utilized. The relative importance of these issues varies by location and the existing water storage capacity of the households. For instance, residents of Dhankuta and Dasharath Chand municipalities are concerned about the drinking water distribution system and the regularity of supply. These municipalities are semi-urban by nature, and most of the houses do not have water storage tanks. Households in these municipalities are more concerned about regular water supply each day as they do not want

Table 23.1 Attributes selected by water users

Attributes	Municipality		
	Dharan	Dhankuta	Dasharath Chand
Water quality	√		√
Water quantity	√		√
Distribution system		√	√
Protection of land from erosion	√	√	
Budget allocation for water source protection		√	

Source Bhatta et al. (2018) and Rai et al. (2017, 2018)

to bear additional costs of purchasing or constructing water tanks for storing water. But in the case of Dharan, where most of the housing units have in-built water storage tanks, the main concern is the quantity/quality of water rather than the regularity.

The users also consider local conditions and the source of water supply while evaluating the water management programmes. Therefore, the selected attributes indicate that in Dharan where it is a part of the watershed, residents, have considered PES as a part of watershed management and the protection of land from erosion and water quality as two separate attributes. Dharan is one of the most affected towns by water-borne diseases due to open defecation in the upstream areas (Pant et al., 2016). The residents feel water quality can be improved through toilet construction to curb open defecation in upstream settlements, reduction in open grazing, and changes in the use of pesticides and fertilizers in the upstream farming. In addition, in Dhankuta, landslides during the rainy season affect the water quality, making protection of land from erosion and budgetary allocation for protection a key requirement. In Dasharath Chand, the water source is in the protected forest. Though the quality of water also gets affected due to sanitation and agriculture-related activities of water source households, the major concern here is that of quality, quantity and regularity of water supply with little concern about conservation of the watershed.

The selected attributes also show how consumers are concerned about the fund allocation for water source management activities. In Dhankuta, the water management committee charges each household NPR[1] 15 per month as watershed management fee, which amounts to around NPR 0.56 million per year. However, Dhankuta water users observed that the money paid to the water source community was being spent on infrastructure development, rather than water source protection. Noting that this may have negative impact on the upstream communities' willingness to undertake activities related to improving the quantity and quality of water the Dhankuta water users' main concern was the allocation of funds, particularly for water source protection. This is important since the conditionality of the PES scheme is to secure the flow of ecosystem services from the service providers (Wunder, 2005).

The preference for water source management also varies with the size of income and family. The importance given to water source management, and WTP increase with income and size of the family. However, the effect of gender on PES schemes is context specific. In Dhankuta, more Female respondents have prioritized improved watershed conditions compared to their male counterparts, while it is the opposite in Dharan and Dasharath Chand. Such context specific heterogeneity in gender response requires further examination.

---

[1] NPR is Nepali currency (the average exchange rate in 2018 was approximately USD 1 = NPR 108).

23.4   Water Source community's Preferences

Since PES is a voluntary agreement between water service users and providers, it is equally important to understand the preferences of the water source community (service providers) while designing a functional mechanism (Nyongesa et al. 2016). The activities identified by the water source community for watershed management should be the basis for determining the budget required to implement watershed management activities. In addition, it is also imperative to assess whether the upstream/water source community's preferences meet the expectations of water users for designing a workable PES mechanism (To et al., 2012). Therefore, the change in the behaviour or activities of the water source communities is related to the maintenance or improvement of the flow of ecosystem services (water supply) as expected by water user communities.

In all three case studies, the upstream/water source communities have identified activities that would support watershed management and also shown their willingness to participate in such activities though the type of activity has been determined by specific local contexts (Table 23.2). Most of the costs in managing the watersheds cost such as toilet construction, and providing piped water to the upstream households would be incurred as initial one-time.

Most of the identified activities are either related to addressing the land degradation due to agriculture or improving sanitation facilities in the upstream/ water source communities. In the study areas, the households in the vicinity of water sources rely heavily on conventional farming practices particularly farming on steep slopes. The farming practices are transitioning towards the use of modern inputs such as chemical fertilizers and improved seeds (Nautiyal et al., 2007). For instance, the use of chemical fertilizer has increased from 37 to 71% within five years in the upstream watershed from where Dhankuta municipality gets its drinking water. Such an increased use of chemical fertilizers around the water sources increases the chance of polluting the drinking water. Therefore, improvement in agricultural practices is one of the strategies that water source communities have proposed, which is acceptable to the water user community. Improvement of agricultural land can have a two-fold advantage—improved livelihood and better natural resource management. Climate resilient agriculture practices can contribute to livelihoods and improvement

Table 23.2 Activities proposed by water source communities

Dharan	Dhankuta	Dasharath Chand
• Grazing management • Toilet construction • Landslide protection • Agriculture improvement • Riparian buffers	• Seeds distribution • Irrigation facilities • Training programmes • Agriculture feeder roads • Land preparation	• Toilet construction • Grazing management • Off-season vegetation farming • Training on non-timber forest products collection/marketing • Provision of piped water in the village

Source Bhatta et al. (2018) and Rai et al. (2017, 2018)

of water quality, while enhancing communities' resilience to climate change (Rai et al., 2019b).

Improved sanitation through toilet construction is another issue identified by the water source communities of Dharan and Dasharath Chand. In Dharan, open defecation is one of the key factors of water contamination since many upstream households do not have toilets. In Dasharath Chand municipality, local tradition prevents women from using family toilets during menstruation compelling them to go outside for defecation and urination and disposal of their used sanitary pads in streams. Similarly, grazing management is also a common issue in the watersheds of these two municipalities.

Upstream/water source households expressed concern about access to the existing community development fund as local elites often tend to control the community development funds. The situation is similar in the PES schemes too and could be counterproductive in the long -run as the marginalized communities who are excluded from access to the funds may not cooperate for maintaining and managing the watersheds (To et al., 2012). The majority of water source households in Dharan and Dhankuta prefer receiving in-kind support to cash to reduce the possibility of elite capture and potential corruption. A PES pilot project in Dasharath Chand shows that when the water source households received cash, and they established a revolving fund and made a provision to provide soft loan rather than using the fund as a grant.

23.5 PES Mechanism Design

In order to make PES financially feasible, the estimated WTP of water users should outweigh the cost of implementing water source management activities. However, the social opportunity costs of water source communities and the WTP amount are context specific. Since PES is considered as a voluntary mechanism, it is important to understand whether PES could be financially feasible at community level. If not, then it is imperative to find another way to motivate the water source community for conserving the watershed and help maintain the quality and quantity of drinking water supply.

There are three types of funding mechanisms in PES: users financed, government financed and compliance (Salzman et al., 2018). Our case studies are focussed on the users' financed PES scheme. However, the estimated WTP and water source management costs indicate that user' financed PES schemes are not financially feasible where the number of users is small (Table 23.3). In this context, small towns (water user communities) require external support from the government to make PES financially feasible. The government support could be in terms of projects rather than cash. For instance, the forest department can support reforestation activities, while the agriculture department can provide agricultural extension (training and technical) support for sustainable and environmentally friendly agricultural practices.

Table 23.3 shows annual WTP and water source management costs in three case study sites. In the case of Dhankuta, the lower limit of users' WTP is the existing

Table 23.3 Annual WTP and cost of water source management (USD)

	Dharan[a]	Dhankuta[b]	Dasharath Chand[b]
WTP of water users	80,000–100,000	5266–224,299	5119
Water source management cost	46,632–49,369	55,514	4505–10,988

Source Bhatta et al. (2018) and Rai et al. (2017, 2018)
Note 1USD = NPR 108[a] and NPR 107[b]

water source protection fee (NPR 15/household/month) and upper limit is their WTP for the improved services. However, the water users are not ready to pay additional fee unless there is an increase in the water supply. On the other hand, the water source community will not invest in watershed management because of their expectation that an increased demand for drinking water in the future would get them a better deal. There is also uncertainty regarding the expected improvement in ecosystem services from the change in land use (Mátyás & Sun, 2014). This is because climate change also affects the flow of ecosystem services (Fu et al., 2017).

This uncertainty about anticipated improvement of ecosystem services and the resulting hesitation of water users to make full payment could be a big obstacle in making an agreement between the two sides. Hence, designing a PES scheme based on the output of ecosystem services may not create a trustworthy environment between service providers and service users. Therefore, payment should be based on the input or activities carried out by water source households (Hejnowicz et al., 2014). We called these schemes as Incentive payment for ecosystem services (IPES).

In this context, a reliable institution is needed to facilitate the negotiation. In the IPES scheme, identifying the right intermediary institution at the initial stage is very crucial for building trust and confidence between service providers and service users (Corbera et al., 2009). In the study sites, the local stakeholders suggested that a tripartite institution led by the local authority of the user community is a requisite to implement the IPES scheme successfully (Bhatta et al., 2018; Rai et al., 2018). The tripartite committee—comprising of service providers, service users and the local authority could facilitate the payment from the users to the service providers and also monitor the activities of the water source community to ensure that the community carries out environment friendly activities as per the agreement (Bhatta et al., 2018). This mechanism increases the trust between ecosystem services users and service provides while improving both horizontal and vertical coordination among multiple stakeholders.

23.6 Conclusion

The three PES schemes for drinking water projects discussed in this chapter suggest that such schemes can be an effective tool in making drinking water projects sustainable where there is a clear distinction between water users and providers. For a sustainable system that avoids conflict, which may arise due to uncertainties resulting from climate change, it is suggested to provide incentive payment for the ecosystem managers based on their activities.

These cases studies clearly indicate that IPES schemes provide both immediate and long-run benefits to both sides. The immediate benefit for water users is the minimization of the potential risk of an obstruction of the water supply by the water source community. In the long-run, such a scheme contributes to uplift the livelihood of water source households through incentive payments which can be invested in income-generating activities and ensures the sustainable supply of the quality and quantity of water to water users. In addition, IPES considered in the three case studies also focuses on improving the sanitation infrastructure (toilets construction and piped water supply) in the water source community, which not only contributes to improve water quality for water users but also improves the health of water source households.

These schemes are clear examples of how IPES can be managed at the community level in a sustainable manner and also generate required resources for providing incentives to the water source community for protecting watersheds. Users' payment is sufficient to cover the IPES scheme in the area, where the size of service users is large. In the case of small communities where service users are fewer, government programmes should be designed to support watershed management activities. Similarly, the experience of these schemes suggests that designing an IPES scheme is a rigorous process as it requires information on biodiversity, land-use pattern, hydrology and economics with intensive dialogue between service providers and users. Therefore, the IPES design should be supported by technical experts (Asquith et al., 2007). The potentially high cost of this service can, however, be minimized if embedded with the initial environmental impact assessment process of the drinking water project. Based on the lessons from the three cases, we recommend designing the payment schemes based on the inputs of watershed management activities to create a trustworthy environment between service providers and consumers.

An additional aspect of IPES is the tripartite institution to implement the scheme at the community level. Since local authorities are responsible for managing drinking water projects and also for the conservation of the local environment, their leadership for coordinating local stakeholders would be readily acceptable. In addition, such a coordinating role of the local authority provides assurance of fund flow from users to service providers; and also, the implementation of environmental friendly activities in the water source area. Last but not least, IPES schemes are context specific since they rely on the preferences of service users and providers and also the hydrology of the watersheds. Therefore, carefully designing the scheme with the active involvement of three sides (local authority, service providers and service users) is a necessary condition to make an IPES scheme implementable and sustainable.

References

Alston, L. J., Andersson, K., & Smith, S. M. (2013). Payment for environmental services: Hypotheses and evidence. *National Bureau of Economic Research, 5*(1), 139–159.

Asian Development Bank. (2011). *South Asia project brief: Nepal Small Towns water supply and sanitation project.* ADB Department, Kathmandu, Nepal.

Asquith, N. M., Vargas, M. T., & Wunder, S. (2007). Selling two environmental services: In-kind payments for bird habitat and watershed protection in Los Negros, Bolivia. *Ecological Economics, 65*(4), 675–684.

Ben-Akiva, M., & Lerman, S. (1985). *Discrete choice analysis: Theory and application to travel demand.* MIT Press.

Bharti, N., Khandekar, N., Sengupta, P., Bhadwal, S., & Kochhar, I. (2020). Dynamics of urban water supply management of two Himalayan towns in India. *Water Policy, 22*(S1), 65–89.

Bhatta, L. D., Khadgi, A., Rai, R. K., Tamang, B., Timalsina, K., & Wahid, S. (2018). Designing community-based payment scheme for ecosystem services: A case from Koshi Hills, Nepal. *Environment, Development and Sustainability, 20*(4), 1831–1848.

Bulte, E. H., Lipper, L., Stringer, R., & Zilberman, D. (2008). Payments for ecosystem services and poverty reduction: Concepts, issues, and empirical perspectives. *Environment and Development Economics, 13*(03), 245–254.

Choi, I.-C., Shin, H.-J., Nguyen, T. T., & Tenhunen, J. (2017). Water policy reforms in South Korea: A historical review and ongoing challenges for sustainable water governance and management. *Water, 9*(9), 717.

Corbera, E., Soberanis, C., & Brown, K. (2009). Institutional dimensions of payments for ecosystem services: An analysis of Mexico's carbon forestry programme. *Ecological Economics, 68*(3), 743–761.

Engel, S., Pagiola, S., & Wunder, S. (2008). Designing payments for environmental services in theory and practice: An overview of the issues. *Ecological Economics, 65*(4), 663–674.

Fu, Q., Li, B., Hou, Y., Bi, X., & Zhang, X. (2017). Effects of land use and climate change on ecosystem services in Central Asia's arid regions: A case study in Altay Prefecture, China. *Science of the Total Environment, 607*, 633–646.

Hejnowicz, A., Raffaelli, D., Rudd, M., & White, P. (2014). Evaluating the outcomes of payments for ecosystem services programmes using a capital asset framework. *Ecosystem Services, 9*, 83–97.

Kattel, R. R., & Nepal, M. (2021). Rainwater harvesting and rural livelihoods in Nepal. In A. K. E. Haque, P. Mukhopadhyay, M. Nepal, & M. R. Shammin (Eds.), *Climate change and community resilience: Insights from South Asia.* Springer Nature.

Kosoy, N., Martinez-Tuna, M., Muradian, R., & Martinez-Alier, J. (2007). Payments for environmental services in watersheds: Insights from a comparative study of three cases in Central America. *Ecological Economics, 61*(2–3), 446–455.

Mátyás, C., & Sun, G. (2014). Forests in a water limited world under climate change. *Environmental Research Letters, 9*(8), 085001.

Nautiyal, S., Kaechele, H., Rao, K. S., Maikhuri, R. K., & Saxena, K. G. (2007). Energy and economic analysis of traditional versus introduced crops cultivation in the mountains of the Indian Himalayas: A case study. *Energy, 32*(12), 2321–2335.

Nyongesa, J. M., Bett, H. K., Lagat, J. K., & Ayuya, O. I. (2016). Estimating farmers' stated willingness to accept pay for ecosystem services: Case of Lake Naivasha watershed payment for ecosystem services scheme-Kenya. *Ecological Processes, 5*(1), 1–15.

Ojha, H., Neupane, K. R., Pandey, C. L., Singh, V., Bajracharya, R., & Dahal, N. (2020). Scarcity Amidst Plenty: Lower Himalayan Cities struggling for water security. *Water, 12*(2), 567.

Pant, N. D., Poudyal, N., & Bhattacharya, S. K. (2016). Bacteriological quality of bottled drinking water versus municipal tap water in Dharan municipality, Nepal. *Journal of Health, Population and Nutrition, 35*(1), 1–6.

Rai, R. K., Bhatta, L. D., Dahal, B., Rai, B. S., & Wahid, S. M. (2019a). Determining community preferences to manage conflicts in small hydropower projects in Nepal. *Sustainable Water Resources Management, 5*(3), 1103–1114.

Rai, R. K., Nepal, M., Bhatta, L. D., Das, S., Khadayat, M. S., Somanathan, E., & Baral, K. (2017). Ensuring water availability to water users through incentive payment for ecosystem services scheme: A case study in a Small Hilly Town of Nepal. *Water Economics and Policy, 05*(04), 1850002.

Rai, R.K., Neupane, K. R., Bajracharya, R. M., Dahal, N., Shrestha, S., & Devkota, K. (2019b). Economics of climate adaptive water management practices in Nepal. *Heliyon, 5*(5), e01668.

Rai, R. K., Shyamsundar, P., Nepal, M., & Bhatta, L. D. (2015). Differences in demand for watershed services: Understanding preferences through a choice experiment in the Koshi Basin of Nepal. *Ecological Economics, 119*(88), 274–283.

Rai, R. K., Shyamsundar, P., Nepal, M., & Bhatta, L. D. (2018). Financing watershed services in the foothills of the Himalayas. *Water, 10*(7), 965.

Rai, S. C., & Rai, S. (2019). Impact of urbanization on portable water in Sikkim, Himalayas. *Political Economy Journal of India, 28*(1–2), 82.

Salzman, J., Bennett, G., Carroll, N., Goldstein, A., & Jenkins, M. (2018). The global status and trends of payments for ecosystem services. *Nature Sustainability, 1*(3), 136–144.

Singh, V., & Pandey, A. (2020). Urban water resilience in Hindu Kush Himalaya: Issues, challenges and way forward. *Water Policy, 22*(S1), 33–45.

Tamang, L., Chhetri, A., & Chhetri, A. (2020). Sustaining local water sources: The need for sustainable water management in the Hill Towns of the Eastern Himalayas. *Water Management in South Asia* (pp. 123–131). Springer.

To, P. X., Dressler, W. H., Mahanty, S., Pham, T. T., & Zingerli, C. (2012). The prospects for payment for ecosystem services (PES) in Vietnam: A look at three payment schemes. *Human Ecology, 40*(2), 237–249.

Wunder, S. (2005). *Payments for environmental services: Some nuts and bolts* (Vol. 42). CIFOR, Jakarta, Indonesia.

Chapter 24
Waste Segregation at Source: A Strategy to Reduce Waterlogging in Sylhet

Muntaha Rakib, Nabila Hye, and A. K. Enamul Haque

Key Messages

- Cities around the world have higher risk of waterlogging due to climate change
- Managing city wastes can help against the threat of waterlogging which would help to build resilience for cities in developing countries
- Increasing awareness by targeting women is the best strategy for ensuring at-home segregation to reduce the volume of solid waste
- There is a need to consider motivational approaches in campaigning that would work for women.

24.1 Introduction

Waste matters not only for its volume—generated due to urbanization, population pressure and economic development—but also for the fact that the rate of generation outstrips the ability of city authorities to manage and recycle (Banerjee & Mitra, 2013; Kien, 2018). The poor waste-management process in urban areas has a harmful impact on all elements of environment and human health (Bhalla et al., 2011; Rathi, 2006). If the waste is not managed properly, then it may contaminate the groundwater aquifer (Vasanthi et al., 2008). Failure to remove waste appropriately

M. Rakib (✉)
Department of Economics, Shahjalal University of Science and Technology, Sylhet, Bangladesh
e-mail: muntaha_rakib@yahoo.com

N. Hye
Asian Center for Development, Sylhet, Bangladesh
e-mail: nabila.hye@gmail.com

A. K. E. Haque
Department of Economics, East West University, Dhaka, Bangladesh
e-mail: akehaque@gmail.com

© The Author(s) 2022
Haque et al. (eds.), *Climate Change and Community Resilience*,
https://doi.org/10.1007/978-981-16-0680-9_24

may lead to waterlogging and increased climate-induced urban flooding (Brand & Spencer, 2020; Lamond et al., 2012; Pervin et al., 2020). For example, un-segregated hazardous waste may cause contamination; medical waste may cause various infections; unattended waste may create health hazards such as water-borne diseases; indiscriminate throwing of waste may cause drainage clogging and increase the intensity of floods and waterlogging (Kalina, 2020; Lamond et al., 2012). On top of this, if a city is vulnerable to climate-induced increased precipitation, it might lead to the failure of the adaptation strategy taken against urban flooding. Adaptation and resilience in urban waste management are recommended to reduce the risk of climate-induced flood and other disasters (Greenwalt et al., 2018; Kalina, 2020; Pervin et al., 2020; Phonphoton & Pharino, 2019). Nepal et al. (2022, Chap. 21 of this volume) have shown households' willingness to pay for better waste management practices. Sharma et al., (2022, Chap. 22 of this volume) have shown possibilities of engaging community-based organizations to make people aware of climate change-related issues in urban areas. Given this, it is important to integrate the waste management process as a part of the adaptation policy.

It is widely acknowledged that public participation inclusive of all major stakeholders is a key prerequisite for efficient waste management (Buckingham et al., 2005; Adebo and Ajewole, 2012; Plavsic, 2013; Vineeshiya & Mahees, 2016; Kien, 2018; Buckingham, 2020). The bulk of municipal solid waste (MSW) in developing countries is kitchen waste from households. Many households dispose of all waste without separating hazardous waste such as, broken glass, electric wastes, cleaning chemicals, batteries, which may be harmful to human health and the environment. Women play the primary role in waste management as part of their daily household chores. However, city managers are mostly men and women, despite being major stakeholders in generating and managing wastes in their homes, are often not consulted, resulting in inefficiency in waste management. Research findings suggest that household waste disposal is seen as a women's responsibility. Studies also find that waste segregation and disposal systems can be made efficient by mainstreaming gender within the system (Al-Khatib et al., 2009; Buckingham, 2020).

Men and women often put different precedence on solid waste management; they also probably conceptualize and produce waste differently within their household (Buckingham, 2020; Plavsic, 2013). Their allocation of time on waste management also varies, and they have different choices of waste disposal and different priorities on waste-management strategies. Moreover, men and women attach different values to the environment (Adebo and Ajewole, 2012). Research has also shown that gender mainstreaming positively affects household recycling behaviour (Babaei et al., 2015). Sustainable city waste management requires involvement and participation of women throughout the decision-making process, inclusion of gender-specific design and approaches in the information, education and awareness materials (dos Muchangos & Vaughter, 2019).

Waste is also a resource if there is a market for recycling or reusing waste. Once it is known, women might be interested in contributing to the family income, and this could be a game-changer in waste management.

Since the waste disposal system requires collection, segregation and disposal at a dumpsite, it is often expensive for city corporations. For example, in Kolkata, one of the largest Indian cities, the Municipal Corporation usually spends 70–75% of its budget on collection, 25–30% on transportation, and 5% on disposal of municipal waste (Chattopadhyay et al., 2009). Most communities in South Asian cities do not have sanitary landfill sites, and so, there is a high risk of groundwater contamination. Improving efficiency in waste management might release financial resources to be used for developing better dumpsites.

An efficient waste-management system requires reusing, recycling and reducing the volume of waste at the source (Otitoju and Seng, 2014; Al-Khateeb et al., 2017). This reduces the burden of carrying the waste to the dumpsites. Cities in developing countries rarely segregate wastes at source. In many developing countries, however, dry waste such as paper, metal, bottles and plastic materials can be sold in the recycling markets. Composting might also allow city corporations to spend more resources on proper disposal of hazardous wastes which are detrimental to the environment. To implement any composting scheme, there is a need to engage the stakeholders who generate waste, and the first strategy is to segregate waste at source (Otitoju and Seng, 2014) into recyclable, reusable and the refuse and segregating hazardous wastes. At each step, women and men have distinct roles to play (Scheinberg et al., 1999).

There are various methods of composting including several mechanisms to initiate a process for recycling and reuse of non-biodegradable waste. The challenge is to create an innovative mechanism to ensure that communities are engaged in the process and come forward to make a change in their usual practices for disposal of waste at home. There is a need to investigate the current practice of waste management to learn about the attitude towards it and the level of awareness. This will help to design a community-engaged and efficient waste management for the city.

In this research, we worked with the Sylhet City Corporation (SCC) to find a solution to urban solid waste management by engaging with women and encouraging them to sort, separate and dispose of their daily household waste in an orderly manner to promote recycling, reuse and composting. The experience we narrate here helped the city corporation learn about a more efficient system of waste management. It also demonstrated that engaging women in the overall architecture of managing waste matters.

Section 24.2 shows the current state of city waste management, the role of women, and their attitude towards waste disposal. Section 24.3 shows how we have designed a strategy of engaging women to segregate waste at source and how it was implemented. Section 24.4 shows the results.

24.2 Waste-Management System in Sylhet City Corporation

Sylhet City with a population of more than half a million has 27 Wards (smaller administrative sub-units of a municipality) which generate between 250 and 300 tonnes of solid waste a day. The Sylhet City Corporation (SCC) is responsible for managing solid waste in the city. Management of solid waste is done in two steps. The door-to-door collection or the primary collection is organized by the local Clubs who are elected by a community of 100–200 houses. Clubs are community-based voluntary organizations, not connected to the city corporation, but the councillors recognize them as an integral part of the community. Households pay a monthly fee for the daily collection of waste, and the waste collectors take the waste to a designated secondary collection centre. The city dump trucks carry the household waste from the secondary collection centre to the dumpsite. This is shown in Fig. 24.1. The dumpsite is located on the outskirts of the city from where rag-pickers collect recyclable or reusable objects. While the process is very similar in other cities in Bangladesh, the difference is that this in Sylhet the collection is organized by local clubs, whereas non-government organizations (NGOs) or civil society organizations (CSOs) manage the door-to-door collection system in other cities.

24.2.1 Role of the Clubs

Homeowners in a lane or several by lanes form a local voluntary association by organizing a committee for two years. The Club provides a manually operated three-wheeler rickshaw van and appoints the waste-collector. Households do not perform any segregation of wastes. However, metal, plastic, glass and paper wastes are bought and sold in the market through *feriwala* (street vendors) who come to the locality to buy these disposable wastes often on a weekly basis. Waste collectors often segregate

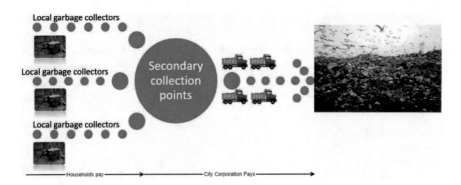

Fig. 24.1 Garbage management system in Sylhet. *Source* Author's conceptualisation

the recyclable items from the wastes and sell this to the *feriwala* while transporting it to the secondary collection point. In many cases, the waste collector takes the waste to a nearby open place beside a canal/drain, takes out the valuables, and then throws the rest into the canal causing a long-term problem for the municipality. The canals get filled up and the clogged drains cause waterlogging during the rainy season.

Estimates show that because of waste mismanagement and improper dumping, and the depth of the canals decreases by between 10 and 20 cm every year. This reduces the drainage flow and during heavy rains nearly 22% of the city area is flooded causing significant damage to properties and assets (Pervin et al., 2020). Clearly, providing un-segregated garbage to the waste collector created the problem as the valuables were the incentive for the collector to deviate from their daily activities. Considering this, researchers assumed that the risk of flooding would greatly reduce if households segregate at source.

24.2.2 A Women-Centric Approach

Significant improvement towards reducing the waterlogging in the city is possible if the current system of waste management in the Sylhet City Corporation—which follows the sequential course of production, collection and segregation and disposal—can be changed to production and segregation followed by collection and disposal. To do so, the role of women or a women-centric approach to waste management needs to be recognized and institutionalized. It is like shuffling the second card from the middle to the front of the deck as shown in Fig. 24.2 implying that the production points (household) also become the segregation points and are connected to the recycling markets, while the collectors are involved in composting the organic wastes.

The general solid waste-management system of Sylhet city is illustrated in A of Fig. 24.2, and the proposed alternative system is shown in B of Fig. 24.2. The

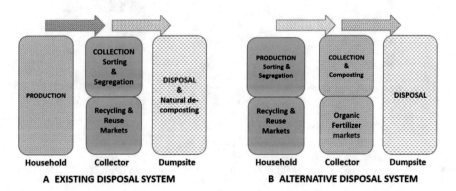

Fig. 24.2 Alternative waste-management systems. *Source* Author's conceptualisation

process, if implemented properly, connects the production and collection stages of household wastes in the city with the recycling and the organic fertilizer markets. It adds value and significantly reduces transportation needs for solid wastes to the dumpsites. The key for success in such transformation is to make the production point at the household into the sorting and segregation point as well.

24.3 Research Design

The research was organized into two phases. Phase I consisted of understanding the current state of waste management in the city including information on current practice and volume of waste generated at the household level. Phase II involved designing several awareness campaigns to influence women to segregate waste before disposal.

While planning solid waste management for cities, the average daily waste per person per day is assumed to be between 0.32 and 0.48 kg (Alamgir & Ahsan, 2007). This was verified using physical measurement of wastes generated per household per day. A small survey of 150 households was conducted for 21 days to measure (a) the composition and (b) volume of daily waste. It shows that daily waste is around 2.24 kg per day per household in traditional homes (one-storey buildings) which is around 0.45 kg per person per day. For multi-storied houses (between 2 and 6 floors), it is around 0.57 kg. However, the result of our study was surprising for the high-rise buildings (above 6 floors) where we observed that per day an apartment generates nearly 4.9 kg of solid waste which is equivalent to 0.98 kg per person per day, nearly double the estimate. Figure 24.3 shows the distribution of waste by different types of houses.

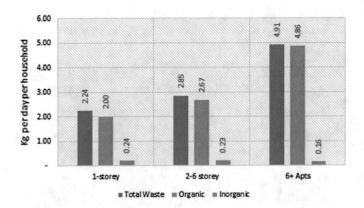

Fig. 24.3 Daily waste per household per day in Sylhet. *Source* Asian Center for Development Survey 2017

The survey, therefore, provides evidence on the volume of solid waste generated daily in the city and shows that the volume is going to rise rapidly as many of the newly constructed houses are high-rise apartments. As such, it also illustrates the fact that managing solid waste in urban areas will be an important strategy in developing countries where the disposal is haphazard and public littering is common leading to clogging of the drainage system and increasing the risk of disasters during any climate-induced event.

The waste from these households was screened and weighed in two parts—(a) compostable and (b) non-compostable. It was found that nearly 92% of the waste is compostable and is mostly organic kitchen waste. As such, the research team decided to study the following strategies for in-house waste management.

- How to promote the idea of the ready market for recyclable waste.
- How to engage household members, especially women to separate kitchen waste which can be composted.
- How to provide community awareness particularly to clubs about segregation and composting including the fact that local composting is a much better option than to transport the solid waste to the dumpsite which is costly and has been ineffective so far.

24.4 The Awareness Programme

To design an effective awareness programme involving women, the research team decided to survey households to understand their current waste disposal behaviour in 2017 and in 2018. Three city wards were selected for the door-to-door campaign on solid waste management. Based on the baseline survey, each household was given two doses of awareness using a door-to-door campaign. The awareness programme is divided into three steps and is presented in Fig. 24.4.

In part I, workshops and seminars with city councillors and club representatives of all 27 wards in the SCC were arranged. The workshop was on the benefits of improved solid waste management to build resilience against flash floods and waterlogging.

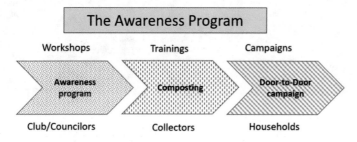

Fig. 24.4 The flowchart of the awareness programme. *Source* Author's conceptualization

In part II, the waste collectors in the three selected wards were trained on composting the waste and on the benefits to the city and their own health. They were trained on how to make compost out of kitchen waste after segregation. Composts produced from the waste were for them to sell in the market as organic fertilizer.

In part III, a door-to-door campaign was conducted for 14 days, in two phases of 7 days each in three blocks of the three wards to motivate women in the house to separate their kitchen waste. During the campaign, volunteers went to every house and explained the motivation for segregation to the household heads and the most prominent women in the house, distributed campaign materials and also showed a small video on diseases caused by mosquitoes due to waterlogging and clogged drainage systems (campaign materials are illustrated in Picture 2). Figure 24.5 shows a few photos from workshops, training and a door-to-door campaign targeted at women.

Figure 24.6a shows health benefits of separating kitchen waste for the city and its potential impact on reducing waterlogging. It also shows two ways of separating household wastes: degradable and non-degradable, and finally gives a pictorial illustration of making composts to motivate the household to separate kitchen waste. Figure 24.6b gives a list of kitchen waste materials which can be easily composted if separated at source. It also explains the benefits of using composts as a substitute for chemical fertilizer.

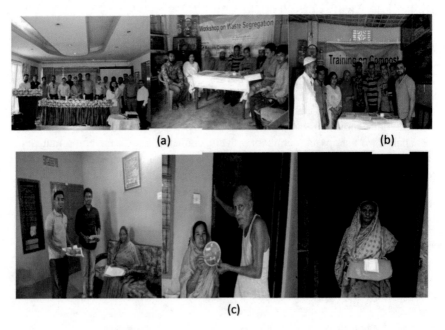

(a)   (b)

(c)

Fig. 24.5 **a** Various workshops and seminars with city councillors and clubs. **b** Training on compost to garbage collectors. **c** Door-to-door awareness campaign to women. *Photo credits* The Asian Center for Development Research Team

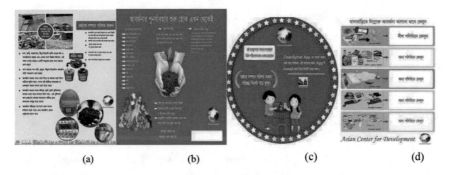

Fig. 24.6 Awareness materials. *Photo credit* Posters produced by the authors for the Asian Center for Development

Furthermore, it also gives a link to a mobile app (produced as part of the campaign) to report littering in the community for the City Corporation to monitor and remove. Figure 24.6c is a sticker distributed to every house for them to stick on their doors to motivate other members in the house to participate in segregation tasks including the link to the app to report littering in the community and to register "Clean Sylhet" campaign in their mind. Figure 24.6d is a sticker for kitchen doors to define segregation of wastes for them. This is to ensure that every member of the house who is in the kitchen remembers how to dispose of their wastes separately. In addition, the volunteers showed a short video on how people are separating household waste in other countries to members present in the house at the time of their visits.

24.4.1 Campaign Game Plan

The campaign designed to promote separation of household wastes at home was targetted at the women of the household and based on four motivational factors: (i) social persuasion; (ii) moral persuasion; (iii) economic incentives and (iv) social recognition.

24.4.2 Mechanism of Social Persuasion

In order to persuade the households to segregate their wastes, we not only sensitized them to separate their kitchen waste, sanitary/medical waste and hazardous waste but also distributed blue coloured W-shaped poly bags of a pack of 60 for two months (each bag was of $12'' \times 9''$ in. in dimension with an average weight carrying capacity of 5 kg). A set of households were also given black-coloured poly bags for separating hazardous wastes (broken glass, battery, electrical and electronic gadget and light bulb).

Since Sylhet is a densely populated urban area, household-based composting seemed to be unfeasible, so we provided community-based compost bins for segregated kitchen wastes and involved local garbage collectors and the club to organize composting.

24.4.3 Mechanism for Moral Persuasion

To motivate households to properly dispose their wastes and to separate kitchen wastes, we (a) developed the "Clean Sylhet" mobile app which can be used by any person in the community to report littering in the community to the City Council; (b) showed them a motivational video on how other communities around the world participate in such activities and separate their wastes and how it benefits the community in terms of reduced health hazards.

24.4.4 Mechanism for Economic Incentives

The research team introduced two mechanisms to develop economic incentives for households and for the garbage collectors. First, by selling the non-degradable wastes in the waste-recycling market households can earn extra income. Garbage collectors could also earn additional income by selling the compost. Our experiment shows that each waste bin can produce nearly 51 kg of good quality compost a month providing the garbage collector with an earning of nearly 15 USD or 1275 taka a month.

24.4.5 Mechanism for Social Recognition

The campaign introduced a process of social recognition by announcing Green Award for different stakeholders in the community. The categories are: "Green Home Award" for three best households, "Green Club Award" for three best clubs and "Clean Sylhet Award" for three waste collectors who actively participated in the activities to make the city clean. The award winners were nominated based on specific criteria that they had to fulfil.

24.5 Lessons Learned

24.5.1 Did the Women-Centric Approach Work?

Our objectives were to understand whether our campaign was effective; whether households became socially responsible and separated their hazardous wastes; and whether they were motivated by the incentives and increased their participation to sell their waste.

Based on the four mechanisms of the campaign to motivate the women in the households, we find the following results (shown in Table 24.1). Comparing the before campaign and immediately after campaign scenarios among the households, Table 24.1 shows that our women-centric approach to pursue households to separate their wastes yielded positive outcomes. First, there has been a 484% increase among participating households in separating kitchen wastes. 60 days long daily monitoring of garbage collection data shows that 72.8% of the households participated regularly and separated their kitchen waste for community-based composting and 6.9% participated in home-based composting. Second, separation of hazardous waste increased by 33% and that of sanitary wastes by 51%. There was no direct benefit to the households for doing this, but it was one of the objectives of the campaign and also used as a criterion for the Green Award declared by the City Mayor. Third, the baseline survey revealed that nearly 59% households used to sell their paper waste, 57% sell their plastic waste, 62% separate and sell their glass waste, 38% sell their metal waste and 58% sell their polybag waste. However, they do so over a long period of

Table 24.1 Waste segregation, reusing, recycling and composting by types of waste

Household activities to manage waste	Before campaign	After campaign	Percentage change (%)	Level of significance
Separate kitchen waste for composting	12.45	72.8	484	***
Dispose sanitary waste separately	30.71	46.32	51	***
Separate Hazardous waste	64.32	85.71	33	***
Reuse plastic/shopping bag	85.06	92.21	8	**
Sell waste	45	47	4	
Compost in compound	6.22	6.93	11	
Sample size (n)	241	231		

Source Field survey 2017 and 2018
Note *indicates the change is statistically significant at 10%, **means 5% and ***means 1%. The sample sizes before and after campaigns are different because, within the span of 2 months during our campaign, 10 households left the locality and moved outside. Numbers indicate per cent of households

time. In our survey, we further asked whether they sold their recyclable wastes in the past two months before and after the campaign. Table 24.1 shows that during the past two months before and after the campaign there was no significant change in their behaviour implying that economic incentives did not change their behaviour significantly. In terms of home composting, the increase was also not significant.

24.5.2 Was There Any Change in the In-House Participation in Garbage Disposal?

All members of the household were present at the time of the campaign by our volunteers. Our base-line survey results show that it is mostly women in the house who participate in the garbage management activities. Table 24.2 shows the role of women within the household in terms of waste management, disposal and compost making before and after the campaign. It shows that as a result of our campaign in the house, it was women who became more motivated than men, and so, there has been a 21% rise in the role of women to manage their waste properly. Women also took more interest in the disposal activities, and there has been a 14% rise in waste disposal activities by women. These are all statistically significant. However, as previously mentioned at-home, composting did not increase significantly.

Table 24.2 Activities of women in waste management

Activities of women	Before the campaign	After the campaign	Percentage change (%)	Level of significance
Waste management in the house by women	62.76	76.19	21	***
Waste disposal in the house by women	65.56	74.89	14	**
Waste to compost in the compound by women	46.67	50	7	
Sample size (n)	241	231		

Source Field survey 2017 and 2018
Note *indicates the change is statistically significant at 10%, **means 5% and ***means 1%. The sample size between before and after campaigns is different because within the span of 2 months during our campaign, 10 households left the locality and moved outside. Numbers indicate per cent of households

24.6 Conclusions

Cities around the world have higher risk of waterlogging due to climate change. Previous research has shown that without developing a proper waste-management policy, investment in drainage infrastructure is likely to fail. There is a need to develop a mechanism to reduce the risk of clogging of the city drainage system. This research was a follow-up to find strategies to increase household-level participation. Based on previous surveys, it was also observed that women are at the centre of waste management in a typical urban house in Bangladesh. Consequently, this research has used motivational approaches to motivate households to participate and dispose of their garbage in an orderly fashion. The team ran a campaign in selected localities for 14 days, two periods of 7 days each, in order to (a) pursue households to separate their kitchen wastes for community level composting, (b) make the city clean and smart given the climate risks.

Four motivational approaches were used: (a) social persuasion, (b) moral persuasion, (c) economic incentives and d) social recognition to motivate the city dwellers to separate their kitchen wastes. All household members present at the time in the household were motivated to participate in the activities, but the campaign also ensured that the main women in the house were present at the time of the campaign. Our results show that social and moral persuasion and the strategy of social recognition worked well, and it is the women members in the house who were more motivated by such campaigns. Women are not particularly attracted by the economic incentives shown in our campaign, maybe because the economic benefit is small compared to the income of the household. In terms of participation in managing household wastes, the campaign resulted in greater motivation in women to manage their wastes responsibly. Finally, in terms of social recognition, it was also evident that all the prize winners were women.

The study draws several conclusions. First, providing awareness by targeting women in the house to manage wastes and to at-home segregation is the best strategy for cities in developing countries to manage city wastes which would help build resilience against the threat of waterlogging. Second, while developing campaigns there is a need to consider motivational approaches that would work for women— these are—motivation to build a better society, motivation to create a safer locality and recognition of their work.

Finally, there is a social dividend to develop such resilience. Nearly, half of the global population lives in cities. Cities in developing countries are generally more vulnerable to climate change, and thus have greater adaptation needs. The study reveals that encouraging women to manage solid waste using community-based composting not only helps the cities to reduce their risks it also reduces the burden on cities to dispose of a huge volume of municipal solid wastes to garbage disposal sites. It makes the cities climate resilient and promotes the concept of sustainable and resilient cities which is one of the sustainable development goals.

Acknowledgements Financial support is gratefully acknowledged from the International Development Research Centre (IDRC), Ottawa, Canada (Grant #08283-001). The Research is jointly collaborated with South Asian Network for Development and Environmental Economics (SANDEE), Asian Center for Development (ACD), Institute of Water Modelling (IWM) and Sylhet City Corporation (SCC). The authors would like to acknowledge the contributions and cooperation of The City Mayor, Chief officials of SCC, Ward Councillors, Club members and all the respondents and enumerators.

References

Adebo G. M., & Ajewole, O. C. (2012). Gender and the Urban environment: Analysis of willingness to pay for waste management disposal in Ekiti-State, Nigeria. *American International Journal of Contemporary Research, 2*(5).

Alamgir, M., & Ahsan, A. (2007). Municipal solid waste and recovery potential: Bangladesh perspective. *Iranian Journal of Environmental Health, Science and Engineering, 4*(2), 4. ISSN: P-ISSN: 1735-1979.

Al-Khateeb, A. J., Al-Sari, M. I., Al-Khatib, I. A., & Anayah, F. (2017). Factors affecting the sustainability of solid waste management system—The case of Palestine. *Environmental Monitoring and Assessment, 189*(2), 93.

Al-Khatib, I. A., Arafat, H. A., Daoud, R., & Shwahneh, H. (2009). Enhanced solid waste management by understanding the effects of gender, income, marital status, and religious convictions on attitudes and practices related to street littering in Nablus-Palestinian territory. *Waste Management, 29*(1), 449–455.

Babaei, A. A., Alavi, N., Goudarzi, G., Teymouri, P., Ahmadi, K., & Rafiee, M. (2015). Household recycling knowledge, attitudes and practices towards solid waste management. *Resources, Conservation and Recycling, 102*, 94–100.

Banerjee, S., & Mitra, S. (2013). Radioactive and hospital waste management: A review. *International Journal of Latest Trends in Engineering and Technology (IJLTET), 3*(1), 275–282.

Bhalla, G., Kumar, A., & Bansal, A. (2011). Assessment of groundwater pollution near municipal solid waste landfill. *Asian Journal of Water, Environment and Pollution, 8*(1), 41–51.

Brand, J. H., & Spencer, K. L. (2020). Will flooding or erosion of historic landfills result in a significant release of soluble contaminants to the coastal zone? *The Science of the Total Environment, 724*, 138150. https://doi.org/10.1016/j.scitotenv.2020.138150

Buckingham, S. (2020). *Gender and environment*. Routledge.

Buckingham, S., Reeves, D., & Batchelor, A. (2005). Wasting women: The environmental justice of including women in municipal waste management. *Local Environment, 10*(4), 427–444. https://doi.org/10.1080/13549830500160974

Chattopadhyay, S., Dutta, A., & Ray, S. (2009). Municipal solid waste management in Kolkata, India—A review. *Waste Management (New York, N.Y.), 29*, 1449–1458. https://doi.org/10.1016/j.wasman.2008.08.030

dos Muchangos, L., & Vaughter, P. (2019). Gender mainstreaming in waste education programs: A conceptual framework. *Urban Science, 3*(29), 1–12. https://collections.unu.edu/view/UNU:7262.

Greenwalt, J., Raasakka, N., & Alverson, K. (2018). Chapter 12—Building urban resilience to address urbanization and climate change. In Z. Zommers & K. Alverson (Eds.), *Resilience* (pp. 151–164). Elsevier. https://doi.org/10.1016/B978-0-12-811891-7.00012-8

Kalina, M. (2020). Waste management in a more unequal world: Centring inequality in our waste and climate change discourse. *Local Environment, 25*(8). https://www.tandfonline.com/doi/abs/10.1080/13549839.2020.1801617

Kien, A. H. (2018). *A gender perspective of municipal solid waste generation and management in the city of Bamenda, Cameroon*. Langaa Rpcig.

Lamond, J., Bhattacharya, N., & Bloch, R. (2012). *The role of solid waste management as a response to urban flood risk in developing countries, a case study analysis* (pp. 193–204). https://doi.org/10.2495/FRIAR120161

Nepal, M., Bharadwaj, B., Nepal, A. K, Khadayat, M. S., Pervin, I. S., Rai, R. K., & Somanathan, E. (2022). Making urban waste management and drainage sustainable in Nepal. In A. K. E. Haque, P. Mukhopadhyay, M. Nepal, M. R. Shammin (Eds.), *Climate change and community resilience: Insights from South Asia* (pp. 325–338). Springer Nature.

Otitoju, T. A., & Seng, L. (2014). Municipal solid waste management: Household waste segregation in Kuching South City, Sarawak, Malaysia. *American Journal of Engineering Research, 10.*

Pervin, I. A., Rahman, S. M. M., Nepal, M., Haque, A. K. E., Karim, H., & Dhakal, G. (2020). Adapting to urban flooding: A case of two cities in South Asia. *Water Policy, 22*(S1), 162–188. https://doi.org/10.2166/wp.2019.174

Phonphoton, N., & Pharino, C. (2019). A system dynamics modeling to evaluate flooding impacts on municipal solid waste management services. *Waste Management, 87*, 525–536. https://doi.org/10.1016/j.wasman.2019.02.036

Plavsic, S. (2013). *An investigation of gender differences in pro-environmental attitudes and behaviors.*

Rathi, S. (2006). Alternative approaches for better municipal solid waste management in Mumbai, India. *Waste Management, 26*(10), 1192–1200.

Scheinberg, A., Muller, M., & Tasheva, E. L. (1999). Gender and waste. *UWEP Working Document, 12.*

Sharma, U., Brahmbhatt, B., & Panchal, H. N. (2022). Do community-based institutions spur climate adaptation in urban informal settlements in India? In A. K. E. Haque, P. Mukhopadhyay, M. Nepal, M. R. Shammin (Eds.), *Climate change and community resilience: Insights from South Asia* (pp. 339–356). Springer Nature.

Vasanthi, P., Kaliappan, S., & Srinivasaraghavan, R. (2008). Impact of poor solid waste management on ground water. *Environmental Monitoring and Assessment, 143*(1–3), 227–238. https://doi.org/10.1007/s10661-007-9971-0

Vineeshiya, M. N., & Mahees, M. T. M. (2016). *Gender perspective of community participation in solid waste management; A case of balangoda urban council*, Sri Lanka.

Part VI
Alternative Livelihood

Chapter 25
Community-Based Tourism as a Strategy for Building Climate Resilience in Bhutan

Ngawang Dendup, Kuenzang Tshering, and Jamyang Choda

Key Messages

- Climate change is affecting the rural households whose primary livelihood is agriculture.
- Community tourism can benefit rural households and provide alternative source of livelihood.
- Households from community that received community tourism program are more likely to have better household wealth (i.e. number of rooms and vehicle ownership).

25.1   Introduction

Bhutan is regarded as one of the most exclusive travel destinations with its reputation for authenticity, remoteness, well-protected cultural heritage and its pristine natural environment (Montes, 2019; RGoB, 2012; Rinzin et al., 2007; TCB, 2019; WWF, 2011). Tourism has been one of the highest revenue generators of the country over the past decades (NSB, 2019; TCB, 2020) and has grown consistently over the years. In addition, the tourism industry is one of the highest foreign currency earners for Bhutan (NCoB, 2016) and according to the annual Tourism Monitor Report 2019,

N. Dendup (✉)
Waseda University, Tokyo, Japan
e-mail: ngawangdendup@gmail.com

K. Tshering
Edith Cowan University, Perth, Australia
e-mail: k.tshering@ecu.edu.au

J. Choda
Perth, Australia
e-mail: jamyang@sherubtse.edu.bt

© The Author(s) 2022
Haque et al. (eds.), *Climate Change and Community Resilience*,
https://doi.org/10.1007/978-981-16-0680-9_25

the gross foreign exchange earnings from international and regional tourists was 88.6 million US dollars.

However, Bhutan's tourism sector is strictly regulated and tourists are required to adhere to the 'daily minimum package' set by the government (Dendup & Tshering, 2018; Montes, 2019; NCoB, 2016; Nyaupane & Timothy, 2010; RGoB, 2012; Rinzin et al., 2007; TCB, 2019). Tourism is only allowed to operate within a limited sphere of government facilities, designated hoteliers and tour operators including community-led tourism facilities such as 'home stays' and 'farm houses' with tourists required to book the trip via a local tour operator and no back-packing allowed. However, it is not evident if communities or households are benefitting from this. This chapter attempts to assess whether the benefits of tourism are reaching households and communities using the latest census data. The financial data related to the tourism industry is available for international tourists, generally categorized into international and regional, the latter comprising of tourists from India, Bangladesh and Maldives who are exempted from the 'minimum daily package'. However, in this chapter, the benefit accruing from tourism is assumed as cumulative from both local and international tourists.

To maintain its status as an exclusive travel destination and to safeguard its heritage, Bhutan strictly adheres to its tourism policy of 'high value-low impact tourism' based on its sustainable development concept of Gross National Happiness (TCB, 2020), despite potential threats to this model from increasing arrivals of both regional (Brunet et al., 2001; Teoh, 2016) and international tourists (Basnet, 2020; Montes & Kafley, 2019). Tourism in Bhutan has been founded on the principles of promoting the environment, preserving local cultural heritage and encouraging economically viable activities (Reinfeld, 2003; RGoB, 2012; TCB, 2019). The carrying capacity of the natural environment and cultural landscape has been a determining factor in shaping tourism policy. For example, even locals are prohibited from visiting many culturally significant sites in the higher altitude of Bhutan during certain times of the year, providing a fragile mountain ecosystem with time to regenerate. Such restrictions imposed across Bhutan under both mainstream conservation laws and customary practices have been very effective in reducing the impacts of tourists on local touristic hotspots like Aja Ney, Singye Dzong and many others. The effectiveness of local customary practices has been acknowledged in reducing pressure on rangelands (Moktan et al., 2008; Tshering et al., 2016) and found to be beneficial in maintaining biodiversity, especially the country's rich faunal diversity (WWF, 2011).

On the other hand, community-based tourism in Bhutan is also being promoted to reduce the vulnerability of the community to effects of climate change and economic shocks. In a subsistence farming economy like rural Bhutan, the majority of farmers are dependent on food crops that are dominated by rainfed agriculture. Every year farming communities lose crops to wind storms, wildlife, pest and erratic rainfall (MoAF, 2017), which could worsen with climate change. Recent policy has therefore ensured that every conservation project in Bhutan implements community-based tourism as a way of enhancing community resilience to reduce vulnerability to climate change. Besides tourism, there are other strategies like migration, alternative farming, crop diversification and also using mobile-based social networks that can also build resilience in rural communities (see similar experiences documented by Ghosh and Roy (2021, Chap. 26 of this volume, Gunathilaka & Samarakoon 2021, Chap. 27 of this volume, Shafeeqa & Abeyrathne 2022, Chap. 9 of this volume and Nazir & Lohano, 2022, Chap. 28 of this volume).

One of the biggest conservation projects, 'Bhutan for Life', which is under implementation and planned to continue till 2032, has identified numerous key performance indicators on promotion of ecotourism and nature-based business models in all the protected areas (BFLS, 2019). Similar approaches are visible in numerous development initiatives, particularly those being implemented within the protected areas of Bhutan, which cover more than 50% of the country. For instance, the National Sheep Breeding Farm has been supporting the farmers' group under Phobjikha and Gangtey (one of the study sites) for production of woollen handicrafts and products as a means of creating alternative income generation from the sale of products to tourists visiting the valley (BTFEC, 2019; The World Bank, 2019). Further, the potential of community-based tourism was also reported by Gurung and Seeland (2011) during an assessment on ecotourism impacts on three communities within the protected areas (Jigme Dorji National Park, Jigme Singye Wangchuck National Park, Phrumsingla National Park) of Bhutan, suggesting a potential for expansion of the ecotourism business model in the country.

The Department of Forest and Park Services, the government agency entrusted with conservation of natural biodiversity of the country, has also embraced the concept in order to promote incentive-based conservation initiatives. The Department promotes ecotourism, as a way of generating alternative income for communities in and around the protected area network system who have often had to forgo numerous economic opportunities under the strict conservation regulations (Penjore, 2008).

The significance of the tourism industry for promoting balanced regional development was also emphasized in the current five-year national development plan—12th Five Year Plan (2018–2023)—where a priority programme has been planned as a 'sustainable tourism development flagship programme' (GNHC, 2019). In general, community-based tourism and ecotourism in the context of Bhutan refer to a platform for the local community to generate economic benefits through offering their products to tourists (both national and international) that ranges from local community lifestyles and natural resources to culture. It is being run as a development programme that is expected to enhance social and cultural benefits to the locals as well as the tourist (Donny & Nor, 2012).

Key tourism products in community-based projects include homestays and farmhouses, traditional hot stone bath, local games like archery and darts, hiking with local guides, and local festivals (e.g. mushroom festival, rhododendron, highland). Some communities in touristic areas also earn income through offering porter and pony services during trekking by international tourists and pilgrimage by local tourists. In certain locations like Khoma village under Bumdeling Wildlife Sanctuary, promotion of the textile industry has been the key income source for the locals. In other areas, community groups have been formed for the management of medicinal plants, black-necked crane conservation, human-wildlife conflict and other issues (BTFEC, 2019; World Bank, 2019).

The managers of the protected area have been instrumental in developing infrastructures such as trekking routes and camping facilities at hot springs and culturally significant touristic sites with funding support from the central government, while the management has been mostly handed over to the local farmers' or youth groups. For example, since 2018, the guesthouse, campsites, and trekking routes built by Bumdeling Wildlife Sanctuary in and around the local pilgrimage hotspot Aja Ney have been handed over to the local community for management. The local farmers group of 16 households manages the facilities and makes a substantive income from it (BTFEC, 2019; World Bank, 2019).

The majority of community-based tourism initiatives in Bhutan are within the protected areas network, which includes parks, wildlife sanctuaries and the network of biological corridors. Any developmental activities are in accordance with the long-term park management plans, which are always within the framework of the government's planning processes. For the communities of Phobjikha and Gangtey sub-districts, which falls within a biological corridor, the community-based tourism was initiated by a local CSO, the Royal Society for Protection of Nature (RSPN) within the framework of overarching management policies set by the Department of Park and Forests Services, the ultimate authority for protected area management in Bhutan (RGoB, 2012).

25.2   Study Area: Bhutan Context

In this study, we consider Gangtey and Phobjikha sub-districts under Wangdiphodrang district, Chhokhor, Tang and Ura sub-districts under Bumthang district and Bumdeling sub-district under Tashiyangtse district as community tourism initiatives (treatment) have been introduced here for some time past and the benefits may not be visible in other areas where the initiatives have been introduced more recently.

The sub-districts that are considered treatment sub-districts are reported in Table 25.1 along with the number of tourist arrivals, district and national park or protected area they fall under.

The main attraction for international tourists in Bumdeling, Phobjika and Gangtey is the presence of the Black-neck crane, a charismatic species with significant spiritual and cultural value for the locals. The marshland of Gangtey and Phobjikha is one of the biggest winter roosting grounds for the migratory bird flying from the Tibetan Autonomous Region of China to Bhutan every winter. The riverine wetlands along the Kholong Chu River under Bumdeling sub-district are another vital winter roosting ground for the crane (RGoB, 2018; RSPN, 2014). These wetlands are also listed as internationally important wetlands and comprise two of the three RAMSAR sites in Bhutan. Other sub-districts like Chhokhor and Tang also receive some cranes in the winter. The engagement of the community, through incentive-based initiatives is vital for meeting conservation and development targets in such critical ecosystems which are vulnerable to climate change.

Table 25.1 Treatment sub-districts with number of tourist arrivals

S. No.	District	Sub-district	Number of international visitors[1] in 2019 in district	Protected area manager
1	Wangdue Phodrang	Phobjikha[2] and Gangtey	30,090	Biological corridor—8
3	Bumthang	Choekhor, Tang and Ura	30,580	Wangchuck Centennial National Park
6	Trashi Yangtse	Bumdeling[3]	1411	Bumdeling Wildlife Sanctuary

[1] These figures shows only international visitors in 2019 as per annual Tourism Monitor Report prepared by Tourism Council of Bhutan (TCB, 2020), thus it doesn't provide any information on number of local tourists. In places Phobjikha and Gangtey, income from local tourist can also be significant part of annual income for home-stays by hosting holidaymakers from capital city Thimphu. Some of the home-stays have also been hosting in-house workshops for smaller groups from numerous offices in Thimphu. Similarly, a significant number of local tourist on pilgrimage to Pemaling and holidaymakers to Bumdeling valley are expected to create significant impact on local income generation.

[2] Managed by Royal Society for Protection of Nature (RSPN), registered CSO in Bhutan. The area is also one of the three RAMSAR sites in Bhutan.

[3] Second RAMSAR site in Bhutan.

The majority of places that are promoted as tourist destinations in Bhutan fall under protected areas or biological corridors. In these areas, the benefits and restrictions are usually well communicated with the key stakeholders even at the level of individual households in order to gain their support. Participatory community planning is the first step, and the park managers actively engage with the local government. The plan is thoroughly discussed with all the key stakeholders, mainly those households willing to participate in ecotourism. These households are then provided technical and material support in terms of health and hygiene and hospitality management, in most cases, for improving basic washroom facilities to enable them to host tourists. The local community is expected to adhere to the restrictions under the Forest and Nature Conservation Act of Bhutan 1995. For example, in Phobjikha and Gangtey sub-districts, households are allowed to practice agriculture, but face restrictions on using pesticides and inorganic fertilizers. Park managers report that such community-led projects have been very successful in the management of the natural environment especially the fragile high altitude ecosystems which are not only often under extreme threat from climate change but also the least monitored due to remoteness. In some areas like Jigme Dorji National Park, Wangchuck Centennial National Park and Bumdeling Wildlife Sanctuary, the attitude of local communities towards conservation has significantly improved through such community-led projects (BTFEC, 2019; The World Bank, 2019). The promotion of sub-districts within national parks or biological corridors as eco-tourist destinations reduces pressure for development of infrastructure like roads and mainstream tourism industry development like hotels.

Key challenges for the management of wetlands of international significance, like Bhutan's RAMSAR sites such as the Phobjikha wetland, include agriculture activities like indiscriminate grazing, reclaiming of wetland areas for agriculture and the heavy use of chemical fertilizers, pesticides and herbicide. Wetlands along the tourist hiking trails were also reported to be under increasing pressure from the increasing number of people and the grazing of horses from the porter services. There were also localized cases of invasive species like water hyacinth taking over the wetland system in a freshwater lake at Samtegang, probably due to increasing sediment flow into the lake and warmer climate due to climate change. In Bumdeling RAMSAR site, the annual flash floods due to erratic monsoon rainfall have washed away huge areas of wetland (RGoB, 2018; Rinzin et al., 2007; RSPN, 2014). The wetlands face imminent threat from climate change, which is the biggest concern for the managers of the protected areas, who often lack appropriate data, knowledge and tools for reducing the vulnerability of the local communities. Anecdotal evidence from the Ecotourism Section under the Nature Conservation Division of Department of Forest and Park Services recommends community-based tourism as one of the most effective tools in promotion of incentive-based conservation in Bhutan and for reducing the economic vulnerability of the local communities to the negative effects of climate change.

25.3 Identification Strategy

In this section, we describe the identification strategy that we use to assess the impact of community-based tourism on household wealth. The households from six sub-districts reported in are treatment sub-districts. The community-based tourism in Gangtey and Phobjikha in the western region was started through the initiative of RSPN, while Chhokhor, Tang and Ura in Bumthang in the central region and Bumdeling in the eastern region were supported through numerous environmental conservation projects funded by the government channelled through the management of Wangchuck Centennial National Park, Phrumsingla National Park and Bumdeling Wildlife Sanctuary. These sites were selected to ensure regional representation of the community-based project sites. Further, community tourism initiatives under these six sub-districts have been promoted to improve community resilience to weather and climate shocks (BFLS, 2019; BTFEC, 2019; The World Bank, 2019). The decision to promote these six sub-districts for community tourism was obviously not random and hence, household wealth from above-treated sub-districts cannot be directly compared with households from other sub-districts that did not receive community tourism programmes. The community tourism initiatives were promoted specifically for conservation. The second source of non-randomness arises at the household level. Once the tourism initiative arrives in the community, it is the decision of households to determine the degree of engagement in participating in the community tourism activity, and this also drives the household income. In order to circumvent non-random assignment of community tourism initiatives in Bhutan, this study adopts propensity score matching. We match households from treated sub-districts with households from control sub-districts that are similar in observed characteristics or that have similar treatment probability, using nearest neighbour matching. Using only those matched households, we estimate the following model

$$w_i = \beta_0 + \tau \text{ Tourism}_i + X_i\theta + \epsilon_i \tag{25.1}$$

where w_i is household wealth indicator and Tourism$_i$ takes value 1 if the household belongs to sub-districts that received community tourism initiatives and zero otherwise. The parameter τ is the effect of receiving community tourism initiatives and vector X_i is a vector of other control variables. We estimate both linear probability model and probit model for Eq. (25.1). Similar identification strategy on matched sample was also implemented by Brucal et al. (2019) and Litzow et al. (2019).

In Eq. (25.1), we only use households that are similar in observed characteristics based on propensity score matching; hence, the unconfoundedness assumption holds. However, propensity score matching does not account for the selection on unobservables or if unobservables in ϵ_i is correlated with community or household-level characteristics that also determines the probability of receiving the tourism initiatives from government, τ may be biased. However, this is of less concern to us, because of the following two reasons. First, Bhutanese society is homogenous unlike other South Asian countries and hence, unobservables should also be similar. Second, vector of controls in Eq. (25.1), includes a host of variables that may have affected the government's decision to implement tourism initiative, and hence, the variables *Tourism* is exogenous after controlling for observed households and individual characteristics. Therefore, we believe that our identification strategy has substantially mitigated concern over non-random assignment of the government tourism initiatives in particular areas.

25.4 Data

This study uses the 2017 Population and Housing Census (PHCB) (NSB, 2018) which was conducted in 2016–2017. It has collected information from entire households in both rural and urban Bhutan. According to the 2017 PHCB, there are about 163,001 households in Bhutan, of which about 63% are classified as rural households. In this study, we use rural households since our objective is to understand the impact of tourism on rural households. PHCB has collected data at both household and individual level.

Based on our definition of treatment households, about 3033 households or about 3% of rural households are treated with community tourism initiatives. From the total of 3033 households, we could find match households from the control group for 1033 households. In this study, control group refers to households from those sub-districts that did not receive community tourism programmes. This reduction in the number of matched households is partly due to inclusion of many confounding household and individual characteristics for the matching purposes or estimating propensity score. The differences in the mean before matching are significant for almost all of the variables. After matching, the mean differences are significant for only seven variables. Therefore, the vector of controls in Eq. (25.1) includes those significant variables as well as demographic characteristics of household heads.

In this study, we examined the effect of tourism on three important wealth indicators in rural settings: numbers of rooms, modern floor and ownership of vehicles. Based on the matched sample, the average number of rooms is about 4 with 8 as the maximum number of rooms. However, when estimating Eq. (25.1) for the number of rooms, we use binary variable indicating 1 if the number of rooms is more than four and zero otherwise. About 9% of households have reported as having a modern floor, and about 36% of households have reported as owning a vehicle which includes a family car or vehicle for commercial purposes.

25.5   Results and Discussion

The results of Eq. (25.1) from the linear probability model are reported in Table 25.2, panel A and the results of probit models and average marginal effects from the probit model are reported in Table 25.2, panel B. Both linear probability models and probit models are estimated using frequency of match as weights in the regression. In linear probability models, the standard errors are clustered at sub-district level. The standard errors for marginal effects are calculated using Delta method. The results from linear probability model and probit models are comparable both in terms of the effect size and significance level.

The results from the linear probability model suggest that households that are exposed to community-based tourism treatment are likely to have 0.5 more rooms compared to control households. Similarly, household size, age and female-headed households are also positively correlated with the number of rooms. On the other hand, households headed by married heads are negatively correlated with the number of rooms, perhaps due to tradeoffs that households face between children expenditure and upgrading the house. Other assets such as ownership of television, livestock and washing machines, are also positively correlated with the number of rooms. Similarly, treated households are about 6% more likely to own a vehicle compared to control households. Household size, marital status, age and level of education are positively correlated with vehicle ownership while female-headed households are negatively correlated with vehicle ownership, perhaps due differences in household needs prioritizations. Similarly, ownership of television, mobile internet and washing machines are also positively correlated with the ownership of vehicles.

The average marginal effect from the probit model, reported in Table 25.2, panel B also shows similar results. The results show that households from communities that have received the community tourism initiative are about 10% more likely to

Table 25.2 Effect of tourism on household wealth

	(1)	(2)	(4)
Variables	Room (0/1)	Modern floor (0/1)	Vehicle (0/1)
Panel A: linear probability model			
Tourism	0.469***	0.008	0.055***
	(0.131)	(0.025)	(0.017)
Controls	Yes	Yes	Yes
Obs	2068	2070	2070
Panel B: probit model			
Tourism	0.0348***	0.077	0.191***
	(0.106)	(0.157)	(0.064)
APE	0.096***	0.012	0.053***
	(0.030)	(0.025)	(0.017)
Controls	Yes	Yes	Yes
Obs	2068	2070	2070

Note The binary variable room is defined as 1 if household has more than four rooms, and zero otherwise. Standard errors reported in parentheses for the linear probability model and probit model are clustered at sub-district level. APE stands for average partial effect (or average marginal effect) and standard error for APE is estimated using Delta method
***, **, * indicate significant at 1, 5 and 10% significance level. Control variables include household head's demographics, education, ownership of assets, distance from road, and access to information

have more than four rooms and about 5% more likely to own a vehicle. The results are highly significant and comparable with the results from the linear probability model. However, for outcome variable modern floor, the results are not significant both in linear probability model and probit model. One of the possible reasons is that there is very limited variation in the outcome variables.

The overall results suggest that there are moderate benefits of community tourism initiatives on rural households. The community tourism initiatives are highly regarded as a successful initiative in Bhutan, and it may be because such initiatives are able to benefit the households directly and hence receive full support from participating communities. Since the majority of these tourism initiatives in Bhutan are linked with conservation of critical ecosystems to combat the adverse effect of climate change, the results suggest that it may help communities in becoming more climate-resilient. The upgrading of houses could enhance earnings from homestay programmes, and these earnings may be invested for income-generating activities such as in the transportation sector through procurement of vehicles. This may also reduce household's dependence on agriculture output, which often is vulnerable to effects of climate change such as erratic rainfall patterns, windstorms, flash floods, emergence of new pests and many others.

25.6 Conclusion

In this chapter, we examined how community tourism initiatives in rural Bhutan enabled households to cope with the effect of changing climate, which is often more exacerbated in fragile mountain ecosystems. Using the household census in Bhutan, we used propensity score matching to examine the effect of community tourism on locally important household wealth indicators. Our results showed that households from the community that received the community tourism initiatives are about 5% more likely to have more rooms and are also about 5% more likely to own a vehicle or car. This improvement in wealth indicators may indicate a reduction in the household's dependence on agriculture and hence, this may be an indication that community tourism initiatives in Bhutan may be contributing towards building household resilience to climate change by diversifying income sources away from climate-threatened agriculture.

However, the external validity of these results may depend on how the tourism initiatives are promoted, necessary support from government and overall tourism policy. Community tourism in Bhutan is promoted with caution and policy impacts such as benefits and restrictions are well communicated with the households during the planning period. In some instances, the government has promoted such initiatives through local non-governmental organizations. However, this chapter contributes towards the larger debate of who benefits the most from tourism and also how the tourism sector can also contribute towards building climate resilience in rainfed agricultural rural communities like Bhutan.

Our results also indicate that the benefits from tourism are not limited to revenue generation for the country, but also one of the ways to tackle the current climate change issues. Promoting tourism in rural villages can help poor households not just in terms of enhancing household income but also provide alternative livelihood options in the poor villages. Further, through such community tourism initiatives, the households' reliance on the natural environment such as forest cover and water bodies may also reduce thus providing the space for the natural environment to regenerate.

References

Basnet, T. R. (2020). Tourism and cultural heritage in Bhutan. *Bhutan Law Network/JSW Law Research Paper Series* (20–04).

BFLS. (2019). *Bhutan for life prospectus*. Thimphu.

Brucal, A., Javorcik, B., & Love, I. (2019). Good for the environment, good for business: Foreign acquisitions and energy intensity. *Journal of International Economics, 121*, 103247.

Brunet, S., Bauer, J., De Lacy, T., & Tshering, K. (2001). Tourism development in Bhutan: Tensions between tradition and modernity. *Journal of Sustainable Tourism, 9*(3), 243–263. https://doi.org/10.1080/09669580108667401

BTFEC. (2019). *Project impact assessment: Sustainable financing for biodiversity conservation and natural resource management*. Thimphu.

Dendup, N., & Tshering, K. (2018). Demand for piped drinking water and a formal sewer system in Bhutan. *Environmental Economics and Policy Studies, 20*(3), 681–703. https://doi.org/10.1007/s10018-018-0211-3

Donny, S., & Nor, N. (2012). Community-based tourism (CBT): Local community perceptions toward social and cultural impacts. In *1st Tourism and Hospitality International Conference*.

Ghosh, S., & Roy, S. (2021). Climate change ecological stress and livelihood choices in Indian Sundarban. In A. K. E. Haque, P. Mukhopadhyay, M. Nepal, & M. R. Shammin (Eds.), *Climate change and community resilience: Insights from South Asia* (pp. 399–413). Springer Nature.

GNHC. (2019). *Twelfth five year plan document (2018–2023)* (Vol. I). Thimphu.

Gunathilaka, R. P. D., & Samarakoon, P. S. M. K. J. (2021). Adaptation by vegetable farmers to climate change in Sri Lanka. In A. K. E. Haque, P. Mukhopadhyay, M. Nepal, & M. R. Shammin (Eds.), *Climate change and community resilience: Insights from South Asia* (pp. 415–429). Springer Nature.

Gurung, D. B., & Seeland, K. (2011). Ecotourism benefits and livelihood improvement for sustainable development in the nature conservation areas of Bhutan. *Sustainable Development, 19*(5), 348–358. https://doi.org/10.1002/sd.443

Litzow, E. L., Pattanayak, S. K., & Thinley, T. (2019). Returns to rural electrification: Evidence from Bhutan. *World Development, 121*, 75–96.

MoAF. (2017). *Agriculture statistics 2017*. Thimphu.

Moktan, M. R., Norbu, L., Nirola, H., Dukpa, K., Rai, T. B., & Dorji, R. (2008). Ecological and social aspects of transhumant herding in Bhutan. *Mountain Research and Development, 28*(1), 41–48.

Montes, J. (2019). *Environmental governance in Bhutan ecotourism*. Wagenigen University.

Montes, J., & Kafley, B. (2019). Ecotourism discourses in Bhutan: Contested perceptions and values. *Tourism Geographies*, 1–24. https://doi.org/10.1080/14616688.2019.1618905

Nazir, A., & Lohano, H.D. (2022). Resilience through Crop diversification in Pakistan. In A. K. E. Haque, P. Mukhopadhyay, M. Nepal, & M. R. Shammin (Eds.), *Climate change and community resilience: Insights from South Asia* (pp. 431–442). Springer Nature.

NCoB. (2016). *Review report on tourism policy and strategies*. Thimphu.

NSB. (2018). *2017 Population and housing census of Bhutan*. Thimphu. ISBN 978-99936-28-50-7

NSB. (2019). *Statistical yearbook of Bhutan 2019*. Thimphu.

Nyaupane, G. P., & Timothy, D. J. (2010). Power, regionalism and tourism policy in Bhutan. *Annals of Tourism Research, 37*(4), 969–988. https://doi.org/10.1016/j.annals.2010.03.006

Penjore, D. (2008). *Is national environment conservation success a rural failure? The other side of Bhutan's conservation story*.

Reinfeld, M. A. (2003). Tourism and the politics of cultural preservation: A case study of Bhutan. *Journal of Public and International Affairs Princeton, 14*, 125–143.

RGoB. (2012). *Ecotourism development in the protected areas network of Bhutan: Guidelines for planning and management*. Thimphu.

RGoB. (2018). *Ramsar COP13 national report*. Thimphu.

Rinzin, C., Vermeulen, W. J. V., & Glasbergen, P. (2007). Ecotourism as a mechanism for sustainable development: The case of Bhutan. *Environmental Sciences, 4*(2), 109–125. https://doi.org/10.1080/15693430701365420

RSPN. (2014). *Study of climate change impact on Wetland ecosystem Phobjikha, West Central Bhutan*. Thimphu.

Shafeeqa, F., & Abeyrathne, R. M. (2022). Climate adaptation by Farmers in three communities in the Maldives. In: A. K. E. Haque, P. Mukhopadhyay, M. Nepal, & M. R. Shammin (Eds.), *Climate change and community resilience: Insights from South Asia* (pp. 129–141). Springer Nature.

TCB. (2019). *Tourism policy of Kingdom of Bhutan*. Thimphu.

TCB. (2020). *Bhutan tourism monitor report*. Thimphu.

Teoh, S. (2016). *"We have become too ambitious, too greedy": Bhutan's gross national happiness (GNH) tourism model*.

The World Bank. (2019). *Implementation completion report (ICR) review of project P127490—Bhutan*. Report Number: ICRR0021742.

Tshering, K., Ning, W., Phuntsho, K., Chhetri, N., Bida, N., & Dema, K. (2016). The fate of traditional rangeland management practices under Bhutan's changing policies and socioeconomic conditions. *Bhutan Journal of Research and Development*, 53–66.

WWF. (2011). *The high ground: Sacred natural sites, bio-cultural diversity and climate change in the Eastern Himalayas*. ISBN 978-2-940443-28-4.

Chapter 26
Climate Change, Ecological Stress and Livelihood Choices in Indian Sundarban

Santadas Ghosh and Sreejit Roy

Key Messages

- The younger generation from Indian Sundarban delta is mostly opting for migrant labour jobs outside the region.
- Individual educational attainment is found to have an inverse relationship with activities like fish and crab collection. Education enables the local youth to gather more information and to move outward.
- A policy of incentivizing and facilitating basic education among youth in these remote rural locations may increase their livelihood resilience and help in reducing ecological stress.

S. Ghosh (✉)
Department of Economics and Politics, Visva-Bharati University, Bolpur, West Bengal, India
e-mail: santadas_ghosh@yahoo.co.in

S. Roy
Department of Economics and Politics, Visva-Bharati University, Bolpur, West Bengal, India
e-mail: sroy106@gmail.com

© The Author(s) 2022
Haque et al. (eds.), *Climate Change and Community Resilience*,
https://doi.org/10.1007/978-981-16-0680-9_26

26.1 Introduction

26.1.1 Sundarban: Geography, Ecology and People

Spread over India and Bangladesh, the Sundarban is the largest single mangrove forest in the world, an ecological hotspot and a UNESCO World Heritage Site for its biodiversity significance. The region is a tidal delta, and its islands are yet to attain matured heights. On the Indian side, out of a total of 102 islands, 48 falls within the Sundarban Reserve Forest (SRF) that is home to the famous Royal Bengal Tiger. The remaining 54 islands are inhabited by over 4.5 million people (WWF, n.d.). The SRF and the settlements are on two mutually exclusive sets of islands. Human settlements were made possible only by earthen embankments all around the islands, which sustain freshwater agriculture amidst surrounding saline waters. In the Indian part, a large area outside the SRF is declared in 1989 as the Sundarban Biosphere Reserve which covers a considerable area outside the deltaic islands. The SRF lies in the eastern corner of India bordering Bangladesh while the populated islands are located along the western periphery of the delta.

This mangrove ecosystem is highly productive in terms of forest biomass and nutrient contribution, especially through detritus-based food webs that support rich biodiversity in this estuarine delta. It is a nursery ground for many kinds of fish and other species and is responsible for maintaining the fish stock and aquatic diversity for a vast area in the northern Indian Ocean (Neogi et al., 2017).

People have started settlements on these erstwhile mangrove-dominated islands within the last two centuries. As population expanded over time, the shortage of productive land led to the clearance of large areas of mangrove for agriculture and aquaculture. Studies have identified the changing salinity profile of the delta due to climate change as well as human intervention. It is, however, indicated that increasing glacial melt from the Himalayas might have decreased the salinity at the mouth of the Ganges in the eastern sector of Sundarban that lies in Bangladesh. At the same time, salinity has increased in the central sector, where the connections to the freshwater sources upriver have been ruptured due to heavy siltation and construction activities. The impact of such changes could be alarming for the ecosystem (Banerjee, 2013).

Sundarban's large human population is exposed to a new set of challenges posed by global climate change. Studies indicate that changes in river discharge, tides, temperature, rainfall and evaporation will affect the wetland nutrient variations, influencing the physiological and ecological processes and hence biodiversity and productivity of Sundarban mangroves. Hydrological changes in wetland ecosystems through increased salinity and cyclones will lower food security and increase human vulnerability to waterborne diseases (Neogi et al., 2017). The ripple effect of these changes will have multi-faceted adverse impacts on the nature-dependent livelihoods of nearby communities. Studies have concluded that elevated health risks, reduced land and labour productivity and increased exposure to storms, floods, droughts and other extreme events will make escape from poverty more difficult for the local communities (Dasgupta et al., 2020).

26.1.2 Climate Change Threats to Local Livelihood

Research interest in this region picked up after climate change predictions identified this large coastal population, including those residing in the Bangladesh part of Sundarban, as one of the most vulnerable communities in the world. While the long-term vulnerability of this region is due to soil salinization and area loss from sea-level rise, the more immediate fallout is the predicted increase in cyclonic storm frequency in the Bay of Bengal. Using data on tropical cyclone frequency over the eastern coast of India from 1891 to 2013, a study concluded that there is indeed a trend of enhanced cyclogenesis during the months of May, October, November and December (Mishra, 2014). Another recent study analysed a geo-referenced panel database of cyclonic storm tracks in Bay of Bengal between 1877 and 2016. Considering the pattern of cyclone landfall history over the entire coastline of Bangladesh and that of the states of Odisha and West Bengal in India, the study found that the median location of cyclones has shifted eastward over time, with the highest-impact zones currently found in northern Odisha and in the Sundarban region of West Bengal in India (Dasgupta et al., 2020). The role that mangroves play by providing protection services has been well-documented by Das (2021, Chap. 17 of this volume) in the context of Odisha, India.

Studies have identified shifts in the distribution of aquatic species as well as in the timing of their reproduction (Neogi et al., 2017). Fisheries constitute an important source of livelihood for this poor remote island population. Studies conducted on the Bangladesh part of Sundarban show that aquatic salinization may have an especially negative impact on poor households in the region by 2050. The estimates indicate that areas with poor populations that lose species are about six times more prevalent than areas gaining species (Dasgupta, 2017). Studies have also assessed respondents' risk perceptions of saline water inundation on such aquaculture through risk assessment tools. It was observed that respondents in low-lying areas of Sundarban considered cyclone and coastal flooding as extreme risk events (Chand et al., 2012).

Studies have also found that land and aquatic salinization would also affect the stock and spatial distribution of mangroves in Sundarban. Such changes are likely to affect the prospects for people's forest-based livelihoods. Salinity-induced mangrove migration is expected to have a strong regressive impact on the value of timber stocks because of the loss of the highest value timber species. In addition, the augmented potential for honey production is likely to increase conflicts between humans and wildlife in the region (Chowdhury et al., 2016; Dasgupta et al., 2017).

26.1.3 Livelihood and Stress on Natural Resources

Agriculture has always been the mainstay of people on these islands and provided local food security. Studies predict progressive salinization of water and soil will result in a decline in rice output by 15.6% in nine sub-districts of the Bangladesh part of Sundarban before 2050 (Dasgupta et al., 2018). Salinity ingression on farmland mostly occurs through storm surges and embankment collapse resulting from extreme weather events. The large-scale impact of such events was amply demonstrated by cyclone Aila in 2009 when it devastated large sections of protective embankments resulting in saltwater inundation of agricultural land and freshwater ponds on most islands, damaging crop productivity and agricultural patterns (Haldar & Debnath, 2014).

River dynamics may have been accentuated in recent times as a result of climate change and sea-level rise, and there is regular river erosion and accretions in different parts of the delta. (Raha et al., 2012). But newly formed mudflats are invariably treated as government land and no one is allowed to settle there. As a result, the delta on the Indian side has a significant share of landless population and marginal farmers. Also, due to lack of irrigation facilities as well as basic infrastructure, agricultural practices are traditional and dependent on seasonal rainfall. People supplement their income by exploiting forest and surrounding aquatic resources (Mahadevia Ghimire & Vikas, 2012). With decreasing land resources, there are more incursions inside the SRF area for fishing, crab catching and honey collection. Fishermen venture into forest creeks for better catch as those places are supposed to be richer in fish concentration. In the Indian part of Sundarban, around 20% of the islands' inhabitants are estimated to be surviving on fishing activities (Ghosh, 2017).

Going inside the reserve forest carries the risk of tiger attacks but the poor continue to do so for their livelihood. A study on the victims (both death and injury) between 1999 and 2014 estimates the risk of tiger attack in the range of 0.11–0.88 for every 10,000 residents of the blocks surrounding SRF. The majority of the victims (68%) were found to be male, aged between 30 and 50 years (Das, 2018).

Another form of natural resource extraction has also caused great harm to the aquatic resource stock. From the late 1980s, international markets started opening up for large-scale export of the locally available shrimp species usually referred as *tiger shrimp (Penaeus monodon)*. Its juveniles were abundantly available in the local rivers and were in high demand by newly established inland shrimp farms. Collection of these juveniles started on a large scale and was a lucrative avenue to earn hard cash for the local poor. Over the next two decades, indiscriminate shrimp seed collection caused serious harm to other fish juveniles. This practice greatly contributed to the gradual decline in the fish populations in Sundarban (Gopal & Chauhan, 2006; World Bank, 2014).

Estimation of yield, exploitation rate and maximum sustainable yield for major shell and finfish species in SRF has signalled overexploitation (Hoque Mozumder et al., 2018). The IUCN has already listed the status of a number of economically important fish species of Sundarban as 'threatened' (Hoq, 2007). Other studies have also identified the northern Bay of Bengal ecosystem as an exploited one as its fish stock is being steadily depleted under huge pressure from fishing (Dutta et al., 2017).

In recent times, the fall in relative price of shrimp-fry as well as government restrictions and environmental awareness programmes have led to some reduction in shrimp juvenile collection. Instead, collection of mud crabs (*Scylla serrata*) from the delta has greatly increased due to better realization of price in the international market (Nandi & Pramanik, 2017). As crab is a keystone species for the mangrove ecosystem (Smith et al., 1991), intensification of its collection from the wild has also accentuated the anthropogenic stress on it.

Pressure on the Sundarban's ecosystem has also resulted from investments requiring clearing of mangrove forests. Studies identified that the Chakoria (Bangladesh) mangrove forest had completely disappeared between 1903 and 2010, primarily to make way for inland shrimp farms. Empirical evidence shows that such interventions resulted in significant livelihood loss for the local communities (Islam, 2014).

26.1.4   Labour Outmigration

Though settlements in Sundarban delta are relatively recent, increasing population and decreasing agricultural land coupled with dwindling fish stock is resulting in large labour outmigration in recent years. Increasing soil salinity associated with extreme weather events has decreased agricultural productivity, which in turn is fuelling such outmigration (Hajra & Ghosh, 2018). In the aftermath of cyclone Aila (2009), which resulted in widespread and prolonged loss of agriculture in the delta, labour outmigration emerged as a predominant coping strategy for the local poor against their agricultural loss (Ghosh, 2013).

The extent of this phenomenon has been investigated in some studies both quantitatively and qualitatively. It is found that almost 75% of migrant labour travel inter-state and work in the western and southern Indian states. Almost 95% of them are male and they mostly work as construction labour and come back to Sundarban in the monsoon period for paddy cultivation (Guha & Roy, 2016; Mistri, 2013).

Much of the empirical literature on Sundarban hovers around climate change implications for local livelihood and its ecological health. The socio-economic profile of poor communities, those who exert direct anthropogenic stress on the ecosystem, is relatively less explored. A study conducted in Bangladesh found that the livelihood condition and education level of the Sundarban fishermen is very poor (Mondal et al., 2018). Similar inference is drawn for its Indian counterparts in another study (Mistri & Das, 2015). However, none of these studies is based on a reasonably large sample and does not detail the livelihood choices and their dynamics in this delta.

On the Indian side, there is a discernible shift in the livelihood mix of the Sundarban people in recent years, caused by population pressure and decreasing land and other natural resources. Migrant labour jobs are being adopted increasingly by the local youth, which is also facilitated by infrastructural development and easier communication through mobile connectivity. These livelihood dynamics have important implications for the local ecosystem as they may help in reducing anthropogenic stress. A study on the pattern of such spontaneous livelihood adjustments is crucial for devising supporting policies. Such a study is conspicuous by its absence in existing literature.

26.2 Objective, Study Area and Data

This study intends to fill the research gap mentioned above. The data is generated from four different sets of household surveys in the region conducted during 2012, 2015, 2017 and 2019. Though the surveyed households were different across years, there is a large overlap in the Gram Panchayat areas that were covered. Altogether, the primary data consists of 2800 households having over 5500 adult earning members in them. It makes this study a rare one that uses such a large primary dataset from Sundarban and lends credence to the empirical evidence.

The set of sample households differed in these surveys as they were conducted with separate study objectives under different research projects.[1] Most of the sampled households are from the remote islands in the Indian part of the Sundarban. The spatial coverage of the four surveys mostly includes Gram Panchayat areas in the periphery of SRF. It should be noted that in spite of the overall cross-section as well as time series characteristics, the aggregate information does not constitute a panel dataset as the primary sample units (households) were different across surveys. The set of all Gram Panchayat areas covered under these surveys is shown in Fig. 26.1.

While the survey rounds differed in their primary study objective, a common module in the survey instrument included individual member profiles including information on age, sex, educational attainment, main occupation and subsidiary occupations. These sets of individual information are rich enough to shed light on the current study objective.

26.3 Findings

26.3.1 Household Size, Earning Members and Number of Activities

The resource stress and low agricultural output per capita imply multiple supplementary earning activities taken up by household members whenever possible. Table 26.1 summarizes the findings in this regard.

---

[1] 2012: *Dynamics of households' adaptation and resilience: Sundarban after cyclone Aila,* sponsored by South Asian Network for Development and Environmental Economics (SANDEE).

2015: *Livelihood impact of electricity and road-links in remote rural villages: a study in Sundarban delta,* sponsored by National Bank for Agricultural and Rural Development (NABARD).

2017: *Developmental interventions, rural transformation and environmental sustainability: A study on Sundarban in India,* sponsored by Indian Council of Social Science Research (ICSSR).

2019: *SAWI-Sundarbans Targeted Environmental Studies,* sponsored by The World Bank.

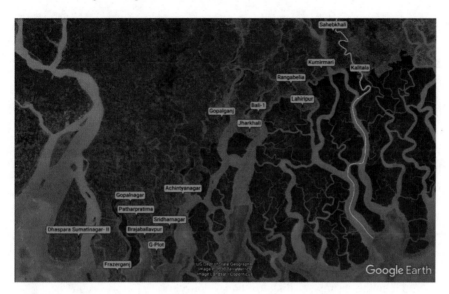

Fig. 26.1 Gram Panchayats covered for primary surveys in the Indian Sundarban. *Source* Google Earth

Table 26.1 Indicators for multiplicity of earning activities

Indicators	First survey (2012)	Second survey (2015)	Third survey (2017)	Fourth survey (2019)
Average household size	4.98	5.45	4.28	4.47
Average number of earning members per household	1.91	1.87	1.72	2.34
Average number of earning activities per earning member	1.57	1.41	1.32	1.70

Source Primary surveys in different years

The highlight of the table is the presence of more than one earning member even though the average household size is small. The earning from the primary occupation was supplemented through other activities.

Though the metrics are provided with their temporal dimension in the table, it would be wrong to conclude anything relating to their time trend. There is spatial concentration of households depending on whether their main activity is based on land or river. Hamlets located close to the rivers and the forest have a relatively larger proportion of households collecting fish and crab. Depending on the specific study objective, the selection of hamlets was done in different years with different shares of these hamlet types. It resulted in noticeable differences across columns in Table

26.1. The main purpose of the table, however, is to provide background information on the sample households and their livelihood intensity. The main findings of this study are summarized in the following sections.

26.3.2   The Livelihood Mix

Information on main and subsidiary occupations across survey years were collected by aggregating all possible earning activities in the following eight mutually exclusive and exhaustive groups:

i.   Agriculture and farm labour: All agriculture related activities including inland fisheries, animal husbandry and farm labour jobs.
ii.   Non-farm labour: All forms of daily wage earning except from farm-sector inside Sundarban region.
iii.   Fish/crab catching: All forms of collections from open rivers and forests including shrimp-seedling and honey collection.
iv.   Migrant labour: All labour jobs taken up in locations outside Sundarban region.
v.   Self-employed professional: All specialized commercial activities requiring skills like private tuition, traditional medicine (quacks), priests etc.
vi.   Business: Any other commercial enterprise requiring capital investment.
vii.   Salaried employment: Regular employment, mostly teachers and staffs in local primary schools, government enrolled village health workers, other staff at local Panchayat office, NGO-workers with regular salary, etc.
viii. Other Miscellaneous: Rent earning, old-age pension, widow pension, disability pension, etc.

Since earning members generally show a multitude of earning activities, it is difficult to identify each of them exclusively with one of the above categories. Hence, all the earning members in different survey rounds are categorized following their self-reported 'primary' occupation only. Figure 26.2 describes the relative share of these occupations over survey years.

The relative shares of activities vary across the survey rounds due to differential choice of hamlets, as described before. But the robustness of finding lies with the consistent high rank of the three livelihood options across the years—agriculture, fishing and migrant labour jobs. The predominance of these three activities, in spite of differences in sample hamlets, endorses the postulate regarding limitation of livelihood options in the delta.

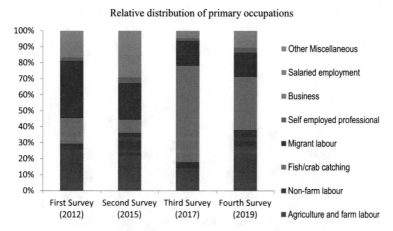

Fig. 26.2   Relative distribution of primary occupations. *Source* Primary survey

26.3.3   Livelihood Across Age and Education

It is interesting to see how these livelihood options are distributed among earning adults when they are categorized across age groups. In the aggregate dataset, age of an earning member showed a range of 10–96 years. Age below 15 years was recorded only for a few members engaged in shrimp juvenile collection. A couple of very old (more than 90 years) earning members were also found who were 'earning' as recipients of old-age pension.

Figure 26.3 describes the average age of earning members across the eight activities over the survey rounds. The figure clearly indicates that migrant labour jobs involve the youngest of the labour force from the delta. The robustness of this observation is also revealed by the figure as this phenomenon is markedly similar across all survey years.

A marked difference in average age across occupations is indicative of the inherent livelihood dynamics across generations. In the beginning, islanders started with agriculture as their main occupation. Gradually people resorted more to fish/crab collection, and later, migrant labour jobs have been replacing both of these activities for the younger generation. The gradual lowering of the height of the bars across these three activities might be interpreted in this light. The average age for 'other miscellaneous' activities is always much higher as it consists of old-age and widow pensioners. All other occupation choices show more or less similar values for their average age, which is also close to the overall average in the sample.

The educational status, indicating the human capital one possesses, is another interesting aspect to look at. For easier understanding, information on individual educational attainments is categorized into four mutually exclusive and exhaustive classes. The overall education profile of the earning members in the sample is shown in Fig. 26.4, which describes the relative shares of the four education classes. It

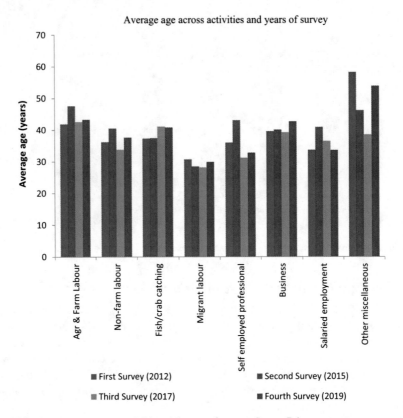

Fig. 26.3 Average age across activities and years of survey. *Source* Primary survey

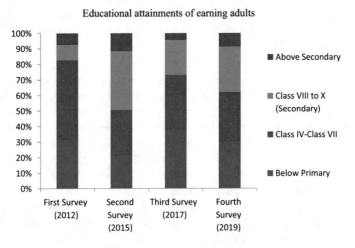

Fig. 26.4 Educational attainments of earning adults. *Source* Primary survey

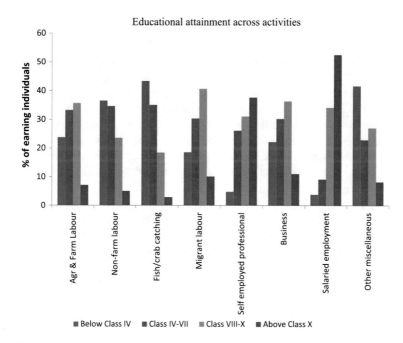

Fig. 26.5 Educational attainment across activities. *Source* Survey in 2019

is clear that the earning adults in the delta are poorly educated, with only a small fraction of them being able to proceed beyond the secondary level.

Further education-activity decomposition gives more interesting insights. To keep the information simple and easy to comprehend, Fig. 26.5 is produced using only the latest survey year (2019). The figure, however, captures the essence of the distribution, which is found to be robust across other survey years as well.

It is observed that the lowest education class (below class IV) forms the modal class for 'Non-farm labourers' and 'Fish/crab catchers'. It is also the modal class for old people earning from old-age and widow pensions (other miscellaneous), who constitute the bulk of the local illiterate population. The spread of more accessible primary and secondary education in these remote islands is a relatively recent phenomenon.

The study finds that only a negligible section of the local people could reach up to graduation level. The majority of these graduates are found to be engaged in private tuition (self-employed professional) and a handful of them could become teachers in local primary schools (salaried employment). This explains the dominance of the highest educational class for these activities.

The most interesting aspect in Fig. 26.5 is the completely opposite educational attainment pattern between 'fish/crab catchers' and 'migrant labourers'. The former category is heavily skewed towards the lowest education class while it is just the opposite for the migrants.

26.4   Conclusions and Policy Implications

The Sundarban is a very important geographical region both for the ecological health of the Bay of Bengal and the millions of poor people in Bangladesh and India that inhabit the area. Its geographical and ecological aspects have been studied over the last several decades. The region caught the fancy of the research community with climate change-related vulnerabilities identified for its large resident population. On the Indian side, the region is also witness to significant dynamics in livelihood owing to availability of mobile communication and transport infrastructure in recent years. Yet, studies on these livelihood dynamics supported by micro-level data are scarce in the literature.

This study bridges this crucial research gap by using information from several rounds of household surveys that were conducted in the Indian part of the Sundarban over the last decade. The majority of the sample observations belong to the peripheral hamlets of Sundarban Reserve Forest. The pooled data of 2800 households and over 5500 adult earning members unfolds an interesting story of people's resilience in the face of climate change-induced loss in their traditional livelihood. The study finds strong empirical evidence suggesting that the younger generation in the delta is coping with the situation by moving out of the region as migrant labour. Such livelihood dynamics might be considered ecologically beneficial as it reduces the anthropogenic stress on the mangrove ecosystem.

The study also finds that a little improvement in individual educational attainments can cause an increase in this labour outflow. A strong positive inducement for this is found to exist for education up to secondary level. Working at a distant location and in a completely different cultural environment might need a level of communication and information gathering skill that the illiterate or very minimally educated people do not possess. This finding has important policy implications. Additional government initiatives in enhancing school enrolment, arresting school dropouts and incentivizing educational attainments up to secondary level can be a policy handle for an ecological goal. It seems that enhancing the basic educational infrastructure in this remote deltaic region goes much beyond its apparent objective of building social capital in general. In an ecological hotspot like the Sundarban, such investments in education would increase community resilience in the face of climate change threats to their traditional livelihood.

26.5   Limitations and Scope for Further Research

This study is intended to shed light on the dynamics of livelihood mix in Indian Sundarban and its implication for local ecology. The empirical estimates reported here should be interpreted with a word of caution. The region is characterized by spatial concentration of fish/crab-based livelihood in the proximity of rivers. For villages or hamlets away from rivers, livelihood is mostly based on agriculture.

The proportion of households depending directly on natural resources would vary depending on the choice of sample hamlets. In this study, hamlet selections were not purely random, and hence, the overall sample might not be directly used for population estimates on livelihood mix. However, this study is not attempting to make any such estimation either. It might be noted that households were selected as stratified random samples from within those hamlets, where stratification was done on the basis of landholding. Deeper investigations on the factors inducing the youth for migrating out, and the marginal effect of such outmigration on local ecological indicators remains an interesting and open research agenda for Sundarban.

Acknowledgements This paper is based on data pooled from four different studies in Indian Sundarban over the decade of 2010–2020. The authors sincerely acknowledge the sponsors of all these studies which include South Asian Network for Development and Environmental Economics (SANDEE, Kathmandu); Department of Economic Analysis and Research, National Bank for Agricultural and Rural Development (DEAR, NABARD, Mumbai); Indian Council of Social Science Research (ICSSR, New Delhi) and the World Bank (Washington, DC). The authors, however, are solely responsible for the conclusions and any possible errors in this study.

References

Banerjee, K. (2013). Decadal change in the surface water salinity profile of Indian sundarbans: A potential indicator of climate change. *Journal of Marine Science: Research & Development, 01*(S11). https://doi.org/10.4172/2155-9910.S11-002

Chand, B. K., Trivedi, R. K., Biswas, A., Dubey, S. K., & Beg, M. M. (2012). Study on impact of saline water inundation on freshwater aquaculture in Sundarban using risk analysis tools. *Exploratory Animal and Medical Research, 2,* 170–178.

Chowdhury, A., Sanyal, P., & Maiti, S. K. (2016). Dynamics of mangrove diversity influenced by climate change and consequent accelerated sea level rise at Indian Sundarbans. *International Journal of Global Warming, 9*(4), 486. https://doi.org/10.1504/IJGW.2016.076333

Das, C. S. (2018). Pattern and characterisation of human casualties in Sundarban by Tiger attacks, India. *Sustainable Forestry, 1*(4). https://doi.org/10.24294/sf.v1i2.873

Das, S. (2021). Valuing the role of mangroves in storm damage reduction in coastal areas of Odisha. In A. K. E. Haque, P. Mukhopadhyay, M. Nepal & M. R. Shammin (Eds.), Climate change and community resilience: Insights from South Asia (pp. 257–273). Springer Nature.

Dasgupta, S. (2017). The impact of aquatic salinization on fish habitats and poor communities in a changing climate: Evidence from Southwest Coastal Bangladesh. *Ecological Economics, 12.*

Dasgupta, S., Hossain, Md. M., Huq, M., & Wheeler, D. (2018). Climate change, salinization and high-yield rice production in coastal Bangladesh. *Agricultural and Resource Economics Review, 47*(1), 66–89. https://doi.org/10.1017/age.2017.14

Dasgupta, S., Sobhan, I., & Wheeler, D. (2017). The impact of climate change and aquatic salinization on mangrove species in the Bangladesh Sundarbans. *Ambio, 46*(6), 680–694.

Dasgupta, S., Wheeler, D., Sobhan, M. I., Bandyopadhyay, S., Nishat, A., & Paul, T. (2020). *Coping with the vulnerability of the sundarban in a changing climate: Lessons from multidisciplinary studies.* World Bank. https://openknowledge.worldbank.org/handle/10986/34770

Dutta, S., Chakraborty, K., & Hazra, S. (2017). Ecosystem structure and trophic dynamics of an exploited ecosystem of Bay of Bengal, Sundarban Estuary, India. *Fisheries Science, 83*(2), 145–159. https://doi.org/10.1007/s12562-016-1060-2

Ghosh, S. (2013). Extreme event, anthropogenic stress and ecological sustainability in Sundarban Islands. *Vidyasagar University Journal of Economics, XVII*, 132–148.

Ghosh, S. (2017). Coping with a natural disaster: Sundarbans after Cyclone Aila. In *Global change, ecosystems, sustainability: Theory, methods, practice* (pp. 116–136). SAGE Publications, Inc.

Gopal, B., & Chauhan, M. (2006). Biodiversity and its conservation in the Sundarban Mangrove ecosystem. *Aquatic Sciences, 68*(3), 338–354. https://doi.org/10.1007/s00027-006-0868-8

Guha, I., & Roy, C. (2016). Climate change, migration and food security: Evidence from Indian Sundarbans. *International Journal of Theoretical and Applied Sciences, 8*(2), 45–49.

Hajra, R., & Ghosh, T. (2018). Agricultural productivity, household poverty and migration in the Indian Sundarban Delta. *Elementa: Science of the Anthropocene, 6*(1).

Haldar, A., & Debnath, A. (2014). Assessment of climate induced soil salinity conditions of Gosaba Island, West Bengal and Its Influence on Local Livelihood. In M. Singh, R. B. Singh, & M. I. Hassan (Eds.), *Climate change and biodiversity* (pp. 27–44). Springer Japan. https://doi.org/10.1007/978-4-431-54838-6_3

Hoq, M. E. (2007). An analysis of fisheries exploitation and management practices in Sundarbans mangrove ecosystem, Bangladesh. *Ocean & Coastal Management, 50*(5), 411–427. https://doi.org/10.1016/j.ocecoaman.2006.11.001

Hoque Mozumder, M. M., Shamsuzzaman, Md. M., Rashed-Un-Nabi, Md., & Karim, E. (2018). Social-ecological dynamics of the small scale fisheries in Sundarban Mangrove Forest, Bangladesh. *Aquaculture and Fisheries, 3*(1), 38–49. https://doi.org/10.1016/j.aaf.2017.12.002

Islam, S. (2014). An analysis of the damages of Chakoria Sundarban mangrove wetlands and consequences on community livelihoods in south east coast of Bangladesh. *International Journal of Environment and Sustainable Development, 13*, 153–171. https://doi.org/10.1504/IJESD.2014.060196

Mahadevia Ghimire, K., & Vikas, M. (2012). Climate change—Impact on the Sundarbans: A case study. *International Scientific Journal, 2*, 7–15.

Mishra, A. (2014). Temperature rise and trend of cyclones over the Eastern Coastal Region of India. *Journal of Earth Science and Climatic Change, 05*(09). https://doi.org/10.4172/2157-7617.1000227

Mistri, A. (2013). Migration and sustainable livelihoods: A study from Sundarban biosphere reserve. *Asia Pacific Journal of Social Sciences, 5*(2), 76–102.

Mistri, A., & Das, B. (2015). Environmental legislations and livelihood conflicts of Fishermen in Sundarban, India. *Asian Profile, 43*, 389–400.

Mondal, M., Islam, M., Islam, M., Barua, S., Hossen, S., Ali, M., & Hossain, M. B. (2018). Pearson's correlation and likert scale based investigation on livelihood status of the Fishermen living around the Sundarban Estuaries, Bangladesh. *Middle East Journal of Scientific Research, 26*. https://doi.org/10.5829/idosi.mejsr.2018.182.190

Nandi, N., & Pramanik, S. (2017). *Livelihood on Mud Crab catchment: A case study of Sundarban Coast, West Bengal, India* (Vol. 1). https://doi.org/10.29011/2577-1493.100003

Neogi, S. B., Dey, M., Kabir, S. L., Masum, S. J. H., Kopprio, G., Yamasaki, S., & Lara, R. (2017). Sundarban mangroves: Diversity, ecosystem services and climate change impacts. *Asian Journal of Medical and Biological Research, 2*(4), 488–507. https://doi.org/10.3329/ajmbr.v2i4.30988

Raha, A., Das, S., Banerjee, K., & Mitra, A. (2012). Climate change impacts on Indian Sunderbans: A time series analysis (1924–2008). *Biodiversity and Conservation, 21*(5), 1289–1307. https://doi.org/10.1007/s10531-012-0260-z

Smith, T. J., Boto, K. G., Frusher, S. D., & Giddins, R. L. (1991). Keystone species and mangrove forest dynamics: The influence of burrowing by crabs on soil nutrient status and forest productivity. *Estuarine, Coastal and Shelf Science, 33*(5), 419–432. https://doi.org/10.1016/0272-7714(91)90081-L

World Bank, W. (2014). *Building resilience for sustainable development of the Sundarbans: Strategy report* (*No. 20116*; *World Bank Other Operational Studies*). The World Bank. https://ideas.repec.org/p/wbk/wboper/20116.html

World Wildlife Fund. (n.d.). *About Sundarbans*. WWF India. https://www.wwfindia.org/about_wwf/critical_regions/sundarbans3/about_sundarbans/. Last accessed December 26, 2020.

Chapter 27
Adaptation to Climate Change by Vegetable Farmers in Sri Lanka

R. P. Dayani Gunathilaka and P. S. M. Kalani J. Samarakoon

Key Messages

- Farmer characteristics such as gender, education level, and willingness to take credit influence resilience to climate change.
- Availability of updated climate information and improved extension services are instrumental in effective adaptation.
- Networking of farmers for initiating an information portal will nurture successful resilience.

27.1 Introduction

A substantial volume of economic and scientific studies have focused on the impact of climate change on agriculture. The agricultural sector's inherent vulnerability to weather and its significant role in developing countries are reasons for this growing concern (Blanc & Reilly, 2017; Chalise et al., 2017; Huong et al., 2019). South Asia is one of the regions most affected by climate change (Cai et al., 2016; Aryal et al.,

---

Disclaimer: The presentation of material and details in maps used in this book does not imply the expression of any opinion whatsoever on the part of the Publisher or Author concerning the legal status of any country, area or territory or of its authorities, or concerning the delimitation of its borders. The depiction and use of boundaries, geographic names and related data shown on maps and included in lists, tables, documents, and databases in this book are not warranted to be error free nor do they necessarily imply official endorsement or acceptance by the Publisher, Editor(s) or Author(s).

R. P. D. Gunathilaka (✉) · P. S. M. K. J. Samarakoon
Department of Export Agriculture, Uva Wellassa University, Badulla, Sri Lanka
e-mail: dayani@uwu.ac.lk

P. S. M. K. J. Samarakoon
e-mail: kalanijaya25@gmail.com

© The Author(s) 2022
Haque et al. (eds.), *Climate Change and Community Resilience*,
https://doi.org/10.1007/978-981-16-0680-9_27

2019). According to the fifth assessment report of the Intergovernmental Panel on Climate Change (IPCC), it is predicted that the temperature will rise by 2 °C on average, hot days will increase in frequency, and unpredictable rainfall patterns will become more frequent and severe with shorter rainfall periods in South Asia by the mid-twenty-first century (Hijioka et al., 2014). In light of the fact that three-fifths of the cropping area of the region is rain-fed, extreme weather events such as floods and droughts will have severe impact on the rural poor and on agriculture (Jat et al., 2016; Esham et al., 2018; Aryal et al., 2019; Ahmed, 2022, Chap. 7 of this volume; Kattel & Nepal, 2021, Chap. 11 of this volume).

Consistent with the projections for South Asia, Sri Lanka is already experiencing climate change and the consequent, erratic rainfall and extreme swings between droughts and intense rainfall (Chithranayana & Punyawardena, 2014; Esham et al., 2018). For example, the island has been exposed to 30 large-scale flooding events between 2000 and 2018 and in 2016–2017 the country experienced a drought which was considered to be the worst such event in the past 40 years, affecting 20 districts and approximately 1.8 million people (Esham et al., 2018; United Nations Office for Disaster Risk Reduction, 2019). Data from the Food and Agriculture Organization (FAO) indicates that severe droughts followed by heavy rainfall have affected cropping areas in Sri Lanka, threatening the food security of some 900,000 people (FAO, 2017). Given that only 34% of the cultivated area in Sri Lanka is irrigated, the consequences of climate extremes have become critical causing serious implications for food production and rural livelihoods. Therefore, comprehensive understanding of the impacts on different food systems and adaptation measures is necessary to minimize the negative impacts and ensure local food security.

There is an expanding body of research to assess and quantify the impact of climate change on the Sri Lankan agriculture sector in general and principal crops such as tea, rubber, coconut, and paddy, in particular (Seo et al., 2005; De Silva et al., 2007; De Costa, 2010; Chithranayana & Punyawardena, 2014; Gunathilaka et al., 2017; Chandrasiri et al., 2020). However, only a few studies have been focused on investigating farmers' adaptation to climate change. For example, Gunathilaka et al. (2018) analyze Sri Lankan tea estate managers' perceptions of climate change impacts, factors which affect their choice of current and potential future adaptation options, and barriers to the adoption of these options. Studies by Menike and Arachchi (2016) and Esham and Garforth (2013) analyze farmers' perception and observations on climate change, actual adaptation at farm level, and factors affecting mainly paddy and vegetable mix farming communities in Sri Lanka. Neither these studies nor other studies examine farmers' responses to minimize the negative impacts of climate change on up-country vegetable farming which provides promising opportunities for employment and an essential source of food and nutrition security in Sri Lanka. Therefore, in this chapter, we investigate farmers' ongoing adaptation to climate change, what factors drive farmers' decision to adapt and barriers to adaptation, based on the case of up-country vegetable farmers in Sri Lanka.

Employing data collected through face-to-face interviews with 150 Up-country vegetable farmers, this study focuses on the following research questions: (i) what are farmers' perceptions of climate change? (ii) which adaptation options are farmers currently adopting? (iii) which factors influence the commonly used adaptation options? (iv) what are the policy implications of commonly used adaptation options for long-term sustainability? The paper is structured as follows: Sect. 27.2 describes the context and climate-related impacts on vegetable farming. Section 27.3 provides the study area and sample characteristics. Section 27.4 presents the methodology. Section 27.5 describes results and discussion along with farmers' observations of the impacts of climate change on vegetable farming. Section 27.6 concludes with policy implications.

27.2   Vegetable Farming: Context and Climate-Related Impacts

Sri Lanka has a tropical climate and virtually no distinct seasons as the temperatures vary, on average between 27 and 30 °C in most regions of the island. The up-country[1] area is at an altitude of about 1000 m above mean sea level where Nuwara Eliya has the coolest and mildest climate with mean temperature of 16 °C. Other upland areas (i.e., Badulla, Bandarawela, and Welimada) also experience a temperate climate that can span between 16 and 20 °C.

Sri Lankan agriculture is divided into two groups namely the plantation and non-plantation sector. Tea, rubber, and coconut together make the plantation crop sector. Non-plantation sector comprises of mainly paddy, vegetables, fruits, and field crops. After paddy, the vegetable subsector is the most prominent and consists of two categories namely low-country and up-country vegetables. Up-country vegetables are cultivated primarily in Nuwara Eliya and Badulla districts of the central highlands of Sri Lanka. Potato, leeks, carrot, beetroot, beans and cabbages are the main vegetables and contribute to Sri Lankan nutritional requirements as a source of vitamins, minerals, and fiber (Perera & Madhujith, 2012; Padmajani et al., 2014).

27.3   Study Area and Sampling

We selected Nuwara Eliya, Welimada, Badulla, and Bandarawela divisional secretariats as the study area (Fig. 27.1). These four areas represent three different agro-ecological regions (AERs)[2] namely up-country wet zone 3 (Nuwara Eliya), up-country intermediate zone 2 (Badulla), and up-country intermediate zone 3 (Bandarawela and Welimada). According to the Department of Census and Statistics as of 2018, 48,960 farmers were involved in vegetable farming in the study area.

---

[1] Similarly it is called hill country encompassing districts of Nuwara Eliya and Badulla.

[2] Agro-ecological regions (AER) are categorized based on rainfall, elevation, and soil type. Sri Lanka comprises 46 AERs.

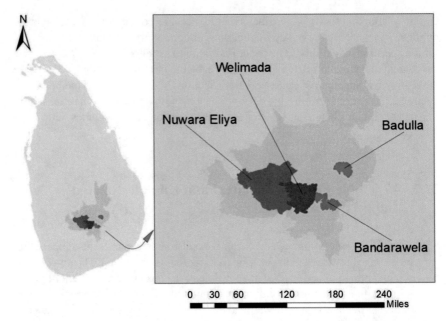

Fig. 27.1 Study area showing the selected divisional secretariats. *Source* Generated by Authors

Using a simple random sampling technique, 150 up-country vegetable farmers were selected for the study. The sampling fraction of farmers from each area was determined on the basis of the farmer population in the respective area. Accordingly, 44% from Nuwara Eliya, 34% from Welimada, 14% from Bandarawela, and 8% from Badulla were included in the sample.

27.4   Methods and Variables

We surveyed the farmers using an in-depth semi-structured interviewing method. The semi-structured questionnaire comprised of both open-ended and close-ended questions and was tested prior to the actual interview to determine validity and appropriateness of the climate change options and driving variables employed in the climate adaptation model here. Table 27.1 shows the variables and types of data collected. The interviews took place between September and November 2019 and obtained data detailing the farmers' age and experience, observations on changes in weather and climate (i.e., temperature, rainfall, drought, wind, and frost patterns) over the course of farming, perceptions of climate change and climate-related risks to their businesses, and responses to climate change and barriers to adaptation. The interviews averaged about 30 min each (Fig. 27.2). All interviews were recorded and then transcribed for subsequent analysis.

Table 27.1 Data types

Variable	Data	Data source
Dependent	No adaptation	The literature and survey
	Early or late planting	
	Switching variety or crop	
	Intensive use of inputs	
Independent	Gender of the farmer	Survey
	Education level	
	Farming experience	
	Income of the farmer	
	Willingness to take credit	
	Extension on crop	
	Information on climate change	
	Farmer-to-farmer extension	
	Total annual rainfall	NRMC
	Average annual temperature	

Source Generated by Authors

(a)   (b)

Fig. 27.2 **a** Interviewing a farmer in Welimada area. **b** Vegetable farm in Nuwara Eliya area. *Photo courtesy* Kalani Samarakoon

Temperature and rainfall data from 2001 to 2018 was obtained from the Natural Resource Management Centre (NRMC) of Sri Lanka. Proxy variable for climate change was constructed taking the percentage change of long-term mean of annual temperature and annual total rainfall. We consider the deviation of long-term mean of the period 2001 to 2009 from that of the period 2010 to 2018.

The study analyzes factors affecting climate change by using a multinomial logit (MNL) model. In this study, MNL model was used based on the previous literature (Belay et al., 2017; Gunathilaka et al., 2018). This model is effective when analyzing decisions across more than two categories (Deressa et al., 2009). Thus, through MNL analysis, this study finds factors influencing the choice of farmers' commonly used adaptation options.

27.5 Results and Discussion

27.5.1 Characteristics of the Sample

The majority of respondents (85%) were male farmers. As the agriculture sector consists of 33.5% female and 65.3% male farmers, this result is as expected. Mean farmland size was 253.02 perches (0.639 ha), and mean household size was four. Among respondents, about 3% had 1–5 years of farming experience while around 20% each had 16–20 years and above 31 years of experience. Nearly four-fifth of the farmers (79%) were over 39 years of age and 92% of farmers had at least senior secondary education.[3]

27.5.2 Farmers' Perceptions on Climate Change

Respondents were from three different agro-ecological regions and they had observed changes in the rainfall pattern, temperature, drought occurrence, and frost conditions. Almost all the farmers (98%) have heard the term "climate change" and 97% of respondents believed climate change is ongoing. In addition, more than 85% of respondents believed that human activities are substantially contributing to climate change and 11% of respondents believed climate change is a natural thing.

Nearly all respondents (99.33%) said that they were experiencing changes in climate.

> In the past, rainy season in Nuwara Eliya was from April to December and the dry period was January to April. Farmers planned their work according to this rainfall pattern. But now this pattern has changed. (Farmer #39, Nuwara Eliya)

Furthermore, in responding to the changes in rainfall pattern, farmers said they alter planting time for different crop types and varieties.

> In the past we did not commence planting potato in June and July. But now, we can plant potato even in June and July. Still we cannot clearly understand this change in the climate.

---

[3] Those who passed grade 10–11 are categorized under the senior secondary level of education in Sri Lanka.

Farmers' observed changes in climate

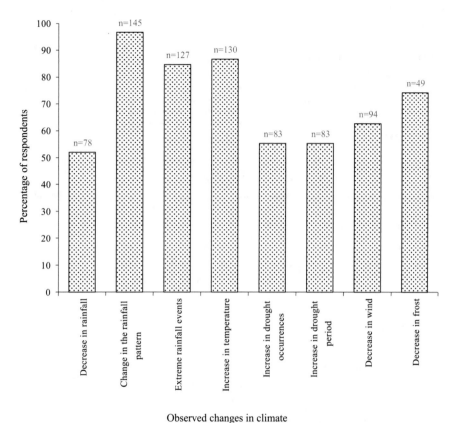

Fig. 27.3 Farmers' observed changes in climate. *Source* Generated by Authors

If we observe a similar climate pattern for three or four consecutive years, we will be able to predict what will happen in next year. (Farmer #41, Nuwara Eliya)

Survey results revealed that farmers have observed numerous changes in climate over the years. Figure 27.3 depicts farmers' observed changes about climate.

27.5.3 Farmers' Observations on Climate Change Impacts

More than 82% of farmers said they were experiencing negative impacts of climate change on their farming while the rest said it had both negative and positive effects. Nearly 75% of the responders said they were trying to minimize impacts of climate change through adaptation methods.

Results revealed that farmers face difficulty in controlling pest attacks and diseases. Consequently, they have had to alter cropping plans and bear yield and income losses.

> Not like in the past, most of times we receive extreme rainfalls and that damages our crops. Also there are lots of pest outbreaks that happen during the extreme rainfall periods. In this area, we can harvest an average of about 20 kg of beans, but today I plucked only 5 kg. I had to throw about 10 kg of beans because of brown spot disease. Heavy rainfall worsens the spread and severity of this disease. I can earn around one and a half million rupees[4] as income if we can sell it for a good market price, but I received only three to four hundred thousand for the last three years as a result of these negative impacts. I lost eight hundred thousand rupees in this season. (Farmer #56, Nuwara Eliya)

Long-lasting fungal diseases, vegetable rotting, and increased cost of production are some of the negative impacts stated by the respondents.

> Rainfall lasting for long period increases the spreading of black rot of cabbage. It is difficult to control potato blight during the period of heavy rains. Weather is unpredictable and it is difficult to plan pest and disease control practices. (Farmer #40, Nuwara Eliya)

According to the farmers, planting of tuber crops should be done avoiding rainy season. However, farmers now face challenges in finding a suitable time for planting since the rainy season is unpredictable. Consequently, they experience huge crop losses.

> Last year my harvest was exactly 50 sacks of carrot by just sowing two tins of seeds. But this year, I received only four sacks. My crop rotted because of rainfall lasted for long period. (Farmer #65, Nuwara Eliya)

Seventy-two percent of farmers consider weather and weather-related information while taking decisions on cultivation and farm management practices. Most of the farmers believe their yield and revenue vary over the years due to climate change.

> I have been cultivating 10 acres of vegetables and sometimes I could earn two million rupees from that cultivation. On the other hand, sometimes I lose one to one and half million rupees. I believe these losses are due to the climate change. (Farmer #39, Nuwara Eliya)

27.5.4   Farmers' Adaptation Options

The results revealed that vegetable farmers have been practicing different adaptation options to minimize climate change impacts. These adaptation options include, early or late planting, switching variety or crop and intensive use of inputs. In this study "no adaptation" was selected as the base category to express relative probabilities in the climate adaptation model.

> Climate during the past twenty years has changed dramatically and forced us to switch the variety or the crop and change the planting time. Some farmers are unable to adopt any adaptation measure mainly because of their financial constraints. (Farmer #39, Nuwara Eliya)

---

[4] LKR 189.84 = USD 1 in December 2020.

Switching variety or crop was adopted by more than 42% farmers as the most common response for adaptation.

> We change the type of crop cultivating and select the crops according to the weather pattern. Cabbage is good for this time since we have adequate rainfall. But it is difficult to protect potato and carrot farms since it is heavy rains. (Farmer #47, Nuwara Eliya)

Early or late planting was adopted by 23% of respondents. Moreover, most of the farmers stated that rainfall seasons[5] have now changed and they are unable to make cropping calendars according to well-known monsoon patterns as they did earlier. As a result, they have to make sudden decisions for their farming and find it hard to plan any farming operations beforehand.

> Earlier we used to start cultivation in October, but now since rain starts earlier, we are planting before October. (Farmer#16, Nuwara Eliya)

Referring to another farmer,

> For blight, in the past we used fungicides at five days interval but now we spray it in three days interval. When rain is heavy, there are times that we use fungicides every other day. (Farmer #24, Bandarawela)

Intensive use of inputs was adopted by 34% of farmers as a commonly used adaptation. Nearly all respondents who have adopted this method revealed that they do not increase the amount of fertilizer. However, they use pesticides, fungicides, and weedicides intensively.

In summary, most of the farmers have started to respond to the impacts of climate change mainly through early or late planting, switching variety or crop and intensive use of inputs.

27.5.5   Drivers of Farmers' Choice of Adaptation

Prospective-independent variables were determined with regard to the literature, availability of data, and pretest. These can be broadly categorized as demographic, socioeconomic, and climate change-related factors. Table 27.2 describes the explanatory variables used in the model.

Different combinations of independent variables were tested out of 14 variables which were collected from semi-structured in-depth interviews. In the preliminary stage, household size and farmland size variables were dropped, as those variables were not significant. Furthermore, independent variables, namely farm income, total income of the farmer, farmer-to-farmer extension, and age of the farmer were tested (model was run with and without those variables) with MNL model. Results indicated that inclusion of farm income and age of the farmer does not make parameter

---

[5] Southwest monsoon season is from May to September and northeast monsoon season prevails from December to February (Department of Meteorology, Sri Lanka).

Table 27.2 Description of explanatory variables used in MNL model

Explanatory variable	Mean	SD	Description
Gender of the farmer	0.85	0.36	Dummy, 1 = male, 0 = otherwise
Education level	2.41	0.79	Categorical (level of education)
Farming experience	4.67	1.71	Categorical (year categories)
Income of the farmer	3.44	1.38	Categorical (income categories)
Willingness to take credit	0.43	0.50	Dummy, 1 = willing, 0 = otherwise
Extension on crop	0.49	0.50	Dummy, 1 = adequate, 0 = otherwise
Information on climate change	0.22	0.42	Dummy, 1 = sufficient, 0 = otherwise
Farmer-to-farmer extension	0.80	0.40	Dummy, 1 = helpful, 0 = otherwise
Total annual rainfall	7.28	3.96	Continuous (percentage change in average annual rainfall between the years 2001–2009 and 2010–2018)
Average annual temperature	0.05	1.09	Continuous (percentage change in average annual temperature between the year 2001–2009 and 2010–2018)

Source Generated by Authors

estimates and marginal effects significant, although total income of the farmer and farmer-to-farmer extension made results significant.

Final combination of explanatory variables was selected based on model fit and statistical significances of marginal effects, as it is the most important result outcome in MNL model.

Model was tested for independence of irrelevant alternatives (IIA) by using seemingly unrelated post-estimation procedure (SUEST) (Deressa et al., 2009; Gunathilaka et al., 2018). SUEST test failed to reject null hypothesis in independence of irrelevant alternatives indicating MNL is appropriate to use for the sample.

27.5.6 Explanation of Variables

The effects of each significant variable on probability of different adaptation options adopted by farmers are explained below.

i.   Gender: Being a male farmer increases the probability of adopting early or late planting by 14.8%.

ii.   Education level: Education of the farmer has a positive effect on switching variety or crop. Well-educated farmers are 11.7% more likely to switch variety or crop as an adaptation to climate change. Education level relates negatively to intensive use of inputs. One unit increase in the level of education would result in 11.8% decrease in the intensive use of inputs. This result is intuitive given farmers with broad knowledge about suitable crops and varieties tend to explore the associated adaptation methods. Similarly, educated farmers recognize harmful effects of intensive input use and reduce usage.

iii. Farming experience: Farming experience has a positive and significant impact on early or late planting. Experienced farmers are 4.5% more likely to practice early or late planting. With experience farmers identify climate change trends and they adjust their cropping calendars according to the changes.

iv. Income: As expected, income has a positive influence on adaptation. Thus, farmers with high-income levels prefer more adaptation. Increase in farmers' income increases the probability of switching variety or crop by 6.8%. The results further emphasize that farmers with higher-income levels can afford adaptation and switching costs.

v. Willingness to use credit: Willingness to use credit has a significant positive effect on adaptation. Farmers with financial problems are willing to use credit as they want capital to minimize the impact of climate through adaptation.

vi. Extension services on crop: As expected, the access to crop extension has a positive and significant effect on adaptation. Precisely, the extension on crop increases the likelihood of early or late planting by 12.3%. Results emphasize the crucial role of disseminating information about climate change and crops. Moreover, farmers receive better knowledge about climate change through extension and are confident to alter planting time.

vii. Information on climate change: Access to climate change information has a significant positive effect on adaptation. Moreover, farmers who receive information on climate change are 19.3% more likely to select switching variety or crop as their commonly used adaptation option.

viii. Farmer-to-farmer extension: Having access to farmer-to-farmer extension has a positive and significant effect on adaptation. The results imply that the adaptation is encouraged by their social network.

ix. Percentage change in average annual rainfall: Farmers who have experienced an increasing percentage of average annual rainfall are more likely to adopt switching variety or crop by 3.9%. This result indicates that farmers switch variety or crop with rainfall changes but do not change the time of planting. As farmers cannot wait until rainfall decreases, they select vigorous crops or varieties to cultivate.

x. Average annual temperature: Unlike increasing percentage of average annual rainfall, higher level of average annual temperature appears to work opposite in terms of adaptation. This may be probably because increasing percentage of average annual rainfall relaxes the implications caused by increasing percentage of average annual temperature on crops and vegetable farming is more sensitive to rainfall and availability of water rather than temperature.

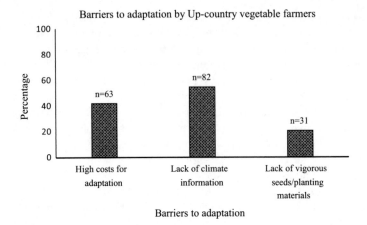

Fig. 27.4 Barriers to adaptation by up-country vegetable farmers. *Source* Generated by Authors

27.5.7 *Barriers to Adaptation*

We employed a descriptive analysis for the qualitative data gathered on barriers to adopt. Lack of climate information, high costs for adaptation and lack of vigorous seed and other planting materials were the three most common barriers stated by the respondents (Fig. 27.4). All non-adapters[6] mentioned insufficient climate change information as a major barrier for adaptation. A disappointing point to note is more than 70% of adopters did not have access to sufficient climate change information. Thus, lack of information was the most cited constraint faced by the farmers (82%).

Some farmers mentioned that though they attempt to identify climate patterns, they could not figure these out as they are not capable of analyzing the data.

> We do not receive near-accurate weather forecasts. Therefore, we have no trust on weather forecasts. If farmers can reach accurate weather information beforehand, it will be a great help for us to plan our cropping practices. (Farmer #72, Welimada)

> When we plan cultivation according to the previous season's rainfall pattern, it will not be the same in the current year. We cannot figure out the pattern of climate change". (Farmer #31, Welimada)

There appears to be no adequate source of information regarding weather patterns through any source of media (i.e., TV, radio, newspapers). Hence, farmers are unable to plan their operations and end up in unprofitable and unsustainable farming. This information barrier was followed by high costs of adaptation (42%). Although farmers want to switch crop or variety, they find it challenging for adaptation since the switching cost is high. Some farmers mentioned that low-income farmers do not wait until suitable environmental conditions arise to commence cultivation and management practices.

---

[6] Non-adapters referred here are the respondent farmers who are not adapted to climate change.

Financially stable farmers plant potatoes in February and harvest in June or July. They wait for harvesting to a better market price. But we cannot wait like them because of financial problems. We have families to feed. These days we are planting short term crops like, Chinese items (Lettuce, coriander) so we can earn some money in short run. (Farmer #50, Nuwara Eliya)

Another major constraint stated by respondents is lack of availability of vigorous seed and other planting materials. Most of the responders have no reliance on new hybrid varieties. They feel hybrid varieties cannot withstand climate change:

There are new blight resistant potato varieties. Although they produce leaves nicely, tubes do not grow adequately. We face problems in selecting proper varieties. (Farmer #55, Nuwara Eliya)

Lack of crop insurance policies also obstructs the farmers' adaptation to climate change. Thus, crop insurance programs should be customized according to the needs of vegetable farmers and in particular to the Up-country.

Crop insurance policies should be changed in a way that it will appropriate and benefit to the Up-country vegetable farmers. Current crop insurance procedure is made for the paddy farmers. It has three steps. Fund is released at the third step. But for potato, this procedure is not suitable because most expenses occur during the early stages of potato cultivation. Therefore, crop insurance should be given for the damages that are occurring in early stages (1st step). Often we receive insurance claim after about nine months of application, so it is too late. (Farmer #40, Nuwara Eliya)

27.6 Conclusions

Farmer adaptation practices are influenced by many factors. Understanding these factors is important in developing effective and appropriate policies and programs related to climate change adaptation. This study investigates climate change adaptation and factors driving the adaptation of vegetable farmers in Sri Lanka. There are three commonly used adaptation options which are early or late planting, switching variety or crop and intensive use of inputs.

By assessing the drivers of adaptation by the farmers, we identify that several characteristics such as gender of the farmer, education level, farming experience, income of the farmer, and willingness to take credit are important when encouraging farmers toward adaptation. The drivers related to information availability for the farmers, e.g., extension on crop, information on climate change, farmer-to-farmer extension, and climate variables themselves are affecting adaptation. Among these factors, education level, willingness to take credit, extension on crops, and information on climate change are highly likely to influence farmers' adaptation through policy interventions.

Information constraints emerge as a major barrier followed by high costs for adaptation and lack of vigorous seeds and other planting materials. Creating well-defined crop extension service and initiating convenient access for reliable and updated climate information could leverage sustainable vegetable farming systems.

Moreover, designing and implementation of a systematic credit system may further promote adaptation. Disseminating climate education through agricultural societies and extension services would require restructuring since farmers are in a dire need of information such as selection of suitable varieties and access to planting materials of these varieties. Educating the farming community, providing comprehensive climate information and improved crop extension service can be underlined as practices which need governmental support. Furthermore, regulating vegetable seed importation, implementing policies based on new research findings, and incentivizing studies on climate change is pivotal. Networking of farmers to initiate an information portal will nurture climate change adaptation among up-country vegetable farmers. All in all, the adaptation strategies incorporated with local knowledge and practices may feasibly result in effective and sustainable adaptation to climate change. This could enhance the up-country vegetable farming community's resilience for climate change consequences.

References

Ahmed, A. (2022). Autonomous adaptation to flooding by farmers in Pakistan. In A. K. E. Haque, P. Mukhopadhyay, M. Nepal & M. R. Shammin (Eds.), *Climate change and community resilience: Insights from South Asia* (pp. 101–112). Springer Nature.

Aryal, J. P., Sapkota, T. B., Khurana, R., Khatri-Chhetri, A., & Jat, M. (2019). Climate change and agriculture in South Asia: Adaptation options in smallholder production systems. *Environment Development and Sustainability, 22*(6), 5045–5075.

Belay, A., Recha, J. W., Woldeamanuel, T., & Morton, J. (2017). Smallholder farmers' adaptation to climate change and determinants of their adaptation decisions in the Central Rift Valley of Ethiopia. *Agriculture and Food Security, 6*(1).

Blanc, E., & Reilly, J. (2017). Approaches to assessing climate change impacts on agriculture: An overview of the debate. *Review of Environmental Economics, 11*(2), 247–257.

Cai, Y., Bandara, J. S., & Newth, D. (2016). A framework for integrated assessment of food production economics in South Asia under climate change. *Environmental Modelling and Software, 75*, 459–497.

Chalise, S., Naranpanawa, A., Bandara, J. S., & Sarker, T. (2017). A general equilibrium assessment of climate change–induced loss of agricultural productivity in Nepal. *Economic Modelling, 62*, 43–50.

Chandrasiri, S., Galagedara, L., & Mowjood, M. (2020). Impacts of rainfall variability on paddy production: A case from Bayawa minor irrigation tank in Sri Lanka. *Paddy and Water Environment, 18*, 443–454.

Chithranayana, R. D., & Punyawardena, B. V. R. (2014). Adaptation to the vulnerability of paddy cultivation to climate change based on seasonal rainfall characteristics. *Journal of the National Science Foundation of Sri Lanka, 42*(2), 119–127.

De Costa, W. (2010). Adaptation of agricultural crop production to climate change: A policy framework for Sri Lanka. *Journal of the National Science Foundation of Sri Lanka, 38*(2), 79–89.

De Silva, C., Weatherhead, E., Knox, J. W., & Rodriguez-Diaz, J. (2007). Predicting the impacts of climate change—A case study of paddy irrigation water requirements in Sri Lanka. *Agricultural Water Management, 93*(1–2), 19–29.

Deressa, T. T., Hassan, R. M., Ringler, C., Alemu, T., & Yesuf, M. (2009). Determinants of farmers' choice of adaptation methods to climate change in the Nile Basin of Ethiopia. *Global Environmental Change, 19*(2), 248–255.

Esham, M., & Garforth, C. (2013). Agricultural adaptation to climate change: Insights from a farming community in Sri Lanka. *Mitigation and Adaptation Strategies for Global Change, 18*(5), 535–549.

Esham, M., Jacobs, B., Rosairo, H. S. R., & Siddighi, B. B. (2018). Climate change and food security: A Sri Lankan perspective. *Environment, Development and Sustainability, 20*(3), 1017–1036.

Gunathilaka, R., Smart, J., & Fleming, C. (2017). The impact of changing climate on perennial crops: The case of tea production in Sri Lanka. *Climatic Change, 140*(3–4), 577–592.

Gunathilaka, R. P. D., Smart, J. C. R., & Fleming, C. M. (2018). Adaptation to climate change in perennial cropping systems: Options, barriers and policy implications. *Environmental Science and Policy, 82*, 108–116.

FAO (2017) Special report on FAO/WFP crop and food security assessment mission to Sri Lanka. Retrieved from http://www.fao.org/3/i7450e/i7450e.pdf

Hijioka, Y., Lin, E., Pereira, J. J., Corlett, R. T., Cui, X., Insarov, G. E., Lasco, R. D., Lindgren, E. & Surjan, A.(2014). Asia. In *Working group II contribution to the IPCC fifth assessment report climate change* (pp. 1327–1370).

Huong, N. T. L., Bo, Y. S., & Fahad, S. (2019). Economic impact of climate change on agriculture using Ricardian approach: A case of northwest Vietnam. *Journal of the Saudi Society of Agricultural Sciences, 18*(4), 449–457.

Jat, M. L., Dagar, J. C., Sapkota, T. B., Govaerts, B., Ridaura, S. L., Saharawat, Y. S., Sharma, R. K., Tetarwal, J. P., Jat, R. K., Hobbs, H., & Stirling, C. (2016). Climate change and agriculture: Adaptation strategies and mitigation opportunities for food security in South Asia and Latin America. *Advances in Agronomy, 137*, 127–235.

Kattel, R. R., & Nepal, M. (2021). Rainwater harvesting and rural livelihoods in Nepal. In A. K. E. Haque, P. Mukhopadhyay, M. Nepal, & M. R. Shammin (Eds.), *Climate change and community resilience: Insights from South Asia* (pp. 159–173). Springer Nature.

Menike, L. M. C. S., & Arachchi, K. A. G. P. K. (2016). Adaptation to climate change by smallholder farmers in rural communities: Evidence from Sri Lanka. *Procedia Food Science, 6*, 288–292.

Padmajani, M. T., Aheeyar, M. M. M., & Bandara, M. A. C. S. (2014). *Assessment of pesticide usage in Up-country vegetable farming in Sri Lanka.* Hector Kobbekaduwa Agrarian Research and Training Institute.

Perera, T., & Madhujith, T. (2012). The pattern of consumption of fruits and vegetables by undergraduate students: A case study. *Tropical Agricultural Research, 23*(3), 261–271.

Samarakoon, P. S. M. K. J., & Gunathilaka, R. P. D. (2020). Adaptation to climate change by Up-country vegetable farmers. In *Proceedings of the International Research Conference of Uva Wellassa University*, July 29–30, 2020.

Seo, S. N. N., Mendelsohn, R., & Munasinghe, M. (2005). Climate change and agriculture in Sri Lanka: A Ricardian valuation. *Environment and Development Economics, 10*(5), 581–596.

United Nations Office for Disaster Risk Reduction. (2019). *Disaster risk reduction in Sri Lanka, Status Report for 2019.* Retrieved from Bangkok, Thailand, United Nations Office for Disaster Risk Reduction (UNDRR), Regional Office for Asia and the Pacific. https://www.unisdr.org/files/68230_10srilankadrmstatusreport.pdf

Chapter 28
Resilience Through Crop Diversification in Pakistan

Adnan Nazir and Heman Das Lohano

Key Messages

- Crop diversification is a potential strategy to enhance community resilience to climate change impacts.
- Farm size, farmer's risk attitude, and the previous exposure to flood or excessive rainfall influence farmer's decision on crop diversification.
- Location-specific factors, such as soil quality, climatic, and agro-ecological condition, determine the extent of crop diversification.

28.1 Introduction

Pakistan was the fifth most affected country globally in terms of the impact of extreme weather events during the period 1999–2018 according to the Global Climate Risk Index 2020 report (Eckstein et al., 2019). The agriculture sector is the most vulnerable to climatic changes as precipitation and temperature are direct inputs into agriculture

---

Disclaimer: The presentation of material and details in maps used in this book does not imply the expression of any opinion whatsoever on the part of the Publisher or Author concerning the legal status of any country, area or territory or of its authorities, or concerning the delimitation of its borders. The depiction and use of boundaries, geographic names and related data shown on maps and included in lists, tables, documents, and databases in this book are not warranted to be error-free nor do they necessarily imply official endorsement or acceptance by the Publisher, Editor(s), or Author(s).

A. Nazir (✉)
Department of Economics, University College of Zhob (UCoZ), BUITEMS, Zhob 85200, Pakistan
e-mail: adnannazir@gmail.com

H. Das Lohano
Department of Economics, Institute of Business Administration, Karachi 75270, Pakistan
e-mail: hlohano@iba.edu.pk

© The Author(s) 2022   431
Haque et al. (eds.), *Climate Change and Community Resilience*,
https://doi.org/10.1007/978-981-16-0680-9_28

production (Deschênes & Greenstone, 2007) and flooding is quite common (Ahmed, 2022, Chap. 7 of this volume). The agriculture sector faces a variety of risks including extreme weather events, pest attacks, and market price fluctuation among others. The existence of these risks has resulted in the development of a number of agricultural risk management instruments and strategies (Velandia et al., 2009).

Information dissemination such as seasonal climate forecast (Manjula, Rengalakshmi, & Devaraj, 2021, Chap. 18 of this volume) and crop diversification including the traditional varieties (Tshotsho, 2022, Chap. 6 of this volume) are recognized as effective risk management strategies (McCord et al., 2015) that can serve to buffer farm businesses from risks including yield risk associated with climatic conditions and price risk associated with commodity markets (Hardaker, Huirne, & Anderson, 1997). It can improve community resilience by creating better ability to suppress pest outbreaks and dampen pathogen transmission, which may worsen under future climate scenarios (Lakhran, Sandeep, & Bajiya, 2017). Adaptation to climate change through climate-smart agricultural practices can transform marginal lands from environmental burdens into productive lands. Some crops such as oil seeds, legumes, cereals, medicinal plants, and some kinds of fruit can adapt in marginal environments such as salinity, waterlogging and drought (Hussain et al., 2020). Thus, crop diversification has been identified as a potential coping strategy for climate change impacts (Bradshaw, Dolan, & Smit, 2004). The National Climate Change Policy of Pakistan also recognizes crop diversification as a method of adaptation to climate change (Government of Pakistan, 2012).

Various studies have been conducted to examine the determinants of crop diversification (Chen, 2007; Kasem & Thapa, 2011; Kumar, Nayak, & Pradhan, 2020; Lin, 2011). Previous studies have examined the determinants of crop diversification such as socioeconomic characteristics of farmers and farms (Burchfield & Poterie, 2018). Likewise, Roesch-McNally, Arbuckle, and Tyndall (2018) examined factors that may influence farmers' decisions to use more diversified crop rotations in the American Corn Belt. Similarly, Shahbaz, Boz, and Haq (2017) investigated the factors influencing the behavior of farmers toward crop diversification at the farm level in Punjab province of Pakistan.

Previous exposure to extreme weather events may influence farmer's decision of crop diversification as the farmer can learn from past experiences and use crop diversification as a coping strategy to climate change. Huang (2014) examined the determinants of crop diversification and found that the decision to diversify crops is influenced by past experience to extreme weather events in China. However, this study did not include the risk attitude of farmers, which is an important determinant of crop diversification (Sarwosri & Mußhoff, 2020). The relationship between crop diversification behavior and exposure to extreme weather events has hardly been examined, especially in Pakistan. This study examines the determinants of crop diversification and investigates how past exposure to extreme weather events, risk attitude, and other farmer and farm characteristics affect the decision on crop diversification in Sindh province of Pakistan.

Sindh is located in the intense heat zone, and rise in temperatures due to climate change is further aggravating the conditions. Furthermore, it is located in the southern part of the Indus River and thus stands to suffer not only from the local climatic and weather changes but also from the climate in the upstream Indus area and from the coastal environment (Rasul et al., 2012). Floods in Sindh have mostly been associated with precipitation and excess flow of water from the upper part of the Indus River. Similarly, the effects of shortage of water and droughts in Sindh are aggravated by decrease in precipitation and reduced flow of water from the upper part of Indus River. In addition, the coastal areas of Sindh are affected by sea water intrusion and the rising sea level in the Arabian Sea (Rasul et al., 2012). These factors may affect both crop yields and returns to farmers.

28.2 Methodology

28.2.1 Study Area

Sindh province in southeastern Pakistan spreads over 44,016 miles^2 (17.7%) out of 307,376 miles^2 of Pakistan's total area (Government of Pakistan, 2019a). The weather in the province is usually dry and hot, which puts it in the arid subtropical zone. The temperature can reach over 50 °C in summer and fall as low as 6 °C in winter. In the last two decades, the province has experienced major floods in 2003, 2006, 2007, 2010, 2011, and 2012 (Kazi, 2014).

Figure 28.1 shows the study area. For this study, seven districts of Sindh province—Badin, Tando Allahyar, Mirpurkhas, Umarkot, Sanghar, Benazirabad, and Sujawal—were selected.

28.2.2 Sampling Strategy and Data Collection

Farm-level data was collected from 480 farmers using multistage stratified random sampling technique. In the first stage, seven districts listed in the earlier section were selected where there was an existing surface drainage system along with other parts of the district without a proper drainage system. In the second stage, two subdistricts (*Talukas*) were selected from each district, with one subdistrict from the drainage area and the other from the area without a proper drainage system. In the third stage, using systematic random sampling, 40 farmers each were selected from 10 subdistricts which are Tando Adam and Shahdadpur from Sanghar district, Digri and Jhuddo from Mirpur Khas district, Pithoro and Samaro from Umarkot district, Jhando Marri and Chamber from Tando Allahyar district, Badin and Golarchi from Badin district. Similarly, 20 farmers each were selected from the remaining four subdistricts which are Daur and Nawabshah from Benazirabad district and Shah Bander and Jati from Sujawal districts. Face-to-face interviews were conducted for collecting data from May to September 2016. Before starting the data collection,

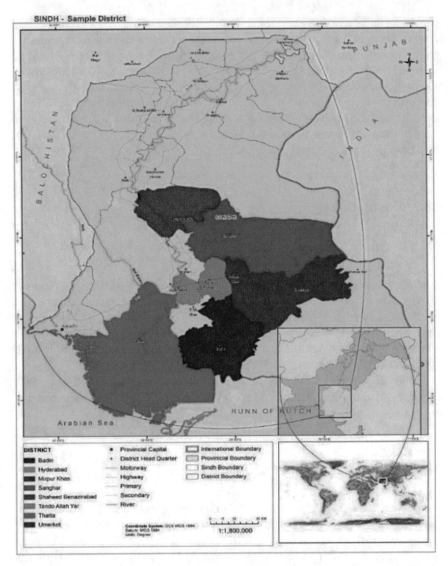

Fig. 28.1 Study area in Sindh province, Pakistan. *Source* Developed by authors using ArcGIS (ESRI 2016)

enumerators were trained in-field and off-field about the study. Questionnaire and data collection methods were properly explained to the enumerators. Further, the questionnaire was pretested in the field for improving the quality of survey data and to avoid missing any important data.

28.2.3 Determinants of Crop Diversification

Crop diversification is the dependent variable in this study. In Pakistan, there are mainly two cropping seasons, namely Rabi (from November to April) and Kharif (from May to October). In the study area, crops cultivated in the Rabi season include wheat, rapeseed, sunflower, sugarcane, tomato, and onion, and crops cultivated in Kharif season include cotton, rice, chillies, and banana. Crop diversification is a count variable and is measured as the number of crops grown by a farmer on his/her farm in the past one year. The determinants of crop diversification include farm characteristics, farmer characteristics, previous exposure to flood or excessive rainfall, community- and location-specific characteristics.

Farm characteristics that may affect crop diversification include farm size and distance from farm to nearest main city. Farm size is an important determinant to examine whether a relatively large farmer grows more crops on the farm or focuses on fewer crops. Moreover, the distance from farm to nearest main city indicates the accessibility to markets and information, and thus may also influence crop diversification decisions. Main city was defined as the city where farmers can purchase farm inputs and sell the produce.

Farmer characteristics that may influence crop diversification decisions include farmer's age, education level, and attitude toward risk. Farmer's age is an important factor to examine whether crop diversification is adopted by relatively older farmers who may have more farming experience or younger farmers who may be willing to take more tasks. Education can increase the knowledge of farmers and access to information and influence farmer's decision on crop diversification. Farmer's risk attitude measures the risk preferences and determines whether the farmer is risk averse or not (risk neutral or risk prone), which was estimated using equally likely certainty equivalent (ELCE) method (Anderson, Dillon, & Hardaker, 1977; Hardaker et al., 2015). Following this method, each farmer was given two risky outcomes (0 and his/her maximum monthly income with 50–50 probability) and was asked the amount of certainty equivalent. These income levels were assigned utility values using a scale from 0 to 1. Next, the farmer was given two other risky outcomes (0 and certainty equivalent in the first round) and was asked the amount of certainty equivalent. This procedure was repeated seven times, and then a cubic utility function was estimated using the data from the responses of the farmer. Using estimated utility function, absolute risk aversion was estimated for each farmer at the average monthly income. If the absolute risk aversion is greater than 0, the farmer is risk averse, otherwise he/she is risk neutral or risk prone. In this study, a dummy variable was created and was equal to 1 if the farmer is risk averse and 0 otherwise. This approach of measuring risk attitude is commonly used in the previous studies (e.g., Saqib et al., 2016).

Previous exposure to flood or excessive rainfall may influence farmer's decision of crop diversification as the farmer can learn from past experiences and can use crop diversification as a coping strategy for climate change. This factor has been used in previous study by Huang et al. (2014) for examining the determinants of crop diversification. In this study, we measure the previous exposure to flood or excessive rainfall by estimating the monetary value of crop losses during the previous event of flood or excessive rainfall in 2011–2012. The monetary value of crop loss was estimated by collecting information on the expected yield and actual yield received by the farmer during the disaster year.

Community and location-specific fixed effects account for the unobservable factors that are specific to the communities and location such as knowledge and preferences of community, soil quality, and climatic and agro-ecological conditions. In this study, we use subdistrict fixed effects to control for these unobservable factors.

28.2.4 Estimation Methods

As the dependent variable is a count variable, we employ the Poisson regression model to examine the determinants of crop diversification. The Poisson model's probability density function is specified as follows (Greene, 2018):

$$Pr(x_i = n) = \frac{e^{-\lambda_i}\lambda_i^n}{n!}, \quad n = 1, 2, 3, \ldots \tag{28.1}$$

It is worth noting that the parameter λ_i is both the mean and the variance under the Poisson distribution. The Poisson model is appropriate as there is no over-dispersion in the data of crop diversification variable, that is, the variance is not greater than the mean value of the variable. It is common to specify the parameter as an exponential function since it is necessary that $\lambda_i > 0$:

$$\lambda_i = \exp(z_i\beta) \tag{28.2}$$

The Poisson model is estimated using maximum likelihood estimation method. For interpreting the results, we compute the marginal effect for each explanatory variable. In the results, we also report the regression results using the ordinary least squares (OLS) method for comparison.

28.3 Results and Discussion

28.3.1 Descriptive Statistics

Summary statistics of variables used in the analysis are reported in Table 28.1. The extent of crop diversification varies from farm to farm, with minimum two crops and maximum eight crops in the study area. On an average, farmers cultivated 3.76 crops in a year. Similar findings were also reported by Huang et al. (2014) who found three

Table 28.1 Descriptive Statistics

Variables	Mean	Standard deviation	Min	Max
Dependent variable				
Crop diversification (Number of crops grown in a year)	3.76	1.35	2	8
Independent variables				
Farm characteristics				
Farm size (acres)	43.09	85.21	2	900
Small farms (1 if farm size <12.5 acres)	0.34	0.47	0	1
Medium farms (1 if farm size from 12.5 to 50 acres)	0.26	0.44	0	1
Large farms (1 if farm size > 50 acres)	0.41	0.49	0	1
Distance from nearest main city (km)	11.44	4.77	3	50
Farmer's characteristics				
Age (years)	46.23	9.45	21	74
Age squared	2,226	896	441	5,476
Education (years)	8.50	3.31	0	16
Risk aversion (1 if risk averse, 0 otherwise)	0.79	0.41	0	1
Previous exposure to flood or excessive rainfall				
Value of crop loss in PKR (USD)	416,482 (3,978)	825,895 (7,888)	614 (5.86)	10,388,377 (99,220)

Source Authors' computations based on survey data. *Note* Value of crop loss in USD, reported in parenthesis, is computed using exchange rate PKR 104.7 per USD during 2016–17 (Government of Pakistan, 2019b)

crops as the average number of crops cultivated by households in China. On average, the farm size in the study area was 43 acres. In the sample, 34% of the farms were small farms (with farm size less than 12.5 acres), 26% were medium farms (12.5–50 acres), and 41 percent were large farms (greater than 50 acres). Average distance from the farm to the nearest main city was 11.44 kms. Average age of farmers was 46 years while their average level of formal education was 8 years. Majority of the farmers (79%) in the study area were risk averse in nature as these farmers tend to avoid risk when they face risky circumstances. We measured the previous exposure to flood or excessive rainfall by estimating the monetary value of crop losses during the previous event of flood or excessive rainfall. The average losses during the previous event were Pakistani rupees 416,482 (USD 3978).

28.3.2   Regression Results

Table 28.2 presents the results of ordinary least square (OLS) regression and Poisson regression to examine the factors that affect crop diversification by a farmer. The parameter estimates from OLS regression directly provides the marginal effects. For the Poisson model, we compute the marginal effects using the coefficient estimates.

Results show that the farm size has statistically significant and positive effects on crop diversification. The extent of crop diversification was higher on medium farms by around 1 crop relative to small farms. Furthermore, crop diversification on large farms was higher by 1.13 crops as compared to small farms. We find that the risk-averse attitude of farmers has a statically significant and positive effect on crop diversification. The extent of crop diversification was higher by 0.43 in farmers who were risk averse relative to farmers who were risk neutral or risk prone.

Results further show that the value of crop losses during the previous exposure to flood or excessive rainfall has statistically significant and positive impact on adopting crop diversification as an ex-ante risk-absorbing instrument. The farmers who faced higher crop losses in a previous event of flood or excessive rainfall had higher levels of crop diversification. These findings are consistent with Huang et al. (2014).

The community- and location-specific fixed effects are statistically significant. This indicates that the unobserved factors, such as soil quality, climatic and agro-ecological conditions, influence the farmer's decision of crop diversification. For example, soil quality and agro-ecological conditions in Digri, Jhudo, Pithoro, and Samaro subdistricts are suitable for producing different types of crops such as wheat, cotton, tomatoes, chillies, vegetables, and other condiments, whereas Badin and Golarchi subdistricts have saline soils and thus farmers in these districts have limited choices for crop selection, such as wheat, rice, oilseeds, and sugarcane.

28.4   Conclusion

Agriculture sector in Sindh province of Pakistan is very vulnerable to climatic changes and extreme events as it is located in the southern part of the Indus River and thus stands to suffer not only directly from the local climatic and weather changes but also from the climate in the upstream Indus and from the coastal environments. Crop diversification is recognized as a potential strategy to enhance resilience to climate change impacts. This study examines how a farmer's decision of crop diversification is affected by past exposure to extreme weather events, risk attitude, and other farmers' and farm characteristics in Sindh. Using farm-level data from 480 farmers, findings of this study show that crop diversification is affected by farm size, farmer's risk attitude, and the previous exposure to flood or excessive rainfall. Furthermore, other location-specific factors, such as soil quality, climatic and agro-ecological condition, determine the extent of crop diversification.

Table 28.2 Regression results

	OLS regression		Poisson regression Marginal effects	
	Coefficient estimates	Standard errors	Coefficient estimates	Standard errors
Dependent variable: crop diversification (Number of crops grown in a year)				
Intercept	1.212	1.082	–	–
Farm characteristics				
Medium farms	0.899***	0.142	0.993***	0.248
Large farms	1.036***	0.138	1.130***	0.235
Distance from main city (km)	0.007	0.012	0.008	0.019
Farmers characteristics				
Age (years)	0.040	0.044	0.041	0.075
Age squared	0.000	0.000	0.000	0.001
Education (years)	-0.013	0.018	−0.012	0.030
Risk aversion attitude	0.404***	0.149	0.433*	0.256
Previous exposure to flood or excessive rainfall				
Value of crop loss (Million PKR)	0.229***	0.072	0.171*	0.098
Community- and location-specific fixed effects (Dummy variable for each subdistrict)				
Shahdadpur	0.650*	0.262	0.723	0.469
Digri	0.512*	0.263	0.588	0.469
Jhudo	1.298***	0.264	1.313***	0.451
Jhando Marri	1.249***	0.263	1.317***	0.456
Chamber	0.842***	0.262	0.927**	0.460
Pithoro	0.715***	0.262	0.793*	0.462
Samaro	1.128***	0.263	1.156***	0.450
Badin	0.182	0.265	0.201	0.488
Golarchi	0.422	0.264	0.470	0.477
Daur	0.834***	0.322	0.923*	0.559
Nawabshah	0.706**	0.320	0.779	0.563
Shah Bander	1.118***	0.322	1.176**	0.542
Jati	0.935***	0.322	1.016*	0.552
Number of observations	480		480	
R^2	0.291		–	
F-value	0.000		–	
LR chi^2 (12)	–		68.27	

(continued)

Table 28.2 (continued)

	OLS regression		Poisson regression Marginal effects	
	Coefficient estimates	Standard errors	Coefficient estimates	Standard errors
Prob > chi^2	–		0.0000	
Pseudo R^2	–		0.0393	
Log likelihood	–		−834.683	

Source Authors' computations based on survey data. Notes: ∗ ∗ ∗, ∗ ∗, ∗ Indicate significance at the 1, 5, and 10% level, respectively. Small farms (less than 12.5 acres) are a reference category in farm size categories. Tando Adam is a reference subdistrict in fixed effects for subdistricts

Crop diversification can improve resilience by enhancing the capacity and ability of farmers to cultivate different crops with different levels of tolerance to excess water or droughts. Furthermore, it can improve resilience by creating a better ability to suppress pest outbreaks and dampen pathogen transmission, which may worsen under future climate scenarios.

References

Ahmed, A. (2022). Autonomous Adaptation to Flooding by Farmers in Pakistan. In A. K. E. Haque, P. Mukhopadhyay, M. Nepal, & M. R. Shammin (Eds.), Climate Change and Community Resilience: Insights from South Asia (pp. 101–112). Springer Nature.

Anderson, J. R., Dillon, J. L., & Hardaker, J. B. (1977). *Agricultural decision analysis.* Iowa State University Press.

Bradshaw, B., Dolan, H., & Smit, B. (2004). Farm-level adaptation to climatic variability and change: Crop diversification in the Canadian prairies. *Climatic Change, 67*(1), 119–141. https://doi.org/10.1007/s10584-004-0710-z

Burchfield, E. K., & de la Poterie, A. T. (2018). Determinants of crop diversification in rice-dominated Sri Lankan agricultural systems. *Journal of Rural Studies, 61*, 206–215. https://doi.org/10.1016/j.jrurstud.2018.05.010

Chen, C. B. (2007). Farmers' diversified behavior: An empirical analysis. *J Agrotech Econ, 1*, 48–54.

Deschênes, O., & Greenstone, M. (2007). The economic impacts of climate change: Evidence from agricultural output and random fluctuations in weather. *American Economic Review, 97*(1), 354–385.

Eckstein, D., Künzel, V., Schäfer, L., & Winges, M. (2019). Global Climate Risk Index 2020. Germanwatch.

ESRI (Environmental Systems Research Institute) (2016). ArcGIS. Available at https://www.esri.com/. Map developed on November 20, 2016.

Government of Pakistan. (2012). *National Climate Change Policy* (Issue September). http://www.mocc.gov.pk/moclc/userfiles1/file/Moclc/Policy/NationalClimate Change Policy of Pakistan (2).pdf

Government of Pakistan. (2019a). *Agricultural Statistics of Pakistan 2017–18.* Ministry of National Food Security and Research, Economic Wing.

Government of Pakistan. (2019b). Pakistan Economic Survey 2019–20. Economic Advisory Wing, Finance Division. Islamabad.

Greene, W. H. (2018). *Econometric analysis.* Pearson Education.

Hardaker, J B, Huirne, R. B. M., & Anderson, J. R. (1997). *Coping with risk in agriculture.* CAB International.

Hardaker, J Brian, Lien, G., Anderson, J. R., & Huirne, R. B. M. (2015). *Coping with risk in agriculture: Applied decision analysis.* CABI.

Huang, J., Jiang, J., Wang, J., & Hou, L. (2014). Crop diversification in coping with extreme weather events in China. *Journal of Integrative Agriculture, 13*(4), 677–686. https://doi.org/10.1016/S2095-3119(13)60700-5

Hussain, M. I., Farooq, M., Muscolo, A., & Rehman, A. (2020). Crop diversification and saline water irrigation as potential strategies to save freshwater resources and reclamation of marginal soils—A review. *Environmental Science and Pollution Research, 27*(23), 28695–28729. https://doi.org/10.1007/s11356-020-09111-6

Kasem, S., & Thapa, G. B. (2011). Crop diversification in Thailand: Status, determinants, and effects on income and use of inputs. *Land Use Policy, 28*(3), 618–628. https://doi.org/10.1016/j.landusepol.2010.12.001

Kazi, A. (2014). A review of the assessment and mitigation of floods in Sindh, Pakistan. *Natural Hazards., 70*, 839–864. https://doi.org/10.1007/s11069-013-0850-4

Kumar, C. R., Nayak, C., & Pradhan, A. K. (2020). What determines crop diversification in North-East zone of India? *Journal of Public Affairs.* https://doi.org/10.1002/pa.2450

Lakhran, H., Sandeep, K., & Bajiya, R. (2017). Crop diversification: An option for climate change resilience. *Trends in Biosciences, 10*(2), 516–518.

Lin, B. B. (2011). Resilience in agriculture through crop diversification: Adaptive management for environmental change. *BioScience, 61*(3), 183–193. https://doi.org/10.1525/bio.2011.61.3.4

Manjula, M., Rengalakshmi, R., & Devaraj, M. (2021). Using climate information for building small holder resilience in India. In A. K. E. Haque, P. Mukhopadhyay, M. Nepal, & M. R. Shammin (Eds.), *Climate change and community resilience: Insights from South Asia* (pp. 275–289). Springer Nature.

McCord, P. F., Cox, M., Schmitt-Harsh, M., & Evans, T. (2015). Crop diversification as a small-holder livelihood strategy within semi-arid agricultural systems near Mount Kenya. *Land Use Policy, 42*(2015), 738–750. https://doi.org/10.1016/j.landusepol.2014.10.012

Rasul, G., Mahmood, A., Sadiq, A., & Khan, S. I. (2012). Vulnerability of the Indus delta to climate change in Pakistan. *Pakistan Journal of Meteorology, 8*(16), 89–107.

Roesch-McNally, G. E., Arbuckle, J. G., & Tyndall, J. C. (2018). Barriers to implementing climate resilient agricultural strategies: The case of crop diversification in the U.S. Corn Belt. *Global Environmental Change, 48*, 206–215. https://doi.org/10.1016/j.gloenvcha.2017.12.002

Saqib, S. E., Ahmad, M. M., Panezai, S., & Rana, I. A. (2016). An empirical assessment of farmers' risk attitudes in flood-prone areas of Pakistan. *International Journal of Disaster Risk Reduction, 18*, 107–114. https://doi.org/10.1016/j.ijdrr.2016.06.007

Sarwosri, A. W., & Mußhoff, O. (2020). Are risk attitudes and time preferences crucial factors for crop diversification by smallholder farmers? *Journal of International Development, 32*(6), 922–942.

Shahbaz, P., Boz, I., & Haq, S. U. (2017). Determinants of crop diversification in mixed cropping zone of Punjab Pakistan. *Direct Research Journal Agricultural Food Science, 5*(11), 360–366.

Tshotsho. (2022). Indigenous practices of paddy growers in Bhutan: A safety net against climate change. In A. K. E. Haque, P. Mukhopadhyay, M. Nepal, & M. R. Shammin (Eds.), *Climate change and community resilience: Insights from South Asia* (pp. 87–100). Springer Nature.

Velandia, M., Rejesus, R. M., Knight, T. O., & Sherrick, B. J. (2009). Factors affecting farmers' utilization of agricultural risk management tools: The case of crop insurance, forward contracting, and spreading sales. *Journal of Agricultural and Applied Economics, 41*(01), 107–123. https://doi.org/10.1017/S1074070800002583

Part VII
Moving Forward

Chapter 29
Communities, Climate Change Adaptation and Win–Win Solutions

A. K. Enamul Haque, Pranab Mukhopadhyay, Mani Nepal,
and Md Rumi Shammin

Key Messages

- The spatial, social, economic and historical diversity of South Asia suggests that there is no single mechanism to ensure success in adaptation.
- This part of the world provides examples of both success stories and failures in community adaptation.
- Traditional knowledge, new innovations, sensitive state intervention and market incentives all play a role in ensuring that community adaptation is optimal.

29.1 Introduction

The Intergovernmental Panel on Climate Change (IPCC) has played a stellar role in sensitising all stakeholders including the academic community, policymakers and local communities on the current state of knowledge on climate change and its potential impacts. The panel is expected to release its sixth assessment report in 2021, but

A. K. E. Haque
Department of Economics, East West University, Dhaka, Bangladesh
e-mail: akehaque@gmail.com

Pranab Mukhopadhyay (✉)
Goa Business School, Goa University, Taleigao Plateau, Goa, India
e-mail: pm@unigoa.ac.in

M. Nepal
South Asian Network for Development and Environmental Economics (SANDEE), International Centre for Integrated Mountain Development (ICIMOD), Kathmandu, Nepal
e-mail: mani.nepal@icimod.org

M. R. Shammin
Environmental Studies Program, Oberlin College, Oberlin, OH, USA
e-mail: rumi.shammin@oberlin.edu

© The Author(s) 2022
Haque et al. (eds.), *Climate Change and Community Resilience*,
https://doi.org/10.1007/978-981-16-0680-9_29

there is likely to be little to cheer about for those who have been or are expected to be adversely affected by rising temperatures and resulting extreme events. There is no country that is immune to some adverse effects of climate change but development economics ignores the environment (Dasgupta & Mäler, 1990).

The Paris Agreement of 2015 gave some hope as most countries committed to limit climate change. While mitigation is of immediate concern, adaptation is expected to be an ongoing challenge. The state must play its role, but it will have to work within the bounds of responsiveness and cooperation from individual consumers and producers. There is some evidence from South Asia that non-government stakeholders are willing to come on board to help fulfil NDC pledges, but some would prefer incentive-based policies rather than command and control policies (Haque et al., 2019).

29.2   Thinking Global, Acting Local

While climate change is a global problem, there is wide acknowledgement that adaptation to climate change will be a local challenge (Yohe & Moss, 2000). This implies that communities will play a significant role in managing the adaptation. They are capable of creating and adopting innovative strategies that are locally suitable, integrative, cost-effective, resource-efficient and culturally appropriate (Shammin, Firoz, & Hasan, 2021a; Chap. 2 of this volume). There is growing evidence and documentation of these successful efforts from across the global South (Shammin, Haque, & Faisal, 2022b, Chap. 3 of this volume).

29.3   Agricultural Adaptation

There are persistent knowledge gaps, for example, on factors that determine farmer adaptation. These include both climatic factors as well as non-climatic factors like demographic and economic characteristics. Most often, farmers' adaptations are driven by market forces, and there exists a significant gap between perceived and actual changes in climatic variables in South Asia (Budhathoki & Zander, 2020; Javed et al., 2020). More studies pertaining to perception and risk attitude behaviour could help evolve suitable policies for better adaptation (Bahinipati & Patnaik, 2021, Chap. 4 of this volume).

Communities are vulnerable not just due to the social hierarchies and inequalities; some are more vulnerable because of where they live and their level of access to technologies and knowledge. The Gangetic delta in South Asia is shared by India and Bangladesh. The communities living in this deltaic region are affected annually by extreme climate events, the severity, and frequency of which is increasing. These mudflats were inhabited about a century and half ago by new settlers. They cleared the mangroves and built an intricate labyrinth of earthen embankments to sustain their agriculture and protect their homes. Studies reveal that they are responding to climate change in two different ways. In one instance, it was found that out-migration, especially by the educated youth, was being used as a coping mechanism by the settlers (Ghosh & Roy, 2022, Chap. 26 of this volume). Apart from building social capital, larger investments in education could also provide greater resilience to vulnerable populations. Amin and Shammin (2022, Chap. 5 of this volume) find that adaptation initiatives took the form of mangrove rehabilitation, freshwater availability, agricultural productivity, and women's empowerment. In another instance, Bari, Haque, and Khan (2022, Chap. 14 of this volume) find that farmers are using traditional knowledge to innovate and produce crops using floating beds (*Baira*) in waterlogged areas of coastal Bangladesh. This technique could be replicated and used as an effective adaptation strategy to overcome vulnerability to waterlogging and provide food security simultaneously.

South Asian agriculture is characterised by subsistence farming by smallholders who are one of the most vulnerable groups. Although advances in science and technology help in adaptation, in some instances, reliance on traditional knowledge systems have helped communities to better adapt to climate change in this part of the world. In Bhutan, where more than 57% of the rice farmers are of this type, the communities are using traditional knowledge and experience to build resilience towards climate change. Tshotsho (2022, Chap. 6 of this volume) finds that the use of traditional rice varieties has helped farmers become more resilient to water scarcity compared to farmers using high-yielding varieties. Farmers in Pakistan are responding to climate change induced floods by crop diversification that helps them spread the risks of crop failure (Nazir & Lohano, 2022, Chap. 28 of this volume).

29.4 Role of External Support

Despite the community's best efforts, in some cases, the outcomes could be improved with external support. In the flood-affected Nowshera district of Pakistan, Ahmed (2021, Chap. 7 of this volume) finds that farmers could improve their adaptation outcomes if there was external technical support from the government. This is also evident in Kerala, India where Devi et al., (2022, Chap. 8 of this volume) document how farmer cooperatives with strong support from state agencies helped increase resilience in a post-flood situation. Collective action with state support helps farmers build on existing social networks that provide resilience during climate threats like sea level rise (Shafeeqa & Abeyrathne, 2022, Chap. 9 of this volume).

We find similar stories from mountain communities of Nepal where traditional technology for rainwater harvesting (RWH) has been re-designed with the help of new material and technology for climate change adaptation. The development and spread of RWH among farmers with external financial support and extension services (mainly providing training to the farmers) have helped farmers diversify their crop from subsistence-level cereal production to commercial high-value vegetables that has resulted in higher net income to the farmers (Kattel & Nepal, 2021, Chap. 11

of this volume). Since the up-front cost of adopting RWH technology is high (30% of farmers' annual income), its adoption rate is low despite a very short payback period of two years for the technology, suggesting that external financial support is needed to increase the adoption rate. In areas like Samanalawewa catchment in Sri Lanka, where the land use has changed over time, soil erosion can be a major concern (Udayakumara, 2022, Chap. 19 of this volume). State support by way of training of farmers in soil and water conservation measures and extension support can help farmers reduce risks due to soil erosion.

29.5 Energy Adaptation

In South Asia, over 60% of the households rely on solid biomass for cooking and heating, mainly using traditional stoves. In the absence of grid-electricity, low-cost solar home systems have been used in many rural communities, which provides clean lighting options (Shammin & Haque, 2022, Chap. 14 of this volume). But, for cooking, the problem is worse in rural areas where there is an absence of alternative sources of cooking energy and improved stoves for burning fuelwood are rarely available. The excessive use of fuelwood in traditional stoves poses the triple challenge of carbon emission, health hazards and impact on the local environment (deforestation). Governments across different countries in the region have been making attempts to reduce fuelwood use and increase alternate fuel use. In India, a series of initiatives including subsidies to the rural poor have been provided to shift dependence on firewood. The extent of adoption of clean cooking energy (refill of LPG cylinders) has differed across states and districts and was found to be dependent on multiple factors, most importantly on the extent of state support for rural employment generation programmes such as the Mahatma Gandhi National Rural Employment Guarantee Act (Thomas et al., 2022, Chap. 13 of this volume). In Bangladesh, Bari et al., (2021, Chap. 14 of this volume) found that the creation of appropriate market chains for LPG in rural areas could reduce forest dependence, with households voluntarily adopting LPG instead of using firewood for cooking.

29.6 Resilience to Extreme Events

Rural communities have dealt with drought and floods by innovating on spatial characteristics of the region they live in. The Small Tank Cascade Systems (STCS) of Sri Lanka (Vidanage, Kotagama, & Dunusinghe, 2021, Chap. 15 of this volume) and rainwater harvesting technology in mountain villages in Nepal (Kattel & Nepal, 2021, Chap. 11 of this volume) are some of the prominent examples. Once the lifeline for survival, these STCS in Sri Lanka became neglected over a period of time due to various reasons. However, with climate change, their revival is being seen as key to adaptation and resilience of rural communities. The cascades that connect water reservoirs at different levels not only fulfil the need for flood control but store water for the dry months. State support for these systems could help significantly in their revival and provide protection against climate change in future.

Just as climate change has brought in its wake increased frequency as well as intensity of extreme weather events, communities and policymakers are learning how to respond to disasters that follow such events. Shammin, Wang and Sosland (2022c, Chap. 16 of this volume) analyse the disaster management frameworks in Bangladesh which have developed a rich matrix of linked institutions between and across different levels. They combine field level observation with expert responses and find decentralised, community-based practices in some locations that could be replicated in other coastal communities with similar circumstances.

In coastal zones, apart from human-made efforts at minimising the damages caused by tidal waves, surges and cyclones, nature itself provides protection services through mangroves. Das (2021, Chap. 17 of this volume) and Mahmud et al., (2021, Chap. 20 of this volume) examine extreme events in India and Bangladesh. The first one looks at the super cyclone in 1999 (that hit Odisha, India), and the second one looks at *cyclone Sidr* in 2007 and *cyclone Roanu in* 2016 (that hit Bangladesh). The 1999 super cyclone showed how important the mangroves were in providing storm protection (to the extent of USD 68,586 per km width and USD 4335 per ha of mangroves) for those in the vicinity of the mangrove forest. Bangladesh too is regularly hit by cyclones. From the individual's perspective, households' investment in housing is driven by location-specific information. People are now inclined to build storm-resistant homes and locate themselves closer to government-sponsored embankments, mangroves, and cyclone shelters apart from wishing to be in the vicinity of a vehicular road, and primary school. The takeaway for policymakers here is to target relief in the aftermath of an extreme event for households living outside embankments and located further away from the mangrove forest.

Both studies find that protection of the mangrove forest is critical for building natural resilience to extreme events for local populations. This would help avoid the large-scale migration that was noticed by Ghosh and Roy (2022, Chap. 26 of this volume) in the aftermath of the severe cyclone *Aila* that hit India and Bangladesh.

29.7 Impact of Information

Communication of weather-related information is critical for reducing risk and building resilience. This is especially true for farmers whose vulnerability and risk-bearing capability is low. Gunathilaka and Samarakoon (2021, Chap. 27 of this volume) and Menon et al. (2021, Chap. 18 of this volume) argue that reliable climate forecasts and weather-related information work well if they are localised, accurate, easy and low-cost to access. Institutions can play a decisive role to enhance resilience if they can help farmers translate this information at the field level.

29.8   The Urban Challenge

Nearly 5 billion people are expected to be living in urban areas by 2030, and half of this increase is expected to be in Asia with China and India absorbing 55% of the regional urban growth (Seto, Güneralp, & Hutyra, 2012), suggesting that South Asia will face challenges in making our urban spaces more resilient. Well-known problems in the global South for urban sustainability include the lack of infrastructure, management capability of local governments and fast changing land use both at the centre and periphery of the urban landscape that leads to disintegration of existing institutions and requires the emergence of new ones. If successfully managed it could lead to greater resilience.

Three of the problems that both large and small urban centres in South Asia are facing relate to supply of safe potable water, recurrent flooding due to lack of drainage, and management of municipal waste. Most cities in the region lack sanitary landfill sites and recycling ability. With municipal waste being dumped in open spaces, forests and riverbanks, the accumulated waste blocks the natural flow of stormwater, resulting in waterlogging and urban flooding (Rahmasary et al., 2019). Urban centres have not increased their drainage commensurate with the city's expansion. While floods in big cities receive global attention, there are many smaller urban centres that face similar problems on a regular basis. The challenge for policymakers is how to finance drainage infrastructure.

Nepal et al. (2021, Chap. 21 of this volume) discuss a similar situation in one of the cities in Nepal (Bharatpur) where the lack of drainage and poor solid waste management has increased the risk of waterlogging and urban flooding. However, they find that households would be willing to pay additional fees for better solid waste management if the city provides waste collection services on a given date and time, so that there is a synchronised disposal and collection of waste by the households and the waste collectors. They also find that at-source segregation of waste helps recycling plastics, papers and metal items that generate a significant amount of revenue for the cities. Cleaner cities have a large price premium for the home owners, suggesting that efficient management of municipal waste generates a win–win situation for the city and its residents. In Sylhet, Bangladesh, Rakib et al., (2022, Chap. 24 of this volume) showed that with little effort at the community level, it was possible to persuade women in the household to segregate their kitchen waste and promote local composting instead of filling up the dumpsites.

Drinking water supply for urban consumption will increasingly be a challenge with climate change. Traditional sources of supply like natural springs in the hills and the mountains and wells in the lowlands are drying up, getting damaged or overdrawn due to poor management. While urban planners have focussed investment on infrastructure for water distribution, the management of water source has often remained unattended. Rai & Nepal (2022, Chap. 23 of this volume) examine the possibility of creating sustainable water supply solutions using an Incentive Payment for Ecosystem Service mechanism between upstream and downstream communities. Local institutions could play the role of intermediary for addressing the missing market that helps providing uninterrupted drinking water supply to urban residents. If designed appropriately, the payment from the water user community would provide incentives to the upstream community for protecting the water source, so that the downstream community can enjoy uninterrupted water supply.

29.9 Resilience and the Role of Women

Women play an important role in the adaptation process. Sharma et al., (2022, Chap. 22 of this volume) examine women's collective action in Ahmedabad and other cities to show how India's urban slum areas are adapting to three types of climatic shocks—heat wave, flooding and vector-borne diseases. These groups are able to increase awareness about the effects of climate change and the adaptation choices that these communities can make. They are also well placed to facilitate these urban changes and increase the resilience of the urban poor. Even in aspects of waste management in urban centres, women can play a significant role by creating awareness and promoting home-based segregation that can reduce the pressure on the city's waste managers (Rakib et al., 2022, Chap. 24 of this volume).

29.10 Innovative Experiments

One of the fastest growing sectors in the service economy is tourism, especially nature-based tourism (Balmford et al., 2009). It has provided alternate livelihood opportunities in many regions of the world. Bhutan has managed its tourism in a responsible manner by controlling the ecological footprint of the sector. Dendup et al., (2021, Chap. 25 of this volume) examine the experiment of community tourism as an adaptation mechanism. They find that community-based tourism helps improve the living conditions and assets of participating households, including the poor.

29.11 Challenges to Adaptation

It is also true that communities are not always able to undertake collective action. In the absence of either collective action or effective state intervention, the outcomes could be unsustainable, and households become more vulnerable. Devi et al., (2022, Chap. 8 of this volume) draw attention to the situation of water scarcity in Kerala where there was neither any visible collective action by communities nor state presence. This has led to unsustainable levels of resource depletion. In fact, long-term trends indicate that in parts of South Asia, the water table has been declining rapidly. This is driven partly by climatic factors (rising temperature, erratic rainfall and drought), but also government policy that subsidises electricity for agriculture. In this situation, state subsidy has not been beneficial for sustainable water use. Where groundwater depletion rate is already high, it could trigger unsustainable withdrawal of water. In the long term, this could adversely affect farmers especially small holders. This would pose a bigger challenge if the well density increases in an area where water scarcity is high or the ground water table is already depleted (Balasubramanian & Saravanakumar, 2021, Chap. 10 of this volume). As the impact of climate change begins to adversely affect human wellbeing especially in developing countries, economists will have to deal with issues of the environment as a core thematic in the subject (Dasgupta, 2021, Foreword of this volume).

References

Ahmed, A. (2021). Autonomous adaptation to flooding by farmers in Pakistan. In A. K. E. Haque, P. Mukhopadhyay, M. Nepal, & M. R. Shammin (Eds.), *Climate change and community resilience: Insights from South Asia* (pp. 101–112). Springer Nature.

Amin, R., & Shammin, M. R. (2022). A resilience framework for climate adaptation: The Shyamnagar experience. In A. K. E. Haque, P. Mukhopadhyay, M. Nepal, & M. R. Shammin (Eds.), *Climate change and community resilience: Insights from South Asia* (pp. 69–84). Springer Nature.

Bahinipati, C. S., & Patnaik, U. (2021). What motivates farm level adaptation in India? A systematic review. In A. K. E. Haque, P. Mukhopadhyay, M. Nepal, & M. R. Shammin (Eds.), *Climate change and community resilience: Insights from South Asia* (pp. 49–68). Springer Nature.

Balasubramanian, R., & Saravanakumar, V. (2021). Climate sensitivity of groundwater systems in South India: Does It matter for agricultural income? In A. K. E. Haque, P. Mukhopadhyay, M. Nepal, & M. R. Shammin (Eds.). *Climate change and community resilience: Insights from South Asia* (pp. 143–156). Springer Nature.

Balmford, A., Beresford, J., Green, J., Naidoo, R., Walpole, M., & Manica, A. (2009). *a global perspective on trends in nature-based tourism. PLoS Biology, 7*(6), e1000144.

Bari. E, Haque, A.K.E., & Khan, Z.K. (2022). Local strategies to build climate resilient communities in Bangladesh. In A. K. E. Haque, P. Mukhopadhyay, M. Nepal, & M. R. Shammin (Eds.). *Climate change and community resilience: Insights from South Asia* (pp. 175–189). Springer Nature.

Budhathoki, N. K., & Zander, K. K. (2020). Nepalese farmers' climate change perceptions, reality and farming strategies. *Climate and Development, 12*(3), 204–215.

Das, S. (2021). Valuing the role of mangroves in storm damage reduction in coastal areas of Odisha. In A. K. E. Haque, P. Mukhopadhyay, M. Nepal, & M. R. Shammin (Eds), *Climate change and community resilience: Insights from South Asia* (pp. 257–273). Springer Nature.

Dasgupta, P. (2021). Foreword. In A. K. E. Haque, P. Mukhopadhyay, M. Nepal, & M. R. Shammin (Eds). *Climate change and community resilience: Insights from South Asia* (pp. xi–xiv). Springer Nature.

Dasgupta, P., & Mäler, K. G. (1990). The environment and emerging development issues. *The World Bank Economic Review, 4*(suppl 1), 101–132.

Devi, P. I., Sam, A. S., & Sathyan, A. R. (2022). Resilience to climate stresses in South India: Conservation responses and exploitative reactions. In A. K. E. Haque, P. Mukhopadhyay, M. Nepal, & M. R. Shammin (Eds.), *Climate change and community resilience: Insights from South Asia* (pp. 113–127). Springer Nature.

Dendup, N., Tshering, K., & Choida, J. (2021). Community based tourism as a strategy for building climate resilience in Bhutan. In A. K. E. Haque, P. Mukhopadhyay, M. Nepal, & M. R. Shammin (Eds.), *Climate change and community resilience: Insights from South Asia* (pp. 387–398). Springer Nature.

Ghosh, S., & Roy, S. (2022). Climate change, ecological stress and livelihood choices in indian, Sundarban. In A. K. E. Haque, P. Mukhopadhyay, M. Nepal, & M. R. Shammin (Eds.). *Climate change and community resilience: Insights from South Asia* (pp. 399–413). Springer Nature.

Gunathilaka, R. P. D., & Samarakoon, P. S. M. K. J. (2021). Adaptation by Vegetable Farmers to Climate Change in Sri Lanka. In A. K. E. Haque, P. Mukhopadhyay, M. Nepal, & M. R. Shammin (Eds.). *Climate change and community resilience: Insights from South Asia* (pp. 415–429). Springer Nature.

Haque, A. E., Lohano, H. D., Mukhopadhyay, P., Nepal, M., Shafeeqa, F., & Vidanage, S. P. (2019). NDC pledges of South Asia: Are the stakeholders onboard? *Climatic Change, 155*(2), 237–244.

Haque, A. K. E., Mukhopadhyay, P., Nepal, M., & Shammin, M. R. (2022a). South Asian stories of climate resilience. In A. K. E. Haque, P. Mukhopadhyay, M. Nepal, & M. R. Shammin (Eds.). *Climate change and community resilience: Insights from South Asia* (pp. 1–7). Springer Nature. https://doi.org/10.1007/978-981-16-0680-9_1.

Haque, A. K. E., Mukhopadhyay, P., Nepal, M., & Shammin, M. R. (2022b). Communities, climate change adaptation and win-win solutions. In A. K. E. Haque, P. Mukhopadhyay, M. Nepal, & M. R. Shammin (Eds.), *Climate change and community resilience: Insights from South Asia* (pp. 445–454). Springer Nature.

Javed, S. A., Haider, A., & Nawaz, M. (2020). How agricultural practices managing market risk get attributed to climate change? Quasi-experiment evidence. *Journal of Rural Studies, 73*, 46–55.

Kattel, R. R., & Nepal, M. (2021). Rainwater harvesting and rural livelihoods in Nepal. In A. K. E. Haque, P. Mukhopadhyay, M. Nepal, & M. R. Shammin (Eds.), *Climate change and community resilience: Insights from South Asia* (pp. 159–173). Springer Nature.

Mahmud, S., Haque, A. K. E., & De Costa, K. (2021). Climate resiliency and location specific learnings from coastal Bangladesh. In A. K. E. Haque, P. Mukhopadhyay, M. Nepal, & M. R. Shammin (Eds). *Climate change and community resilience: Insights from South Asia* (pp. 309–321). Springer Nature.

Manjula, M., Rengalakshmi, R., & Devaraj, M. (2021). Using climate information for building small holder resilience in India. In A. K. E. Haque, P. Mukhopadhyay, M. Nepal, & M. R. Shammin (eds.), *Climate change and community resilience: Insights from South Asia* (pp. 275–289). Springer Nature.

Nazir, A., & Lohano, H.D. (2022). Resilience through crop diversification in Pakistan. In A. K. E. Haque, P. Mukhopadhyay, M. Nepal, & M. R. Shammin (Eds.), *Climate change and community resilience: Insights from South Asia* (pp. 431–442). Springer Nature.

Nepal, M., Bharadwaj, B., Karki Nepal, A., Khadayat, M. S., Pervin, I. S., Rai, R. K., & Somanathan, E. (2021). Making urban waste management and drainage sustainable in Nepal. In A. K. E. Haque, P. Mukhopadhyay, M. Nepal, & M. R. Shammin (Eds.), *Climate change and community resilience: Insights from South Asia* (pp. 325–338). Springer Nature.

Rakib, M., Hye,N., & Haque, A.K.E. (2022) Waste segregation at source: A strategy to reduce waterlogging in sylhet. In A. K. E. Haque, P. Mukhopadhyay, M. Nepal, & M. R. Shammin (Eds.), *Climate change and community resilience: Insights from South Asia* (pp. 369–383). Springer Nature.

Rai, R. K., & Nepal, M. (2022). A tale of three Himalayan Towns: Would payment for ecosystem services make drinking water supply sustainable? In A. K. E. Haque, P. Mukhopadhyay, M. Nepal, & M. R. Shammin (Eds.), *Climate change and community resilience: Insights from South Asia* (pp. 357–367). Springer Nature.

Rahmasary, A. N., Robert, S., Chang, I.-S., Jing, W., Park, J., Bluemling, B., Koop, S., & van Leeuwen, K. (2019). Overcoming the challenges of water, waste and climate change in Asian Cities. *Environmental Management, 63*(4), 520–535.

Seto, K. C., Güneralp, B., & Hutyra, L. R. (2012). Global forecasts of urban expansion to 2030 and direct impacts on biodiversity and carbon pools. *Proceedings of the National Academy of Sciences, 109*(40), 16083–16088.

Shafeeqa, F., & Abeyrathne, R. M. (2022). Climate adaptation by farmers in three communities in the Maldives. In A. K. E. Haque, P. Mukhopadhyay, M. Nepal, & M. R. Shammin (Eds.), *Climate change and community resilience: Insights from South Asia* (pp. 129–141). Springer Nature.

Shammin, M. R., Firoz, R,. & Hasan, R. (2021). Frameworks, stories and lessons from disaster management in Bangladesh. In A. K. E. Haque, P. Mukhopadhyay, M. Nepal, & M. R. Shammin

(Eds), *Climate change and community resilience: Insights from South Asia* (pp. 239–256). Springer Nature.

Shammin, M. R., & Haque, A. K. E. (2022). Small-scale solar solutions for energy resilience in Bangladesh. In A. K. E. Haque, P. Mukhopadhyay, M. Nepal, & M. R. Shammin (Eds). *Climate change and community resilience: Insights from South Asia* (pp. 205–224). Springer Nature.

Shammin, M. R., Haque, A. K. E., & Faisal, I. M. (2022). A framework for climate resilient community-based adaptation. In A. K. E. Haque, P. Mukhopadhyay, M. Nepal, & M. R. Shammin (Eds), *Climate change and community resilience: Insights from South Asia* (pp. 11–30). Springer Nature.

Shammin, M., Wang, A., & Sosland, M. (2022). A survey of community-based adaptation in developing countries. In A. K. E. Haque, P. Mukhopadhyay, M. Nepal, & M. R. Shammin (Eds), *Climate change and community resilience: Insights from South Asia* (pp. 31–47). Springer Nature.

Sharma, U., Brahmbhatt, B., & Panchal, H. N. (2022). Do Community-based institutions spur climate adaptation in urban informal settlements in India? In A. K. E. Haque, P. Mukhopadhyay, M. Nepal, & M. R. Shammin (Eds), *Climate change and community resilience: Insights from South Asia* (pp. 339–356). Springer Nature.

Thomas, L., Balakrishna, R., Chaturvedi, R., Mukhopadhyay, P., & Ghate, R. (2022). What influences rural poor in India to refill their LPG? In A. K. E. Haque, P. Mukhopadhyay, M. Nepal, & M. R. Shammin (Eds.), *Climate change and community resilience: Insights from South Asia* (pp. 191–203). Springer Nature.

Tshotsho. (2022). Indigenous practices of paddy growers in Bhutan: A safety net against climate change. In A. K. E. Haque, P. Mukhopadhyay, M. Nepal, & M. R. Shammin (Eds.), *Climate change and community resilience: Insights from South Asia* (pp. 87–100). Springer Nature.

Udayakumara, E.P.N. (2022). Farmer adaptation to climate variability and soil erosion in Samanalawewa Catchment, Sri Lanka. In A. K. E. Haque, P. Mukhopadhyay, M. Nepal, & M. R. Shammin (Eds.), *Climate change and community resilience: Insights from South Asia* (pp. 291–307). Springer Nature.

Vidanage, S. P., Kotagama, H. B., & Dunusinghe, P. M. (2021). Sri Lanka's small tank cascade systems: Building agricultural resilience in the dry zone. In A. K. E. Haque, P. Mukhopadhyay, M. Nepal, & M. R. Shammin (Eds.), *Climate change and community resilience: Insights from South Asia* (pp. 225–235). Springer Nature.

Yohe, G., & Moss, R. (2000). Economic sustainability, indicators and climate change. In *Proceedings of the IPCC expert meeting on development, equity and sustainability*. Colombo, Sri Lanka. Geneva: Intergovernmental Panel on Climate Change and World Meteorological Organization.

Correction to: Climate Change and Community Resilience

A. K. Enamul Haque⑩, Pranab Mukhopadhyay⑩, Mani Nepal⑩, and Md Rumi Shammin⑩

Correction to:
Haque et al. (eds.),
Climate Change and Community Resilience,
https://doi.org/10.1007/978-981-16-0680-9

This book was inadvertently published with the incorrect book editor's name in citations and references as "A. K. Enamul Haque" instead of "A. K. E. Haque" in the whole book now has been updated and in chapter 21 chapter authors names are listed in order. This has now been amended throughout the book.

The Open Access license and logo has been changed from "This chapter is licensed under the terms of the Creative Commons Attribution 4.0 International License (http://creativecommons.org/licenses/by/4.0/)" to "This chapter is licensed under the terms of the Creative Commons Attribution-NonCommercial-NoDerivatives 4.0 InternationalLicense (http://creativecommons.org/licenses/bync-nd/4.0/), This has now been amended throughout the book.

The updated version of the book can be found at
https://doi.org/10.1007/978-981-16-0680-9
https://doi.org/10.1007/978-981-16-0680-9_21

Index

Printed in the United States
by Baker & Taylor Publisher Services